# ChemCom
## Chemistry in the Community

A Project of the American Chemical Society

***Teacher's Guide***
Third Edition

KENDALL/HUNT PUBLISHING COMPANY
4050 Westmark Drive P.O. Box 1840 Dubuque, Iowa 52004-1840

## ChemCom

# Third Edition Revision Team

**Chief Editor:**
Conrad L. Stanitski

**Editor of Teacher's Guide:**
Patricia J. Smith

**Third-Edition Editorial Advisory Board:**
Joseph Breen, I. Dwaine Eubanks, Lucy Pryde Eubanks, Regis Goode, Henry Heikkinen (*chair*), Mary Kochansky, Ivan Legg, Steven Long, Nina McClelland, Carlo Parravano, Robert Patrizi, Max Rodel, K. Michael Shea, Patricia Smith, Susan Snyder, Conrad L.Stanitski, Sylvia Ware (*ex officio*)

**ACS Staff:**
Janet M. Boese, Ted Dresie, Martha Turckes Lester, Rebecca Mason Simmons, Michael J. Tinnesand

**Teacher Reviewers of Second Edition:**
Michael Clemente, Carlson High School, Gibraltar, MI; Steven Long, Rogers High School, Rogers, AR; William Penker, Neillsville High School, Neillsville, WI; Audrey Mandel, Connetquot High School, Bohemia, NY; Barbara Sitzman, Chatsworth High School, Chatsworth, CA; Kathleen Voorhees, Shore Regional High School, West Long Branch, NJ; Debra Compton, Cy-Fair High School, Houston, TX; Christ Forte, York Community High School, Elmhurst, IL; Gwyneth D. Sharp, Cape Henlopen High School, Lewes, DE; Louis Dittami, Dover-Sherborn High School, Dover, MA; Sandra Mueller, John Burroughs School, St. Louis, MO; Kirk Soule, Sunset High School, Beaverton, OR; Sigrid Wilkinson, Athens Area High School, Athens, PA; Millie McDowell, Clayton High School, Clayton, MO; Leslie A. Roughley, Steward School, Richmond, VA; Robert Houle, Bacon Academy, Colchester, CT; Robert Storch, Bishop Ireton High School, Alexandria, VA; Michael Smolarek, Neenah High School, Neenah, WI; Fred Nozawa, Timpview High School, Provo, UT; Michael Sixtus, Mar Vista High School, Imperial Beach, CA

**Production Editor:**
Marcia Vogel

**Production Assistant:**
Jody Koenig

**Safety Consultant:**
William H. Breazeale, Jr.

**Editorial:**
The Stone Cottage

**Design:**
P.C. & F., Inc.

**Design of Teacher's Guide:**
Peggy Corrigan, Manager, Graphic Services
Cornithia Harris, Designer

**Teacher's Guide Editorial Services:**
Carol Edwards

This material is based upon work supported by the National Science Foundation under Grant No. SED-88115424 and Grant No. MDR-89470104. Any opinions, findings, and conclusions or recommendations expressed in this publication are those of the authors and do not necessarily reflect the views of the National Science Foundation. Any mention of trade names does not imply endorsement by the National Science Foundation.



Thanks to the faculty, administration, and students of Hinsdale Central High School (Illinois) and the faculty and administration of Suitland High School (Maryland) for use of their chemistry laboratory facilities in photographs depicting laboratory setups and procedures in this textbook.



ISBN 0-7872-0561-3



Printed in the United States of America
10 9 8 7 6 5 4 3 2 1

# TABLE OF CONTENTS

# IMPORTANT NOTICE

*ChemCom* is intended for use by high school students in the classroom laboratory under the direct supervision of a qualified chemistry teacher. The experiments described in this book involve substances that may be harmful if they are misused or if the procedures described are not followed. Read cautions carefully and follow all directions. Do not use or combine any substances or materials not specifically called for in carrying out experiments. Other substances are mentioned for educational purposes only and should not be used by students unless the instructions specifically so indicate.

The materials, safety information, and procedures contained in this book are believed to be reliable. This information and these procedures should serve only as a starting point for good laboratory practices, and they do not purport to specify minimal legal standards or to represent the policy of the American Chemical Society. No warranty, guarantee, or representation is made by the American Chemical Society as to the accuracy or specificity of the information contained herein, and the American Chemical Society assumes no responsibility in connection therewith. The added safety information is intended to provide basic guidelines for safe practices. It cannot be assumed that all necessary warnings and precautionary measures are contained in the document and that other additional information and measures may not be required.

# PREFACE

*Chemistry in the Community (ChemCom)* represents a major effort to enhance science literacy through a high school chemistry course that emphasizes chemistry's impact on society. Preliminary work on *ChemCom* began in 1980 with the formation of a steering committee, staff, and groups of writing teams consisting of high school, college, and university teachers, assisted by chemists from industry and government. Developed by the American Chemical Society (ACS) with financial support from the National Science Foundation and several ACS funding sources, the themes of *ChemCom* were developed, emphasizing the application of chemistry to societal issues and utilizing student-centered activities. During the summers of 1982 and 1983 the first drafts of the eight *ChemCom* units were written and carefully evaluated by content specialists and social science consultants, and field-tested in local communities. In 1984, the overall syllabus was developed, and the national field test version was written by a team of writers under the direction of Henry Heikkinen, then at the University of Maryland.

In 1995, the field test involved 61 teachers and 2900 students located at 13 sites around the country. In several locations the teachers met monthly to share ideas and iron out difficulties, proposing alterations in the text and laboratory material. The field test was very successful, as indicated by the increase in requests for textbooks by field test teachers, and after final revision the book was published by Kendall/Hunt Publishing Company in 1988. Since then, *ChemCom* has been successfully implemented by chemistry teachers in thousands of classrooms. Many teachers report that the program offers a motivational, engaging approach to the study of chemistry for a remarkably wide range of students. The second edition was released in 1993.

Briefly, *ChemCom* is designed to help students

- realize the important roles that chemistry will play in their personal and professional lives.
- use chemistry knowledge to think through and make informed decisions about issues involving science and technology.
- develop a lifelong awareness of the potential and limitations of science and technology.

Each of *ChemCom's* eight units focuses on a chemistry-related technological issue currently confronting our society and the world. The issue serves as a basis for introducing the chemistry needed to understand and analyze it.

The major *ChemCom* topics are *Supplying Our Water Needs; Conserving Chemical Resources; Petroleum: To Burn? To Build?; Understanding Food; Nuclear Chemistry in Our World; Living in a Sea of Air: Chemistry and the Atmosphere; Personal Chemistry and Choices; and The Chemical Industry: Promise and Challenge.*

The eight units include the major concepts, vocabulary, thinking skills, and laboratory techniques expected in any introductory chemistry course. However, the program contains a greater number and variety of student-oriented activities than is customary. In addition to numerous laboratory exercises, including many developed especially for the *ChemCom*, each unit contains three levels of decision-making activities and several types of problem-solving exercises.

The first four *ChemCom* units are designed to be studied in sequence. Each new unit builds on the previous ones. For example, considering water as a resource leads to consideration of resources in general, which leads to information about another very special resource, petroleum. The two uses of petroleum, for energy production and building petrochemicals, mimic the two main uses of food for energy production and building components. The first four units also involve systematic development of basic chemical concepts, problem-solving skills, and decision-making abilities. By contrast, the last four units have been designed independently, so that they can be studied in any sequence, based on the interests and needs of your students.

It is important for teachers to realize that it is not necessary to teach all chemical concepts to mastery as soon as they are introduced in the text. Important concepts and skills will reappear in subsequent modules in different contexts. For example, teachers should not attempt to extend the early sections on stoichiometry to a multi-week exercise. This is not necessary, and will interfere with completion of much of the text. Topics with which teachers are familiar through traditional courses may be treated in a completely different fashion in *ChemCom* or even eliminated. This is deliberate!

The third edition of *ChemCom*, while maintaining the overall structure and approach of the first and second editions, provides updated information on many *ChemCom* topics as well as detailed improvements based on suggestions from classroom experience. A new feature, appropriately named Chemistry at Work, describes how individuals in widely diverse careers find uses for chemistry in their daily work. All third-edition changes are intended to make *ChemCom* even more "user-friendly" for both teachers and students.

Dozens of professionals from all segments of the chemistry community have contributed their talents and energies to create *ChemCom*. Their hope is that its impact will be substantial and lasting and that those who study *ChemCom* will find chemistry interesting, captivating, and useful.

# CREDITS

*ChemCom* is the product of teamwork involving individuals from all over the United States over the past decade. The American Chemical Society is pleased to recognize all who contributed to *ChemCom*.

The team responsible for the third edition of *ChemCom* is listed on the copyright page. Individuals who contributed to the initial development of *ChemCom*, to the first edition in 1988, and to the second edition in 1993 are listed below.

**Principal Investigator:**
W. T. Lippincott

**Project Manager:**
Sylvia Ware

**Chief Editor:**
Henry Heikkinen

**Contributing Editor:**
Mary Castellion

**Assistant to Contributing Editor:**
Arnold Diamond

**Editor of Teacher's Guide:**
Thomas O'Brien

**Revision Team:**
Diane Bunce, Gregory Crosby, David Holzman, Thomas O'Brien, Joan Senyk, Thomas Wysocki

**Editorial Advisory Board:**
Glenn Crosby, James DeRose, Dwaine Eubanks, W. T. Lippincott *(ex officio)*, Lucy McCorkle, Jeanne Vaughn, Sylvia Ware *(ex officio)*

**Writing Team:**
Rosa Balaco, James Banks, Joan Beardsley, William Bleam, Kenneth Brody, Ronald Brown, Diane Bunce, Becky Chambers, Alan DeGennaro, Patricia Eckfeldt, Dwaine Eubanks (dir.), Henry Heikkinen (dir.), Bruce Jarvis (dir.), Dan Kallus, Jerry Kent, Grace McGuffie, David Newton (dir.), Thomas O'Brien, Andrew Pogan, David Robson, Amado Sandoval, Joseph Schmuckler (dir.), Richard Shelly, Patricia Smith, Tamar Susskind, Joseph Tarello, Thomas Warren, Robert Wistort, Thomas Wysocki

**Steering Committee:**
Alan Cairncross, William Cook, Derek Davenport, James DeRose, Anna Harrison (ch.), W. T. Lippincott *(ex officio)*, Lucy McCorkle, Donald McCurdy, William Mooney, Moses Passer, Martha Sager, Glenn Seaborg, John Truxall, Jeanne Vaughn

**Consultants:**
Alan Cairncross, Michael Doyle, Donald Fenton, Conrad Fernelius, Victor Fratalli, Peter Girardot, Glen Gordon, Dudley Herron, John Hill, John Holman, Chester Holmlund, Kenneth Kolb, E. N. Kresge, David Lavallee, Charles Lewis, Wayne Marchant, Richard Millis, Joseph Moore, Kenneth Mossman, Herschel Porter, Glenn Seaborg, Victor Viola, William West, John Whitaker

**Synthesis Committee:**
Diane Bunce, Dwaine Eubanks, Anna Harrison, Henry Heikkinen, John Hill, Stanley Kirschner, W. T. Lippincott *(ex officio)*, Lucy McCorkle, Thomas O'Brien, Ronald Perkins, Sylvia Ware *(ex officio)*, Thomas Wysocki

**Evaluation Team:**
Ronald Anderson, Matthew Bruce, Frank Sutman (dir.)

**Field Test Coordinator:**
Sylvia Ware

**Field Test Workshops:**
Dwaine Eubanks

**Field Test Directors:**
Keith Berry, Fitzgerald Bramwell, Mamie Moy, William Nevill, Michael Pavelich, Lucy Pryde, Conrad Stanitski

**Pilot Test Teachers:**
Howard Baldwin, Donald Belanger, Navarro Bharat, Ellen Byrne, Eugene Cashour, Karen Cotter, Joseph Deangelis, Virginia Denney, Diane Doepken, Donald Fritz, Andrew Gettes, Mary Gromko, Robert Haigler, Anna Helms, Allen Hummel, Charlotte Hutton, Elaine Kilbourne, Joseph Linker, Larry Lipton, Grace McGuffie, Nancy Miller, Gloria Mumford, Beverly Nelson, Kathy Nirei, Elliott Nires, Polly Parent, Mary Parker, Dicie Petree, Ellen Pitts, Ruth Rand, Kathy Ravano, Steven Rischling, Charles Ross, Jr., David Roudebush, Joseph Rozaik, Susan Rutherland, George Smeller, Cheryl Snyder, Jade Snyder, Samuel Taylor, Ronald Tempest, Thomas Van Egeren, Gabrielle Vereecke, Howard White, Thomas Wysocki, Joseph Zisk

**Field Test Teachers:**
Vincent Bono, Allison Booth, Naomi Brodsky, Mary D. Brower, Lydia Brown, George Bulovsky, Kay Burrough, Gene Cashour, Frank Cox, Bobbie Craven, Pat Criswell, Jim Davis, Nancy Dickman, Dave W. Gammon, Frank Gibson, Grace Giglio, Theodis Gorre, Margaret Guess, Yvette Hayes, Lu Hensen, Kenn Heydrick, Don Holderread, Gary Hurst, Michael Ironsmith, Lucy Jache, Larry Jerdal, Ed Johnson, Grant Johnson, Robert Kennedy, Anne Kenney, Joyce Knox, Leanne Kogler, Dave Kolquist, Sherman Kopelson, Jon Malmin, Douglas Mandt, Jay Maness, Patricia Martin, Mary Monroe, Mike Morris, Phyllis Murray, Larry Nelson, Silas Nelson, Bill Rademaker, Willie Reed, Jay Rubin, Bill Rudd, David Ruscus, Richard Scheele, Paul Shank, Dawn Smith, John Southworth, Mitzi Swift, Steve Ufer, Bob Van Zant, Daniel Vandercar, Bob Volzer, Terri Wahlberg, Tammy Weatherly, Lee Weaver, Joyce Willis, Belinda Wolfe

**Field Test Schools:**
California: Chula Vista High, Chula Vista; Gompers Secondary School, San Diego; Montgomery High, San Diego; Point Loma High, San Diego; Serra Junior-Senior High, San Diego; Southwest High, San Diego. Colorado: Bear Creek Senior High, Lakewood; Evergreen Senior High, Evergreen; Green Mountain Senior High, Lakewood; Golden Senior High, Golden; Lakewood Senior High, Lakewood; Wheat Ridge Senior High, Wheat Ridge. Hawaii: University of Hawaii Laboratory School, Honolulu. Illinois: Project Individual Education High, Oak Lawn. Iowa: Linn-Mar High, Marion. Louisiana: Booker T. Washington High, Shreveport; Byrd High, Shreveport; Caddo Magnet High, Shreveport; Captain Shreve High, Shreveport; Fair Park High, Shreveport; Green Oaks High, Shreveport; Huntington High, Shreveport; North Caddo High, Vivian; Northwood High, Shreveport. Maryland: Charles Smith Jewish Day School, Rockville; Owings Mills Junior-Senior High, Owings Mills; Parkville High, Baltimore; Sparrows Point Middle-Senior High, Baltimore; Woodlawn High, Baltimore. New Jersey: School No. 10, Patterson. New York: New Dorp High, Staten Island. Texas: Clements High, Sugar Land; Cy-Fair High, Houston. Virginia: Armstrong High, Richmond; Freeman High, Richmond; Henrico High, Richmond; Highland Springs High, Highland Springs; Marymount School, Richmond; Midlothian High, Midlothian; St. Gertrude's High, Richmond; Thomas Dale High, Chester; Thomas Jefferson High, Richmond; Tucker High, Richmond; Varina High, Richmond. Wisconsin: James Madison High, Madison; Thomas More High, Milwaukee. Washington: Bethel High, Spanaway; Chief Sealth High, Seattle; Clover Park High, Tacoma; Foss Senior High, Tacoma; Hazen High, Renton; Lakes High, Tacoma; Peninsula High, Gig Harbor; Rogers High, Puyallup; Sumner Senior High, Sumner; Washington High, Tacoma; Wilson High, Tacoma

**Safety Consultant:**
Stanley Pine

**Social Science Consultants:**
Ross Eshelman, Judith Gillespie

**Art:**
Rabina Fisher, Pat Hoetmer, Alan Kahan (dir.), Kelley Richard, Sharon Wolfgang

**Copy Editor:**
Martha Polkey

**Administrative Assistant:**
Carolyn Avery

**Student Aides:**
Paul Drago, Stephanie French, Patricia Teleska

ACS also offers thanks to the National Science Foundation for its support of the initial development of *ChemCom*, and to NSF project officers Mary Ann Ryan and John Thorpe for their comments, suggestions, and unfailing support.

## Second Edition Revision Team

**Project Manager:**
Keith Michael Shea

**Chief Editor:**
Henry Heikkinen

**Assistant to Chief Editor:**
Wilbur Bergquist

**Editor of Teacher's Guide:**
Jon Malmin

**Second Edition Editorial Advisory Board:**
Diane Bunce, Henry Heikkinen *(ex officio)*, S. Allen Heininger, Donald Jones (chair), Jon Malmin, Paul Mazzocchi, Bradley Moore, Carolyn Morse, Keith Michael Shea *(ex officio)*, Sylvia Ware *(ex officio)*

**Teacher Reviewers of First Edition:**
Vincent Bono, New Dorp High School, New York; Charles Butterfield, Brattle Union High School, Vermont; Regis Goode, Spring Valley High School, South Carolina; George Gross, Union High School, New Jersey; C. Leonard Himes, Edgewater High School, Florida; Gary Hurst, Standley Lake High School, Colorado; Jon Malmin, Peninsula High School, Washington; Maureen Murphy, Essex Junction Educational Center, Vermont; Keith Michael Shea, Hinsdale Central High School, Illinois; Betsy Ross Uhing, Grand Island Senior High School, Nebraska; Jane Voth-Palisi, Concord High School, New Hampshire; Terri Wahlberg, Golden High School, Colorado.

**Safety Consultant:**
Stanley Pine

**Editorial:**
The Stone Cottage

**Design:**
Bonnie Baumann

**Art:**
Additional art for this edition by Seda Sookias Maurer

# SAFETY IN THE LABORATORY

In *ChemCom* you will frequently perform laboratory activities. While no human activity is completely risk free, if you use common sense as well as chemical sense, and follow the rules of laboratory safety, you should encounter no problems. Chemical sense is an extension of common sense. Sensible laboratory conduct won't happen by memorizing a list of rules, any more than a perfect score on a written driver's test ensures an excellent driving record. The true "driver's test" of chemical sense is your actual conduct in the laboratory.

The following safety pointers apply to all laboratory activities. For your personal safety and that of your classmates, make following these guidelines second nature in the laboratory. Your teacher will point out any special safety guidelines that apply to each activity.

If you understand the reasons behind them, these safety rules will be easy to remember and to follow. So, for each listed safety guideline,

- Identify a similar rule or precaution that applies in everyday life—for example, in cooking, repairing or driving a car, or playing a sport;
- Briefly describe possible harmful results if the rule is not followed.

## Rules of Laboratory Conduct

1. Perform laboratory work only when your teacher is present. Unauthorized or unsupervised laboratory experimenting is not allowed.
2. Your concern for safety should begin even before the first activity. Always read and think about each laboratory assignment before starting the laboratory activity.
3. Know the location and use of all safety equipment in your laboratory. These should include the safety shower, eye wash, first-aid kit, fire extinguisher, blanket, exits, and evacuation routes.
4. Wear a laboratory coat or apron and impact/splash proof goggles for all laboratory work. Wear solid shoes (rather than sandals or open-toed shoes) and tie back loose hair. Shorts or short skirts must not be worn.
5. Clear your benchtop of all unnecessary material such as books and clothing before starting your work.
6. Check chemical labels twice to make sure you have the correct substance and the correct concentration of a solution. Some chemical formulas and names may differ by only a letter or a number.
7. You may be asked to transfer some laboratory chemicals from a common bottle or jar to your own test tube or beaker. Do not return any excess material to its original container unless authorized by your teacher.
8. Avoid unnecessary movement and talk in the laboratory.
9. Never taste laboratory materials. Do not bring gum, food, or drinks into the laboratory. Do not put fingers, pens or pencils in your mouth while in the laboratory.
10. If you are instructed to smell something, do so by fanning some of the vapor toward your nose. Do not place your nose near the opening of the container. Your teacher will show the correct technique.
11. Never look directly down into a test tube; view the contents from the side. Never point the open end of a test tube toward yourself or your neighbor. Never heat a test tube *directly* in a Bunsen burner flame.
12. Any laboratory accident, however small, should be reported immediately to your teacher.
13. In case of a chemical spill on your skin or clothing, rinse the affected area with plenty of water. If the eyes are affected, water-washing must begin immediately and continue for at least 10 to 15 minutes. Professional assistance must be obtained.
14. Minor skin burns should be placed under cold, running water.
15. When discarding or disposing of used materials, carefully follow the instructions provided.
16. Return equipment, chemicals, aprons, and protective goggles to their designated locations.
17. Before leaving the laboratory, make sure that gas lines and water faucets are shut off.
18. Wash your hands before leaving the laboratory.
19. If in doubt, ask!

# COMBINED EXPENDABLE ITEMS SUPPLY LIST

Quantities listed are for a class of 24 students working in pairs and completing all recommended *ChemCom Laboratory Activities*. For information concerning the actual quantities needed for specific activities, see the individual supply lists found at the beginning of each unit of the Teacher's Guide.

**Code:** W: → Supplying Our Water Needs
R: → Conserving Chemical Resources
P: → Petroleum: To Burn? To Build?
F: → Understanding Food
N: → Nuclear Chemistry in Our World
A: → Living in a Sea of Air: Chemistry and the Atmosphere
C: → Personal Chemistry and Choices
I: → The Chemical Industry: Promise and Challenge

| Item | Unit(s) | Quantity |
|---|---|---|
| Acetate sheets | C | 24 |
| Acetic acid, concentrated | W, P, F | 3 mL |
| Alcohol, 2-propanol (isopropyl) | R, N | 200 mL |
| Ammonia, concentrated (demo) | A | 100 mL |
| Ammonium chloride, crystals | W | 25 g |
| Ammonium hydroxide, household | A | 15 mL |
| Ammonium molybdate reagent | I | 325 mL |
| Ammonium nitrate | I | 10 g |
| Amylase enzyme powder | C | 3 g |
| Apples | A, C | 3 |
| Ascorbic acid, vitamin C, crystals | F,I | 21 g |
| Asphalt, pieces | P | 50 g |
| Balloons, small (demo) | F | 10 |
| Balloons, large (demo) | A | 10 |
| Barium chloride dihydrate | W | 2 g |
| Benedict's qualitative solution | C | 300 mL |
| Beverage samples | F | 400 mL |
| Boiling chips | P, C, F | 200 g |
| Bottle, empty soft drink (demo) | A | 1 |
| Bottle, plastic, with lid (demo) | A | 1 |
| Boxes, cigar or shoe | N | 12 |
| Boxes, cardboard with lids | N | 12 |
| Bromthymol blue indicator | A | 25 mL |
| Calcium chloride | W | 3 g |
| Calgon | W | 1 box |
| Can, aluminum, soft drink | P, A | 13 |
| Candies, M & M's | F | 1 pkg |
| Candies, Skittles | F | 1 pkg |
| Candles, birthday (demo) | A | 24 |
| Candles, paraffin | P, A | 35 |
| Capillary melting point tubes | A | 24 |
| Carbon, electrodes | R, I | 24 |
| Carbon, pencil lead | R | 5 g |
| Carbon dioxide gas, cylinder (optional) | A | 1 |
| Cardboard or wax paper pieces | N,A | 17 |
| Cation-exchange resin | W | 100 g |

| Item | Unit(s) | Quantity |
|---|---|---|
| Charcoal, decolorizing | W | 200 g |
| Chromatography paper, 1-inch wide | F | 6 feet |
| Clay or cornstarch | W | 200 g |
| Cloth, pieces | N | 12 |
| Coffee grounds, used | W | 1/2 cup |
| Copper foil approximately 0.1 mm thick | R, I | 150 $cm^2$ |
| Copper wire, 24 gauge | R, I | 200 cm |
| Copper wire, 18 gauge | A | 100 cm |
| Copper(II) chloride dihydrate | R | 8 g |
| Copper(II) nitrate trihydrate | R, I | 5 g |
| Copper(II) sulfate pentahydrate | W/I | 35 g |
| Cotton balls | W, F, A | 100 |
| Ethanol, denatured | W, P, F | 230 mL |
| Filter paper, Whatman No. 1, | W, F | 70 sheets |
| Food samples | F | 5 kinds |
| Garlic powder | W | 1 tsp |
| Gloves, disposable | N, A | 36 pairs |
| Graph paper | N | 50 sheets |
| Gravel, fine grind | W | 2 kg |
| Hair strands | C | 500 |
| Helium gas, cylinder (demo) | A | 1 |
| Hexane | W, P | 900 mL |
| Hydrion paper AB, full scale | A, C | 3 packages |
| Hydrochloric acid, concentrated | R, F, A, C, I | 100 mL |
| Hydrogen gas, cylinder (demo) | A | 1 |
| Hydrogen peroxide, 3% solution | A, C | 2 L |
| Iodine crystals | A | 30 g |
| Iron, filings or thin wire | R | 5 g |
| Iron strips | I | 150 $cm^2$ |
| Iron(II) sulfate heptahydrate | W, I | 3 g |
| Iron(III) chloride hexahydrate | A, W, F | 15 g |
| Iron(III) nitrate nonahydrate | W, I | 12 g |
| Kerosene | P | 450 mL |
| Limewater | A | 25 mL |
| Litmus paper, red and blue | A,I | 3 packages each |
| Liver (raw) | C | 2 pieces |
| Magnesium, ribbon | R, F, A, I | 100 cm |
| Magnesium nitrate hexahydrate | R, I | 6 g |
| Marble chips, calcium carbonate | A | 24 pieces |
| Marshmallows, large (demo) | A | 2 |
| Methanol | P | 24 mL |
| Milk, whole | W | 50 mL |
| Naphthalene flakes | W | 25 g |
| Nichrome or platinum wire | I | 120 cm |
| Nitric acid, concentrated | R, A | 140 mL |
| Nitrogen gas, cylinder (demo) | A | 1 |

| Item | Unit(s) | Quantity |
|---|---|---|
| Oil, automotive | P | 450 mL |
| Oil, corn | A | 300 mL |
| Oil, household lubricating | P | 450 mL |
| Oil, mineral (light) | P | 450 mL |
| Oil, vegetable | W, A | 250 mL |
| Paper bags, small brown | N | 24 |
| Paraffin wax | P | 50 g |
| Paramecium (or mixed) culture | A | 1 |
| Perfume (demo) | A | 1 bottle |
| Permanent wave neutralizer | C | 90 mL |
| Permanent wave solution | C | 180 mL |
| Phenolphthalein (demo) | A | 5 mL |
| Plastic bags, self-sealing, pint | A | 12 |
| Potassium iodate | F | 1 g |
| Potassium iodide | F | 1 g |
| Potassium monohydrogen phosphate | I | 2 g |
| Potassium nitrate | I | 1 g |
| Potassium thiocyanate | W, F | 5 g |
| Potato, raw | C | 1 |
| Rubber bands, orthodontist's type | C | 225 |
| Rubber bands, 10-cm diameter | P | 20 |
| Rubber bands, small | A | 324 |
| Salt (table), sodium chloride | W | 50 g |
| Salt, low-sodium (Lite™) | N | 1 box |
| Sand | W | 1 kg |
| Silicon, lump | R | 5 g |
| Silver nitrate | W, R | 4 g |
| Soap, Ivory (liquid) | W | 1 bottle |
| Sodium carbonate monohydrate | W | 2 g |
| Sodium hydrogen carbonate | C | 1 g |
| Sodium hydroxide | C, I | 7 g |
| Sodium nitrate | I | 5 g |
| Sodium oxalate monohydrate | W | 1 g |
| Sodium phosphate dodecahydrate | I | 20 g |
| Sodium sulfate decahydrate | I | 2 g |
| Sodium sulfite | A, F | 12 g |
| Splints, wooden | A, C | 72 |
| Starch, water soluble | F, C | 3 g |
| Steel wool pads | R, A, I | 150 g |
| Straws, drinking | C | 24 |
| Sulfur hexafluoride gas, cylinder (demo) | A | 1 |
| Sulfur, lump roll | R | 5 g |
| Sulfuric acid, concentrated | P, F, I | 100 mL |
| Sun-sensitive paper | N, C | 14 pieces |
| Suntan oil (range of SPFs) | C | 1 bottle of each |
| Tin, mossy or foil | R | 5 g |
| Trisodium phosphate | C | 50 g |
| Universal indicator solution | A, C | 375 mL |
| Zinc, finely granulated | R | 55 g |
| Zinc, strips (0.25 mm thick) | I | 150 $cm^2$ |
| Zinc nitrate hexahydrate | R, I | 115 g |

# COMBINED NONEXPENDABLE ITEMS SUPPLY LIST

## Standard Glassware and Hardware for a Class of 24 (Working in Pairs)

| Item | Quantity |
|---|---|
| Balances | |
| Beakers | |
| 50 mL | 12 |
| 100 mL | 24 |
| 150 mL | 24 |
| 250 mL | 24 |
| 400 mL | 36 |
| Beral pipets | 120 |
| Bottles, gas collecting | 12 |
| Bottles, wash | 24 |
| Burets | 12 |
| Burners | 24 |
| Clamps | 48 |
| Condensers | 12 |
| Corks, assorted | |
| Crucibles | 24 |
| Culture tubes, 36 mL | 60 |
| Dropping bottles | 18 |
| Flasks, Erlenmeyer, 250 mL | 12 |
| Flasks, Florence, 125 mL | 12 |
| Forceps | 12 |
| Funnels, conical | 12 |
| Graduated cylinders | |
| 10 mL | 24 |
| 25 mL | 12 |
| 100 mL | 12 |
| Marking pencils | 2 |
| Pasteur pipets | 48 |
| Ring stands | 24 |
| Spatulas | 12 |
| Stirring rods, 20 cm | 24 |
| Test tubes | |
| 13 × 100 | 24 |
| 15 × 125 | 12 |
| 18 × 150 | 48 |
| Test tube brushes | 12 |
| Test tube holders | 12 |
| Test tube racks | 24 |
| Thermometers, spirit-filled, -20 °C to 110 °C | 24 |
| Tongs | 12 |
| Troughs, pneumatic | 12 |
| Tubing, glass | 6 m |
| Tubing, rubber | 720 cm |
| Vials | 84 |
| Wellplates, 24-well | 48 |
| Wire screens | 12 |

## Other Nonexpendable Items

| Item | Quantity | Unit used |
|---|---|---|
| Alligator clips with leads | 12 | I |
| Assorted small objects for black box | 12 of each | N |
| Batteries, 9-V | 12 | R, I |
| Battery connectors | 12 | R, I |
| Beads, 7-mm metal | 72 | P |
| Cloud chamber kits | 12 | N |
| Coins, assorted | 12 | N |
| Dominoes | 12 sets | N |
| Glass plates, 5 cm × 5 cm × 0.1 cm | 144 | N |
| Hammers | several | R |
| Lead plates, 5 cm × 5 cm × 0.1 cm | 48 | N |
| Light sources, intense | 12 | N |
| Pennies (pre-1982, post-1982) | 1400 | N |
| Radiation sources (alpha, beta, gamma) | 12 sets | N |
| Ratemeter | 12 | N |
| Sand paper | 12 pieces | R |
| Timers or stopwatches | 12 | P |

# CHEMICALS AND SCIENTIFIC EQUIPMENT SUPPLIERS

The following representative firms provide general supplies and equipment for all areas of science teaching as well as specific items for chemistry teaching.

Aldrich Chemical Company, P.O. Box 355, Milwaukee, WI 53201. 414-273-3850, 800-558-9160.

Beckley-Cardy Group, 100 Paragon Parkway, Mansfield, OH 44903. 419-589-1900, 800-225-FREY.

Carolina Biological Supply Company, 2700 York Road., Burlington, NC 27215. 910-584-0381, 800-334-5551. Request physical sciences and biological sciences catalogs.

Central Scientific Company, 11222 Melrose Avenue, Franklin Park, IL 60131-1364. 708-451-0150, 800-262-3626.

Connecticut Valley Biological Supply Company, Inc., 82 Valley Road., P.O. Box 326, Southampton, MA 01073. 413-527-4030, 800-628-7748.

Edmund Scientific, 101 East Gloucester Pike, Barrington, NJ 08007. 609-573-6250.

Fisher Scientific Company, Educational Materials Division, 485 South Frontage Road, Burr Ridge, IL 60521. 630-655-4410.

Flinn Scientific Inc., P.O. Box 219, Batavia, IL 60510. 630-879-6900, 800-452-1261. The *Flinn Chemical Catalog* also provides information on chemical safety, storage, and disposal.

Hach Company, P.O. Box 608, Loveland, CO 80539-0608. 800-227-4224. Test kits for environmental studies.

Lab Safety Supply, P.O. Box 1368, Janesville, WI 53547-1368. 608-754-2345. Specializes in safety equipment and supplies.

LaMotte Company, P.O. Box 329, Chestertown, MD 21620-0329. 410-778-3100, 800-344-3100. Test kits for environmental studies.

Micro Mole Scientific, 1312 North 15th Street, Pasco, WA 99301. 509-545-4904.

Nalgene Labware Division, P.O. Box 20365, Rochester, NY 14602-0365. 716-586-8800. Specializes in transparent and translucent plastic laboratory equipment.

NASCO, 901 Janesville Avenue, Ft. Atkinson, WI 53538. 414-563-2446, 800-558-9595.

Ohaus Corporation, 29 Hanover Road., Florham Park, NJ 07932. 201-377-9000, 800-672-7722.

Sargent-Welch, 911 Commerce Court, P.O. Box 5229, Buffalo Grove, IL 60089-5229. 847-459-6625, 800-727-4368.

Science Kit & Boreal Laboratories, Inc., 777 E. Park Drive, Tonawanda, NY 14150-6782. 716-874-6020, 800-828-7777.

Wards Natural Science Establishment, Inc. 5100 West Henrietta Road., P.O. Box 92912, Rochester, NY 14692-9012. 716-359-2502, 800-962-2660.

# CHEMICAL SAFETY, STORAGE, AND DISPOSAL REFERENCES

While no laboratory exercise is without some potential risk, most accidents can be prevented by basing laboratory practices and procedures on established chemical safety guidelines. The following references offer valuable background information. Since risk assessments are subject to change with new studies, it is important to keep up to date by reading professional journals (the *Journal of Chemical Education, The Science Teacher,* etc.) and learning about any guidelines and regulations issued in your local school district or state.


*Aldrich Catalog Handbook of Fine Chemicals,* Aldrich Chemical Co., P.O. Box 355, Milwaukee, WI 53201. 414-273-3850; 800-558-9160.

*Chemical Catalog/Reference Manual,* Flinn Scientific, Inc., P.O. Box 219, 770 Raddant Rd., Batavia, IL 60510. 630-879-6900; 800-452-1261.

*Fire Protection Guide on Hazardous Materials,* National Fire Protection Association, 11 Tracy Drive, Avon, MA 02322. 617-770-4543. National safety code as applied to laboratory fire prevention and protection.

*The Merck Index,* 12th edition, Merck & Co., Inc., Whitehouse Station, NJ 08889. Authoritative, brief descriptions of properties (including health hazards) of many substances.

*NIOSH Pocket Guide to Chemical Hazards,* U.S. Department of Health and Human Services, Public Health Service, Centers for Disease Control, National Institute for Occupational Safety and Health, Washington, DC 20201 A pocket summary of hazardous properties of substances regulated by NIOSH.

*School Science Laboratories: A Guide to Some Hazardous Substances,* Council of State Science Supervisors and U.S. Consumer Products Safety Commission, Washington, D.C. 20207.

*Committee on Chemical Safety (ACS), Safety in Academic Chemistry Laboratories,* 6th edition, American Chemical Society, Washington, DC 20036.

*School Science Safety Lecture Series/Secondary,* Flinn Scientific, Inc., Box 219, 770 Raddant Rd., Batavia, IL 60510. 630-879-6900; 800-452-1261. Overhead transparency masters, suggested student activities, student assessments.

Furr, A. Keith. *CRC Handbook of Laboratory Safety,* 4th edition, CRC Press, Boca Raton, FL 33431.

Wood, Claire. *Safety in School Science Labs,* Flinn Scientific, Inc., P.O. Box 219, 770 Raddant Rd., Batavia, IL 60510. 630-879-6900; 800-452-1261.

Committee on Chemical Safety (ACS). *A Model Chemical Hygiene Plan for High Schools,* American Chemical Society, Washington, DC 20036. Product HS70A. 800-209-0423.

*Prudent Practices in the Laboratory: Handling and Disposal of Chemicals,* National Academy Press, Washington, DC, 1995.

*Starting With Safety: An Introduction for the Academic Chemistry Laboratory,* American Chemical Society, Washington, DC 20036. 35-minute VHS videotape and teacher's guide. Product V5100. 800-209-0423.

*Sourcebook,* ChemSource, Inc., College of New Rochelle, New Rochelle, NY 10805. *Sourcebook, Sourceview,* and *Guidebook to Preservice Use of ChemSource* are available from the American Chemical Society, Washington, DC 20036. 800-209-0423.

# COMPOSITION OF CONCENTRATED ACIDS AND BASES

| | Molar mass of reagent, g/mol | Approximate strength of concentrated reagent[a] | Molarity of concentrated reagent (M) | Milliliters of concentrated reagent needed to prepare one liter of 1 M solution[b] |
|---|---|---|---|---|
| Acetic acid, $HC_2H_3O_2$ | 60.5 | 99.8 | 17.4 | 57.5 |
| Hydrochloric acid, HCl | 36.46 | 37.2 | 12.1 | 82.5 |
| Nitric acid, $HNO_3$ | 63.01 | 70.4 | 15.9 | 63.0 |
| Sulfuric acid, $H_2SO_4$ | 98.07 | 96.0 | 18.0 | 55.5 |
| Ammonium hydroxide $NH_3$ (aq) | 17.03 | 28.0 | 14.5 | 69.0 |

[a]Representative value, mass percent
[b]Rounded to nearest 0.5 mL

# INTRODUCTION

The goal of the Teacher's Guide is to provide you with (1) the history, philosophy, and assumptions upon which *ChemCom* is based; (2) day-to-day teaching support; and (3) a variety of optional activities, supplemental references, and resource materials for you and your students to explore.

*ChemCom* is a student-centered, activity-based, issues-oriented chemistry course that encourages small-group learning. Developed by the American Chemical Society, it includes many traditional chemical concepts and laboratory skills, as well as more biochemistry, industrial, organic, and nuclear chemistry than is normally encountered in a standard high school chemistry text. Thus any prior teaching experience with chemistry will prove invaluable.

In terms of level, *ChemCom* meets or exceeds the chemistry presented in the National Physical Science Content Standards (see Section F). In addition, it contains chemistry found in the other Content Standards.

Perhaps the most unique feature of *ChemCom* is that chemistry is taught on a "need-to-know" basis, with societal and technological issues/problems determining the depth and breadth of chemical concepts taught. If your previous background in "real world" chemistry and chemical technology is limited, we invite you to explore the Optional Background Information and Supplemental Readings provided in this guide. Additionally, you'll find that articles about *ChemCom* topics in local newspapers, magazines, and TV news will help you to assist students in seeing chemistry in the context of their own communities.

Most of the complex and perplexing issues and problems facing our nation involve more than scientific concepts. They also involve individual and community values and group decision-making processes. Accordingly, *ChemCom* aims to prepare students for informed, effective citizenship through stimulating and engaging decision-making activities that call on the higher cognitive abilities. Thus, *ChemCom* places much of the responsibility for learning on students themselves as they explore how chemical concepts apply to their everyday lives. The hands-on laboratory activities, most of which incorporate small-scale techniques, constitute about half the course.

The decision-making focus, central to the *ChemCom* approach, requires you to engage in some teaching approaches not typical in traditional chemistry courses. The following material outlines the premises underlying this focus and highlights similarities and differences among various types of decision-making activities, especially the support you will be asked to give, as manager of the classroom learning process. It is our hope that this guide will help make teaching *ChemCom* as enjoyable for you as it will be for your students.

## A: Premises Underlying *ChemCom*'s Decision-Making Focus

1. All students will eventually be required, as citizens, to make decisions on a variety of issues that involve some understanding of chemistry.
2. Academic problem solving as presented in traditional school settings does not model the real-life decision making students will encounter later as voting citizens and professionals.
3. Problem solving and decision making use higher level cognitive skills, which can be improved with appropriate experience and practice.
4. Science curricula should be designed to include opportunities for students to practice real-life decision-making strategies.
5. A science curriculum featuring an issues focus (such as *ChemCom*) can help to develop students' decision-making skills.
6. It is as important for students to be able to apply chemical knowledge as to acquire it.
7. Modern society depends on group cooperation as much as on individual competition. Schools should provide opportunities for students to work together on cooperative ventures.
8. Academic preparation for life is at least as important as preparation for a subsequent course of study.

## B: Characteristics of Problem Solving

| In School | In Real Life |
|---|---|
| One result expected | Multiple alternatives |
| Fully defined | Imperfectly defined |
| Discipline focused | Multidisciplinary |
| Right/wrong answers | Burden/benefits |
| Judged immediately | Judged later |
| Algorithmic | Heuristic |
| Driven by knowledge | Constrained by missing knowledge |
| Objectively oriented | Value-laden |
| No solution is not a solution | No decision is itself a decision (!) |
| Dull (at worst) | Agonizing (at worst) |

Source: Heikkinen, H. "Decision Making and *ChemCom*." *Chemunity*, Vol. 8, No. 2 (Summer 1987), p. 7.

## C: Overview of Types of *ChemCom* Decision-making Activities

**YOUR TURN** — ***ChemQuandary*** — ***You Decide*** — ***Putting It All Together***

| | → | |
|---|---|---|
| 10 minutes | Instructional time | 2 days |
| A "right answer" | Open endedness | Alternatives |
| Low | Small-group learning | High |
| Teacher-directed | Student autonomy | Student-run |
| Academic | Problem solving/decision making | Real life |

### YOUR TURN

**Goal:** to give students practice and reinforcement of basic chemical concepts, skills, and calculations in the context of applied, "real world" chemistry problems.

**Time:** typically assigned as homework, followed by brief (5-15 minutes) class discussion.

**Instructional group:** work performed by individual students—"checking" may be done in a variety of formats, but most commonly through whole-class, teacher-led discussions.

**Teacher's role:** ensure students have acquired prerequisite skills before assignment and to provide subsequent in-class feedback and reinforcement.

**Sample activities:** paper and pencil activities on the metric system, nomenclature, computations, balancing equations, graphing and graphical analysis, etc., typically leading to single "right" answers.

**Typical number per unit:** 10.

### CHEMQUANDARY

**Goal:** to motivate and challenge students to think about chemical applications and societal issues, which are often open-ended and may generate additional questions beyond a specific "right" answer.

**Time:** may be assigned as either homework or in-class start-up activities with brief (10-15 minutes) follow-up discussion.

**Instructional group:** individual or small groups of students (3-5) followed by class discussion.

**Teachers role:** encourage students to use scientific concepts to help frame and answer relevant questions; at times it may be important to avoid giving students "the answer," allowing them to wrestle with and discover answers on their own.

**Sample activities:** discrepant-event type information; indirect water use; benefit and burden analyses; chlorination of water; developing a scenario of life without abundant fuel; comparing and contrasting industrial and automotive pollution control.

**Typical number per unit:** 2-3 (plus other options in the Teacher's Guide).

## YOU DECIDE

**Goal:** an inquiry activity where students are presented aspects of a societal/technological problem, asked to collect and/or analyze data for underlying patterns, and challenged to develop and support or refute hypotheses/solutions based on scientific evidence and clearly stated opinions.

**Time:** one-half to one class period (some involve at-home work).

**Instructional group:** individual and/or small group (3-5 students) assignments leading to class discussions.

**Teacher's role:**

1. teach students prerequisite communication and research skills (library, mathematical, etc.) and conceptual knowledge.
2. hold a short pre-activity briefing (equivalent to pre-lab presentation), forming student groups and clarifying objectives and general procedures of the activity (in some cases this may be a week or more in advance of the actual class activity to allow time for student data gathering).
3. arrange the physical environment to support small-group and/or whole-class communication, as appropriate.
4. monitor and encourage students to:

   use their individual talents and strengths to contribute to group learning

   separate misconception/myth from fact, fact from opinion, and to use scientific concepts to clarify the issue and their own opinions

   work cooperatively, receiving and respecting the opinions of others

   - deal productively with conflict and controversy by evaluating ideas and issues, not people

   -give students the autonomy they need to discover their own solutions, not dominating or obtrusively managing the discussion by trying to "solve" all group-dynamic problems (you should not lead the activity any more or less than you would lead a conventional laboratory activity).
5. evaluate individual and/or group work based on written reports, resource notes, and actual interactions/discussions.
6. provide any needed review and closure.

**Sample activities:** collect and analyze data on personal use of resources; interpret implications of environmental data; survey public understanding of nuclear issues; prepare personal diet and health analyses; interview the elderly on the "good old days."

**Typical number per unit:** 5.

## PUTTING IT ALL TOGETHER

**Goal:** to give students the opportunity to sum up, review, and apply what they have learned in a unit in the context of a hypothetical or real societal/technological, chemically related community issue. Students are expected to develop and defend scientifically sound positions that acknowledge the roles of economics, politics, and personal/social values (this typically involves weighing benefits and burdens and developing workable solutions).

**Time:** two days of class time preceded by individual or group research.

**Instructional group:** individual assignment within small groups (3-5 students) which then interact in a whole-class format (simulation, debate, etc.).

**Teacher's role:** similar to those listed for *You Decide*, but also:
1. provide timely and clear pre-assignments with periodic progress checks as needed.
2. conduct a more extensive debriefing of the activity, ensuring that any scientific misconceptions are corrected and providing feedback to students that encourages improvements in their small-group learning skills.

**Sample activities:** simulation of a town meeting on implications of a fish kill; evaluating the success of air pollution controls; forecasting alternative depletion curves for metal reserves in light of various assumptions; evaluating statements on nuclear issues in terms of validity and underlying values.
**Typical number per unit:** 1 (end of unit activity).

## D: *ChemCom* Problem-Solving Activities

## Laboratory Activities

**Goal:** to motivate student interest in chemistry and its applications; to introduce and develop key chemical concepts; and to provide practice with laboratory skills.
**Time:** 50-minute class period (a few require two days).
**Instructional group:** student pairs
**Teacher's role:** similar to that for conventional chemistry laboratory activities:
1. conduct a pre-lab discussion to clarify objectives, demonstrate new techniques, and emphasize safety procedures.
2. monitor student laboratory work at all times (never leave the students unattended) to:
   -ensure safety regulations are followed;
   -ensure that students are actively engaged in the laboratory procedures;
   -provide needed assistance.
3. direct a post-lab discussion, pulling together and analyzing class data, clearing up any misconceptions, and connecting previous and subsequent learning activities.
4. evaluate student work.
5. encourage students to identify additional everyday applications of the concepts they discover.

**Note:** if you elect to substitute or add some of your own laboratory activities, try to maintain the "real-life" focus by strategies such as using household and grocery store chemicals and supplies.
**Sample activities:** purifying "foul" water; making a petrochemical (ester); simulating half-life; comparing sunscreens; qualitative and quantitative analysis of fertilizers, foods, etc.; electroplating.
**Typical number per unit:** 5.

## SUMMARY QUESTIONS

**Goal:** to help students review important concepts and skills developed in each major part of a unit and to introduce additional practical applications of chemistry.
**Time:** typically assigned as homework with an in-class review of usually no more than one-half a class period.
**Instructional group:** individual and/or group.
**Teacher's role:** provide appropriate class settings (teacher-led recitations, review games, computer databases, posted answer keys, etc.) that enable students to check their responses and receive needed remediation (it may be desirable to use some items for grading purposes).
**Sample activities:** specific items may involve naming compounds, listing key points, performing calculations, balancing equations, answering essay questions, etc. In many cases students are asked to apply their knowledge in new (but still familiar) contexts.
**Typical number per unit:** a typical *ChemCom* unit contains five sections; each section is supported by 3-12 summary questions.

## EXTENDING YOUR KNOWLEDGE

**Goal:** to stimulate and challenge able students and/or arouse interest among all students; to apply chemistry to new situations; to encourage students to seek additional information; may be open ended or not and may involve review, application, and extension of concepts previously learned. These are optional activities to be used at teacher discretion.

**Time:** highly variable (out-of-class time).

**Instructional group:** individual or group.

**Teacher's role:** encourage students to pursue selected issues in greater depth; direct students to appropriate resources; and (if desired) to allow class time for student reports.

**Sample activities:** optional library research on oil spills, origins of chemical symbols, DDT, history of the metric system, alternative automotive fuels and engines, cryogenics, Boyle's law and scuba diving, etc.; optional at-home "laboratory" activities on chemical indicators, acid rain, etc.

**Typical number per unit:** most units contain 1-3 (the Teacher's Guide contains additional project idea).

## E: CHEMICAL CONCEPTS GRID

| | ***ChemCom* Units** | | | | | | | |
|---|---|---|---|---|---|---|---|---|
| **Concepts** | **Water** | **Resources** | **Petroleum** | **Food** | **Nuclear** | **Air** | **Health** | **Industry** |
| Metric (SI measurement) | I | U | E | U | U | U | U | U |
| Scale and order of magnitude | I | U | U | U | U | U | U | U |
| Physical and chemical properties | I | E | E | U | E | E | E | E |
| Solids, liquids, and gases | I | U | E | | U | E | U | U |
| Solutions and solubility | I | E | U | U | U | U | E | U |
| Elements and compounds | I | E | E | E | E | U | U | U |
| Nomenclature | I | E | E | E | E | U | U | U |
| Formula and equation writing | I | E | E | E | U | U | U | U |
| Atomic structure | I | E | E | | E | | | |
| Chemical bonding | I | U | E | E | | U | E | U |
| Shape of molecules | I | | E | U | | | E | U |
| Ionization | I | U | E | | E | E | E | E |
| Periodicity | | I/E/U | | | | | | |
| Mole concept | | I | E | U | E | E | U | U |
| Stoichiometry | | I | | U | | U | U | E |
| Energy relationships | | I | E | E | E | E | E | E |
| Acids, bases, and pH | I | | | E | E | | E | U |
| Oxidation-reduction | | I | | U | | U | U | E |
| Reaction rate/kinetics | | | | I | E | U | E | U |
| Gas laws | | | | | | I/E/U | | |
| Equilibrium | | | | | | | | I/U |
| Chemical analysis | I | E | E | E | | U | U | U |
| Chemical synthesis | | | I | | | U | | E |
| Biochemistry | | | | I | U | | E | |
| Industrial chemistry | I | E | E | E | E | E | E | E |
| Organic chemistry | | | I | E | | | E | |
| Nuclear chemistry | | | | I/E/U | | | | |

I = Introduce; U = Utilize; E = Elaborate

## F: *ChemCom* AND THE NATIONAL SCIENCE EDUCATION CONTENT STANDARDS[1]

The National Science Education Content Standards cover the categories of Science as Inquiry; Physical Science; Life Science; Earth and Space Science; Science and Technology; Science in Personal and Social Perspectives; and the History and Nature of Science. In each category, the statement of the standard is brief. For example, for grades 9-12, Physical Science Standard B states:

*"As a result of their activities in grades 9-12, all students should develop an understanding of*

- *Structure of atoms*
- *Structure and properties of matter*
- *Chemical reactions*
- *Motions and forces*
- *Conservation of energy and increase in disorder*
- *Interactions of energy and matter"*

This standard clearly addresses the intellectual domain of physics as well as chemistry, so it is not to be expected that the standard be fully addressed by a chemistry or a physics course alone.

The Life Science Standard C, while mainly focusing on the intellectual domain of the biological sciences, also includes important chemistry concepts that may be assumed, but not taught, in Biology 1 classes. Similarly, the Earth and Space Science Standard also includes chemistry, although it is not fully defined by chemistry content. In fact, modern chemistry is addressed throughout the seven content standards listed above, a reflection of the central and useful character of chemical knowlege.

In the Standards document, for clarity of intent, each standard is accompanied by both a section on "Developing Student Understanding" as well as a "Guide to the Content Standard." It is this guide that describes the fundamental concepts and principles that underlie each component of each standard. The relationship of *ChemCom* to these underlying concepts and principles is addressed in the following set of tables. For any component of each of the seven content standards, *ChemCom* may:

- include concepts that go beyond those concepts described in the "Guide,"
- address all of the concepts and principles in the "Guide,"
- address most of the concepts and principles in the "Guide,"
- address some of the concepts and principles, or
- not address the concepts and principles at all.

The following tables also briefly describe the concepts and principles that are included in *ChemCom*, and locate the sections in which they are addressed.

Finally, there is also a standard that addresses concepts and processes that provide a connection between and among different scientific disciplines. This Unifying Concepts and Processes Standard states:

*"As a result of activities in grades K-12, all students should develop understanding and abilities aligned with the following concepts and processes:*

- *systems, order, and organization*
- *evidence, models, and explanation*
- *constancy, change, and measurement*
- *evolution and equilibrium*
- *form and function"*

We encourage *ChemCom* teachers to work with other science teachers in their school to ensure that these unifying concepts are addressed across the curricula. They are not explicitly addressed in the *ChemCom* student text, although the text is rich with examples of these concepts as they apply to chemistry. One way to introduce these concepts might be to organize cross-disciplinary seminars that give students the opportunity to discuss each of these unifying concepts in the context of the particular sciences they are studying. The students themselves should be asked to organize and conduct these seminars. We encourage you to consult the National Science Education Standards for additional information on these unifying concepts and processes.

---

[1]National Research Council, 1996. *National Science Education Standards.* Washington, DC: National Academy Press.

## *ChemCom* AND THE STANDARDS

| Standards (grades 9–12) | Fundamental Concepts and Principles (from "Guide to the Standards") | | | | | |
|---|---|---|---|---|---|---|
| A. Inquiry | 1 | 2 | | | | |
| B. Physical science | 1 | 2 | 3 | 4 | 5 | 6 |
| C. Life science | 1 | 2 | 3 | 4 | 5 | 6 |
| D. Earth and space science | 1 | 2 | 3 | 4 | | |
| E. Science and technology | 1 | 2 | | | | |
| F. Personal and social perspectives | 1 | 2 | 3 | 4 | 5 | 6 |
| G. History and nature | 1 | 2 | 3 | | | |

### Legend

Addresses all fundamental concepts and principles in this section of "Guide"

Covers most fundamental concepts and principles

Addresses some concepts and principles

Section concepts not addressed in *ChemCom*

No additional sections

## A: Inquiry Standard

| COMPONENTS | *ChemCom* CONCEPTS AND PRINCIPLES | ACTIVITIES IN *ChemCom* | *ChemCom* UNIT |
|---|---|---|---|
| 1. Abilities necessary to do scientific inquiry | Addresses most concepts: includes investigations, use of technology and math, using logic and evidence, modifying explanations, recognizing alternatives, communicating and defending scientific arguments | Labs, demonstrations, decision-makers, math problems with graphs, word problems, library research | Throughout |
| 2. Understandings about scientific inquiry | Addresses most concepts: includes use of math, logic, methods of scientists, purposes for research, development of science ideas | Labs, demonstrations, decision-makers, math problems with graphs, word problems, library research | Throughout |

## B: Physical Science Standard

| COMPONENTS | *ChemCom* CONCEPTS AND PRINCIPLES | ACTIVITIES IN *ChemCom* | *ChemCom* UNIT |
|---|---|---|---|
| 1. Structure of atoms | Addresses more than section concepts: includes—atomic particles, fission & fusion, radioactive isotopes, radioactive series, nuclear pile. See also standards A, F, G | Labs, demonstrations, decision-makers, math problems with graphs, word problems, library research | Water, resources, nuclear |
| 2. Structure & properties of matter | Addresses more than section concepts: includes periodic table & properties, ionic and covalent bonding, hydrogen bonding, physical & chemical properties, states of matter, organic structures, gas laws. See also standards A, F | Labs, demonstrations, decision-makers, graphs, word problems, library research | Water, resources, petroleum, food, personal chemistry, industry |
| 3. Chemical reactions | Addresses more than section concepts: includes organic, redox, acid/base reactions; reaction rates; catalysis; consumer, industrial & environmental reactions; pH; moles; stoichiometry; enthalpy, activation energy. See also standards A, F | Labs, demonstrations, decision-makers, math problems with graphs, word problems, library research | Water, resources, petroleum, food, air, personal chemistry, industry |
| 4. Motions & forces | **Concepts not addressed in *ChemCom*** | | |
| 5. Conservation of energy & increase in disorder | Addresses most concepts: includes energy transfer, kinetic & potential energy, heat as random motion | Labs, decision-makers, math problems with graphs, graphs, word problems, library research | Petroleum, nuclear, air, industry |
| 6. Interaction of energy & matter | Addresses some concepts related to radiation | Labs, decision-makers, word problems | Air, personal chemistry |

## C: Life Science Standard

| COMPONENTS | *ChemCom* CONCEPTS AND PRINCIPLES | ACTIVITIES IN *ChemCom* | *ChemCom* UNIT |
|---|---|---|---|
| 1. The cell | Addresses some concepts: chemical reactions in cells, proteins, enzymes, photosynthesis. See also standards A, B, D, F | Labs, decision-makers, math problems, word problems, library research | Food, air, personal chemistry |
| 2. Molecular basis of heredity | **Concepts not addressed in *ChemCom*** | | |
| 3. Biological evolution | **Concepts not addressed in *ChemCom*** | | |
| 4. Interdependence of organisms | Addresses some concepts: carbon & nitrogen cycles. See also standards A, B, D, F | Graphs, decision-makers, word problems, library research | Food, air, industry |
| 5. Matter, energy & organization in living systems | Addresses most concepts: includes food for energy & growth, biomolecules, ATP, cycle of energy & matter through living and nonliving systems. See also standards A, B, D, F | Labs, decision-makers, math problems, word problems, library research | Food, air, personal chemistry, industry |
| 6. Behavior of organisms | **Concepts not addressed in *ChemCom*** | | |

## D: Earth and Space Science Standard

| COMPONENTS | *ChemCom* CONCEPTS AND PRINCIPLES | ACTIVITIES IN *ChemCom* | *ChemCom* UNIT |
|---|---|---|---|
| 1. Energy in the earth system | Addresses some concepts: earth's energy balance, atmosphere & climate, greenhouse effect. See also standards A, B, C, F | Decision-makers, graphs, word problems, library research | Air |
| 2. Geochemical cycles | Addresses most concepts: includes nonrenewable sources, conservation, carbon cycle. See also standards A, B, F | Labs, decision-makers, graphs, word problems, library research | Resources, air |
| 3. Origin & evolution of earth system | Addresses some concepts: radioactive isotopes for dating, half-lives. See standards A, B | Lab, math problems, word problems, decision-making, library research | Nuclear |
| 4. Origin & evolution of universe | Addresses some concepts: nuclear fusion reactions. See standard B | Word problems, library research | Nuclear |

## E: Science and Technology Standard

| COMPONENTS | *ChemCom* CONCEPTS AND PRINCIPLES | ACTIVITIES IN *ChemCom* | *ChemCom* UNIT |
|---|---|---|---|
| 1. Abilities of technological design | Addresses most concepts: includes evaluation and redesign of existing technologies. | Decision-making, word problems, library research | Water, resources, petroleum, nuclear, health, air, industry |
| 2. Understandings about science & technology | Addresses some concepts: chemical, petroleum, and nuclear technologies, nature of science and technology | Labs, word problems, library research | Resources, petroleum, nuclear, industry |

## F: Science in Personal and Social Perspectives Standard

| COMPONENTS | *ChemCom* CONCEPTS AND PRINCIPLES | ACTIVITIES IN *ChemCom* | *ChemCom* UNIT |
|---|---|---|---|
| 1. Personal & community health | Addresses some concepts: personal choice, food, chemistry and health | Labs, decision-makers, word problems, library research | Food, personal chemistry |
| 2. Population growth | Concepts not addressed in *ChemCom* | | |
| 3. Natural resources | Addresses most concepts: includes renewable, and nonrenewable resources, uses & distribution of resources, recycling | Labs, decision-makers, graphs, math problems, word problems, library research | Resources, petroleum, industry |
| 4. Environmental quality | Addresses most concepts: includes air and water quality issues and pollution prevention, nuclear wastes | Labs, decision-makers, math problems with graphs, word problems, library research | Water, resources, air, industry |
| 5. Natural & human-induced hazards | Addresses some concepts: risk analysis | Decision-makers, graphs, word problems, library research | Nuclear, air, personal chemistry |
| 6. Science & technology in local, national, & global challenges | Addresses all concepts: science, technology, economics, politics, & social factors, complexity of decision-making | Labs, demonstrations, decision-makers, math with graphs, word problems, library research | Water, resources, petroleum, food, nuclear, air, personal chemistry, industry |

## G: History and Nature of Science Standard

| COMPONENTS | *ChemCom* CONCEPTS AND PRINCIPLES | ACTIVITIES IN CHEMCOM | CHEMCOM UNIT |
|---|---|---|---|
| 1. Science as a human endeavor | Addresses most concepts: includes careers, interactions of scientists with society, work of scientists | Decision-making, word problems, library research | Throughout |
| 2. Nature of scientific knowledge | Addresses some concepts but material tends to be less explicit than described in Standards | Labs, decision-makers, math & word problems | Throughout |
| 3. Historical perspectives | Addresses some concepts: atomic theory, nuclear chemistry, some molecular biology, some medical technology | Labs, demonstrations, word problems, library research | Resources, petroleum, nuclear, air, industry |

# AUDIOVISUAL/COMPUTER SUPPLEMENTS

The following audiovisual and computer software are possible supplements for *ChemCom* units. Citation of a specific product does not imply endorsement. The listings are meant only to suggest types of materials that may prove useful in the classroom. In using such materials, it is important to remind students that learning is an active process. To guard against passivity and to reinforce active learning, utilize such techniques as overviews that tie an audiovisual to previous learning or to potential discussion/debate questions that may arise from the subject matter. As with any supplement, audiovisuals must be previewed by the teacher and a decision made as to appropriateness of use with regard to content and language as well as to age and maturity of students and community morés.

## GENERAL SOURCES

**The American Chemical Society**, Education Division, 1155 Sixteenth Street, NW, Washington, DC 20036. 800-ACS-5558, ext. 2113. *Science Teaching Resources Catalog*.

**The Annenberg/CPB Collection**, P.O. Box 2345, S. Burlington, VT 05407-2345. 800-LEARNER. This is the World of Chemistry Series.

**Bullfrog Video**, P.O. Box 149, Oley, PA 19547. 610-779-8226 or 800-543-3764.

**Bureau of Mines**, Cochrans Mill Road, P.O. Box 18070, Pittsburgh, PA 15236. 412-892-6400 (general) and 412-892-6845 (audiovisual).

**Cambridge Development Laboratory, Inc**., 86 West Street, Waltham, MA 02154. 800-637-0047. Offers a variety of software and interfacing devices.

**Coronet Films and Video**, P.O. Box 2649, Columbus, OH 43216. 800-777-8100. Offers a wide selection of science-related programs on video. Discussion guides are available for most titles.

**Films for the Humanities and Sciences**. P.O. Box 2053, Princeton, NJ 08543-2053. 800-257-5126 or 609-452-1128.

**Guggenheim Productions**. John Collins, 3121 South Street, NW, Washington, DC 20007. 202-337-6900 or fax 202-337-9639. This is for "The Making of Liberty" video.

**Hawkhill Associates, Inc.,** P.O. Box 1029, Madison, WI 53701-1029. 800-422-4292. Offers a variety of science/technology/society programs on video. With every order you get a free year-long subscription to the Hawkhill Science Newsletter for Scientific Literacy.

**International Film Bureau (IFB) Inc.,** 332 South Michigan Avenue, Chicago, IL 60604. 312-427-4545.

**Modern Talking Pictures Service,** 5000 Park Street North, St. Petersburg, FL 33709. 813-541-7571 and 800-243-6877. Videos produced by corporations, civic organizations, and government agencies are available. Videos will be shipped from the nearest library to minimize shipping time and return mail cost.

**National Technical Information Services, NTIS,** Springfield, VA 22161. 703-487-4650.

**NOVA, WGBH Educational Foundation,** Box 2222-TG, South Easton, MA 02375. Weekly STS-oriented 60-minute shows broadcast on public television. Semi-annual teacher's guides (containing a program synopsis, viewing goals, discussion questions, suggested activities, vocabulary, and recommended readings) and program transcripts are available. Include NOVA program title desired, Box 322, 125 Western Avenue, Boston, MA 01234. 617-492-2777 ext. 5400 (programming) and ext. 3848 (educational print and outreach). Inquire with your school audiovisual coordinator concerning fair use and legal videotaping, or rent/purchase the programs through Coronet Films and Video.

**Optical Data Corporation**. ODC, 30 Technology Drive, Warren, NJ 07059. 800-524-2481. "The Living Textbook: Cosmic Chemistry."

**PBS Video**, 1320 Braddock Place, Alexandria, VA 22314-1698. 800-344-3337.

**Project SERAPHIM**, John W. Moore, Director, Department of Chemistry, University of Wisconsin, Madison, WI 53706. 608-263-2837. A National Science Foundation funded clearinghouse for instructional computer information and software in chemistry. Request a current catalog of software titles, the *Teacher's Guide to SERAPHIM Software* (keys software to six popular chemistry books), and *Teaching Tips*.

**TVOntario**. TVO Video, 142 W. Franklin Street, Chapel Hill, NC 27514. 800-331-9566.
**Vestron Video**. P.O. Box 4000, Stamford, CT 06907.

## CAREERS IN SCIENCE/TECHNOLOGY

**Hawkhill Associates, Inc.,** P.O. Box 1029, Madison, WI 53701. 800-422-4292.

"People in Science Today." Focuses on individual scientists from the "Scientists at Work" series. Relevant titles include "A Nuclear Physicist," "An Astrophysicist," "A Toxicologist," "An Industrial Chemist," " A Microbiologist," "A Biotech Entrepreneur," "A Molecular Biologist," "A Biochemist," and "A Geologist."

"Scientists at Work." Titles include "Chemists at Work": Part 1 presents a short history of chemistry, from the alchemists to the present; Part 2 profiles the work of four living chemists; and Part 3 challenges students to consider careers in chemistry and their impact on economic, political, and social issues. Also available are the titles "Physicists at Work," "Biologists at Work," and "Earth Scientists at Work."

"Women in Science." "Part 1. Women in Science Yesterday" and "Part 2. Women in Science Today" highlight five outstanding young women scientists in biology, physics, and chemistry.

**American Chemical Society,** Education Division, 1155 Sixteenth Street, NW, Washington, DC 20036. 800-209-0423 (for orders).

"Chemical Careers in Brief." A view of various career areas within the chemical sciences. Titles include "Agriculture Chemistry" (1994); "Biotechnology" (1994); "Chemical Information" (1994); "Environmental Chemistry" (1994); "Hazardous Waste Management" (1993); "Materials Science" (1994); "Polymer Chemistry" (1994); and "Medicinal Chemistry" (1994). Sold as a set only.

"Chemistry & Your Career: Questions & Answers" (1995). Answers the questions most often asked by high school students about chemistry careers, what chemists do, and how to prepare for future studies. Product C02.

"Chemistry and You: Discovering Your World" (1992) Career poster. Product C01.

"I Know You're a Chemist, But What Do You Do?" (1996). Job descriptions and advice from chemists at all degree levels and in all types of work environments. Product C42.

"Elements of Your Chemistry Career" (1994). This satellite videoconference video highlights careers in the chemical industry. Product C08.

"Employment Outlook". Report reprinted from *Chemical & Engineering News* on the current outlook for career opportunities in chemistry. Product C39.

"Inventing the Future: African-American Contributions to Scientific Discovery and Invention" (1994 videotape with 12-page guide). This 14-minute video highlights selected contributions to scientific discovery and invention in the U.S. from Colonial days to the present.

"Tracing the Path: African-American Contributions to Chemistry in the Life Sciences" (1991 videotape with 12-page guide). This 18-minute video traces selected scientific contributions by pre-colonial Africans in Africa, African-Americans during the late 1800s and early to mid-1900s in the U.S., and modern African-Americans to the field of chemistry as they relate to the life sciences. The two videos are sold as a set. Product PH08.

## SUPPLYING OUR WATER NEEDS

"Lake Study." Project SERAPHIM. Intended to involve students in scientific exploration by allowing them to collect data and identify the pollutant that is killing the fish in a lake. "Pond Study" is a less rigorous version.

"Waqual." Project SERAPHIM. Computer simulation of a wastewater treatment plant. Students choose the fraction of waste that will receive primary, secondary, and tertiary treatment and note the effect on the pollution level in a river.

"Pollute." Cambridge Development Laboratory, Inc. Interactive simulation that examines the impact of various pollutants on typical bodies of water. Students consider the impact on oxygen content and fish survival of water temperature, waste type, waste treatment type, rate of waste dumping, and type of receiving body of water. Generates both tabular and graphical data, and enables students to test various hypotheses.

"Toxic Turmoil: The Silicon Valley Story." Modern Talking Picture Service. Describes how leaks from underground storage facilities contaminated groundwater, tarnished the clean image of the electronics industry, and caused widespread public concern. This well-balanced documentary examines the sources of the contamination, industry's response to the problem, and the potential risks involved. Producer: Clean Water Task Force. Available only in California.

"The New Willamette." Modern Talking Picture Service. Documents the joint efforts of industry, citizens, and government to clean up the once polluted Willamette River in Oregon. It serves as a model for the cleanup task ahead for other rivers of the world. Producer: U.S.Army Corp of Engineers.

"Water—It's What We Make It." Modern Talking Picture Service. Focuses on the need for pure water and filtering technologies that can be used to obtain it. Producer: Water Quality Association.

"Water." Hawkhill Associates, Inc. Introduction to the physics, chemistry, and biology of water. Concepts include the shape of the water molecule, heat of fusion and vaporization, function in erosion, role as the Earth's thermostat and circulatory system, as well as the mystery and beauty of flowing water.

"Water: More Precious Than Oil." PBS. Examines the issues of water management and its conservation on a global basis. Producer: KTCA, Minneapolis-St.Paul, MN.

## CONSERVING CHEMICAL RESOURCES

"Mineral Resources." Project SERAPHIM. A simulation emphasizing energy and entropy. Acquaints students with factors related to the maintenance of mineral resources.

"Copper." Bureau of Mines/U.S.Dept. of Interior. Depicts both the past and present uses of copper, the methods devised for winning it from the Earth, and the importance of recycling.

"The Minerals Challenge." Bureau of Mines/U.S.Dept. of Interior. Shows the key role of minerals in today's society and how the ever increasing need for fuels, metals, and other mineral materials is being met by advanced technologies to find and mine thinner deposits, lower grade ores, and deeper and less accessible veins. Specific sequences include surface and underground mining of metals, nonmetals, and coal; ore preparation and transportation; petroleum and natural gas operations; and research to extract oil from oil shale.

"Wealth Out of Waste." Bureau of Mines/U.S.Dept. of Interior. Identifies the variety of wastes generated in the United States, and tells the story of recycling and waste utilization research aimed at turning the rubbish heaps into valuable resources.

"Mining and the Environment." Modern Talking Picture Service. Describes copper and silver mining in Troy, Montana, and how the ecology of the land and its abundant wildlife are being preserved while providing a livelihood for 350 families. An example of enlightened management working to provide the minerals needed by modern society while preserving the environment. Producer: Asarco Inc.

"Natural Resources: Doomsday or Utopia." Hawkhill Associates, Inc. Program on natural resources and their role in past, present, and future societies. Concepts include resource "value," exploitation, and limitations.

"Population and Pollution." International Film Bureau Inc. Demonstrates how air (industry, planes, and automobiles), water (sewage, detergents, and poisonous chemicals), and land (littering) pollution results from the misuse of the environment and the demands of a constantly growing population. Emphasizes the need for a change in attitudes, and offers specific suggestions for individuals.

"Waste: The Penalty of Affluence." International Film Bureau Inc. Presents the environmental hazards of improper disposal of household, commercial, and industrial waste. Demonstrates sanitary landfills, composting, and incineration as well as techniques for converting waste to raw materials. Producer: Dr. Heinrich Feuter for the World Wildlife Fund.

"Poison in the Rockies." NOVA. Cobalt mining in the Rockie Mountains and its consequences.

"Statue of Liberty." Vestron Video. 1989 Academy award.

"Green Energy." Films for the Humanities and Science. This program examines some renewable alternatives to petroleum products and explains how biological and organic products like wood chips, corn, and garbage can become major sources of energy.

## PETROLEUM: TO BURN? TO BUILD?

"Octane." Project SERAPHIM. Tutorial and game on gasoline, octane rating, engine knock and compression ratio, and gasoline additives. Game involves a simulated cross-country drive for which gasoline must be blended to match characteristics of the car.

"Refiner." Project SERAPHIM. Tutorial: explains the various operations of a modern refinery. Game: the player (an operations manager of a refinery) makes decisions about operating a refinery in order to maximize profits.

"After Oil What?" International Film Bureau Inc. Discusses the limited life span of fossil fuels and considers the advantages and disadvantages of the alternatives of solar, wind, wave, tidal, and geothermal energies. Considers hydroelectric power, nuclear fission, fusion, and breeder reactors.

"Fields of Fuel: The Ethanol Debate." International Film Bureau Inc. Presents evidence from World War II and from present day U.S. and Brazilian farms and research laboratories that demonstrates that alcohol fuels are a practical and economic alternative for vehicles. Producer: Iowa State University.

"Energy—A Family Album." National Audiovisual Center—Energy Research and Development Center. Presents a history of energy use in America and highlights future options.

"Rethinking Tomorrow." National Technical Information Services—Department of Energy. Depicts the growing national concern and mobilization for energy conservation. Flashbacks show how cheap oil fueled the turn of the century industrial revolution. Highlights energy programs of three cities.

"Plastics: The World of Imagination." Modern Talking Picture Service. Emphasizes the role of plastics in aeronautics, architecture, textiles, packaging, transportation, energy conservation, and medicine.

"Refiner." Modern Talking Picture Service. Highlights both the scientific and economic principles behind refining. Extensive animated sequences show how molecules are changed and reformed. Producer: Exxon Corp.

"The Refinery Film." Modern Talking Picture Service. How science works to produce hundreds of useful products from the hydrocarbon soup that is crude oil. Students learn a little about chemistry, a bit of physics, and a lot about what goes on inside the tanks and pipes of a large, modern oil refinery. Producer: Chevron USA.

"What Price Coal?" PBS. Focuses on environmental health and safety concerns associated with increased reliance on coal as a source of energy. Producer: NOVA.

## UNDERSTANDING FOOD

"The Food Processor." Cambridge Development Laboratory, Inc. A dietary analysis program built around a database of over 1500 brand name and generic foods that is capable of analyzing individual foods and whole menus for nutritional content, and comparing these values to RDA recommendations.

"Biotechnology: Unlocking Nature's Secrets." Modern Talking Picture Service. Shows how scientists at Pioneer Hi-Bred International are combining traditional plant breeding procedures with new genetic engineering techniques to create improved crop seeds.

"Food for the Future." Modern Talking Picture Service. Pavilion at EPCOT Center to see Kraft scientists working to discover new ways to grow food.

"Food: Its Science and Your Future." Modern Talking Picture Service. Discusses the various career options (such as product development, food processing, quality control, regulatory compliance, and package design) within the field of food science and technology. Producer: Institute of Food Technologists.

"The Insect Challenge." Modern Talking Picture Service. Uses macrophotography to portray the complexity and sophistication of insects and man's efforts to control them. Producer: Chevron Chemical Co.

"Insects and Agriculture." Modern Talking Picture Service. Explores the scientific research conducted in the struggle humans wage with insects for production of food and fiber. Producer: Chevron Chemical Co.

"Soybeans: Food, Feed and Future." Modern Talking Picture Service. Describes how Pioneer Hi-Bred International breeders are trying to improve the soybean varieties. This title is not available in the states of Arkansas or Hawaii.

## NUCLEAR CHEMISTRY IN OUR WORLD

"Nuclear Chemistry." Project SERAPHIM. Contains two programs: "Nuke 4" is a simulation of first-order radioactive decay involving a dice-throwing experiment, and "Decay" is a drill program on radioactive decay.

"Rutherford." Project SERAPHIM. Part I provides an illustrated overview of the classic alpha particle scattering experiment. Part II is an experiment simulation allowing students to experiment creatively with scattering phenomena.

"Chernobyl: The Taste of Wormwood." Films for the Humanities and Science. Winner of the *Palme d'Or*, 1987 Monte Carlo International Films and Television Festival.

"Radiation: Types and Effects." Films for the Humanities and Science. This program considers three aspects of radiation: what ionizing radiation is, the harm it can cause, and the risks connected with it compared with the risks encountered in everyday living.

"The Atom." Hawkhill Associates, Inc. "Part 1: How We Found Out About Atoms." traces the history of the atomic theory from Democritus to Bohr. "Part 2: What Is an Atom?" presents the modern view of the atom, including the work being done with the supercooled accelerator at Fermilab on quarks and an understanding of its relevance to the ordinary citizen.

"Nuclear Power." Hawkhill Associates, Inc. "Part 1: The History of Nuclear Power" and "Part 2: Nuclear Power—Today and Tomorrow" explains how chain reactions work, the difference between fission and fusion, and the difference between controlled nuclear reactions in a power plant, and explosive ones in a nuclear bomb. Concludes with a summary of the pros and cons of nuclear power and challenges students to debate the issue.

"Radiation." Hawkhill Associates, Inc. "Part 1: How Radiation Was Discovered" traces discoveries from Isaac Newton to Marie Curie. "Part 2: What Radiation Is, and What Radiation Can Do" distinguishes ionizing and nonionizing radiation, as well as electromagnetic radiation and nuclear radiation.

"Conquest of the Atom." International Film Bureau Inc. Recreates the experiments of J.J. Thomson, Ernest Rutherford, Sir John Cockcroft, Sir James Chadwich, and Hahn and Strassman. Points out that, in an atomic pile, the chain reaction is controlled so that a continuous supply of energy can be produced.

"The Discovery of Radioactivity." International Film Bureau Inc. Surveys the work of Roentgen, Becquerel, Curie, Elster, Geitel, and Rutherford.

"The Day Tomorrow Began." National Audiovisual Center—Atomic Energy Commission. Traces the history of the building and testing of the first atomic pile and the team that achieved the first sustained chain reaction on Dec. 2, 1942.

"Suicide Mission to Chernobyl." NOVA. This program accompanies Soviet scientists on a deadly mission inside the massive structure that entombs the Chernobyl nuclear reactor.

## LIVING IN A SEA OF AIR: CHEMISTRY AND THE ATMOSPHERE

"Solids/Liquids/Gases." Project SERAPHIM. Simulations of Charles' Law, Boyle's Law, a ride in a hot air balloon, and a tutorial on"The Earth's Atmosphere." Coronet Films and Video. Discusses the layered structure of the Earth's atmosphere and its interaction with solar radiation.

"The Energy Balance." Coronet Films and Video. Examines Earth's delicate energy balance and how humans consciously or unconsciously affect it.

"The Hole in the Sky." Coronet Films and Video. Discusses the role of chlorofluorocarbons in creating a "hole" in the Earth's ozone layer and the possible effects. Producer: NOVA.

"Air." Hawkhill Associates, Inc. Explores the history of how the Earth came to have an atmosphere, its present composition, and its life-sustaining roles.

"Air is Life." International Film Bureau Inc. Using visuals from around the world, stresses a global perspective and a need for cooperative international efforts as well as prudent individual fuel consumption. Producer: Condor Films.

"The CFC Story." Modern Talking Picture Service. Discusses the uses of chlorofluorocarbons and their possible effects on the stratospheric ozone layer. Producer: DuPont.

"The Climate Factor." National Technical Information Services—National Oceanic and Atmospheric Administration. Discusses how climate has affected both animals and humans over the last 100 million years.

"National Center for Atmospheric Research." PBS. Focuses on the research of scientists and technicians at the NCAR, especially as connected to the efforts of increased carbon dioxide levels. Producer: Creativity with Bill Moyers.

"To Catch a Cloud: A Thoughtful Look at Acid Rain." Modern Talking Picture Service. Seeks to dispel the propaganda and rhetoric surrounding this controversial issue with factual information about the atmosphere and research results. Producer: Alliance for Balanced Environmental Solutions.

## PERSONAL CHEMISTRY AND CHOICES

"Health Awareness Games." Cambridge Development Laboratory, Inc. Five interactive programs: "Coronary Risks," "Why Do You Smoke?" "Exercise and Weight," "Life Expectancy," and "Lifestyle" that relate student personal health and lifestyles to how well and how long they'll live.

"Health Risk Appraisal." Cambridge Development Laboratory, Inc. Using data from the Centers for Disease Control in Atlanta, this program helps students assess their health risks. By answering 40 health-related questions, students can determine morality risks and means of lowering them.

"Disease and Health." Hawkhill Associates, Inc. "Part 1: The Agony and the Pride—a History of Disease and Health" highlights the work of Jenner, Pasteur, Koch, and Nightingale. "Part 2: Avoiding Disease, Achieving Health" emphasizes personal choice issues like smoking, diet, and drugs and presents an overview of the consensus and the disputes within the medical and scientific communities.

"Cancer and the Environment." National Audiovisual Center—National Institutes of Health. Discusses pollution, food additives, smoking, and diet as possible causes of cancer, problems of the industrial environment, drug and chemical testing, prevention, and the mathematics of risk assessment.

"Cancer—What Is It?" National Audiovisual Center—National Institutes of Health. Overview of the nature of cancer, including definitions of biomedical terms and illustrated examples of cancer growth.

"Enzymes: Nature's Magic." Modern Talking Picture Service. Examines the functions and roles of enzymes in nature and their potential use in combating pollution and world hunger. Producer: Novo Industries.

## THE CHEMICAL INDUSTRY: PROMISE AND CHALLENGE

"BCTC." Project SERAPHIM. Allows students to gather data on a fictitious, potential industrial pollutant. It illustrates the contribution science can make to a social problem for which the solution is not obvious.

"Toxic Wastes." Hawkhill Associates, Inc. "Part 1: A History of Toxic Wastes in the Biosphere" points out many striking examples of toxic wastes in the "good old days." "Part 2: Toxic Wastes Today and Tomorrow" discusses the nature of chemicals and their cycling in the biosphere, food chains, tolerance levels, radiation, and environmental controls.

"The Development of Electrochemistry." International Film Bureau Inc. Surveys work on the production of an electric current by chemical action as conducted by Volta, Davy, Oersted, Faraday, Van't Hoff, and contributions of Ostwald, Arrehenius, Heroult, and Hall. Illustrates some modern technological applications of electrochemistry.

"Doing Something." Modern Talking Picture Service. Shows what the chemical industry is doing to insure safe production, transportation, use, and disposal of chemicals. Producer: Chemical Manufacturers' Association.

"The Electrolysis of Brine." Modern Talking Picture Service. Demonstrates the electrolysis of brine into chlorine and sodium hydroxide and the importance of these chemicals in daily life. Producer: PPG Industries, Inc.

"How Much Is Too Much?" Modern Talking Picture Service. Explains how chemical companies (such as Dow) test products to insure that they can be made, transported, and used safely. Concepts such as risk analysis, risk vs. hazard, and toxicity are discussed using everyday examples.

"The Need to Know." Modern Talking Picture Service. Shows what the chemical industry is doing about the issue of hazardous waste disposal. Producer: Chemical Manufacturers' Association.

"The Right Chemistry." Modern Talking Picture Service. Explores the three major divisions of Morton Thiokel, Inc. (salt) and their functions.

# *ChemCom* NEWSLETTER

An important part of the *ChemCom* package is *Chemunity News*, the free newsletter that highlights the educational programs and resources of the American Chemical Society as well as movements and trends in science education including *ChemCom*. *Chemunity News* is published four to five times a year. If you wish to be added to the mailing list, send your name and address to American Chemical Society, *Chemunity News*, Room 831, 1155 Sixteenth Street, NW, Washington, DC 20036.

# TEACHING SCHEDULE

| | TIME BLOCK 1 | TIME BLOCK 2 | TIME BLOCK 3 | TIME BLOCK 4 | TIME BLOCK 5 |
|---|---|---|---|---|---|
| Class Work | Read "A Special Note to Students"<br>Read A-A.1<br>CQ A.1<br>Getting Ready | Discuss A.1<br>Review "Safety in the Laboratory"<br>Prelab LA A.2 | Lab A.2 | Postlab A.2+ questions | YT A.5 |
| Homework | YT A.1 | Read A.4 | YD A.3 (3days) | Read A.5 | Read A.6-A.9 |
| | **TIME BLOCK 6** | **TIME BLOCK 7** | **TIME BLOCK 8** | **TIME BLOCK 9** | **TIME BLOCK 10** |
| Class Work | DiscussA.6-A.7<br>YD A.8 | YD A.9 | Review Part A<br>Quiz Part A<br>Prelab LA B.3<br>Getting Ready | LA B.3 | YT B.5<br>Getting Ready<br>Prelab LAB.8 |
| Homework | | SQ Unit A<br>Read B.1-B.2 | Read B.4-B.7 | YT B.4 | Read B.9-B.10 |
| | **TIME BLOCK 11** | **TIME BLOCK 12** | **TIME BLOCK 13** | **TIME BLOCK 14** | **TIME BLOCK 15** |
| Class Work | Lab B.8 | YD B.9 | Review Part B<br>Quiz Part B | Discuss C.2<br>YT C.2 | YD C.5 |
| Homework | | SQ Part B | Read C.1-C.2<br>YT C.1 | Read C.3-C.4 | Read C.6-C.7 |
| | **TIME BLOCK 16** | **TIME BLOCK 17** | **TIME BLOCK 18** | **TIME BLOCK 19** | **TIME BLOCK 20** |
| Class Work | Discuss C.6-C.7<br>YT C.7 | YD C.10<br>Discuss C.11<br>Prelab LA C.12<br>Getting Ready | LA C.12 | Review Part C<br>Quiz Part C<br>Prelab LA D.2<br>Getting Ready | LA D.2 |
| Homework | Read C.8-C.9, C.11 | SQ Part C | Read D-D.1 | | Read D.3-D.5 |
| | **TIME BLOCK 21** | **TIME BLOCK 22** | **TIME BLOCK 23** | **TIME BLOCK 24** | **TIME BLOCK 25** |
| Class Work | YD D.6 | Review Part D<br>Quiz Part D<br>PIAT<br>"Town Meeting" Preparation | PIAT<br>"Town Meeting" Presentation | Review Water unit | Exam Water unit |
| Homework | SQ Part D<br>Read E-E.2 | | PIAT Letter to the Editor | | |

**LA**=Laboratory Activity; **CQ**=ChemQuandary; **YT**= Your Turn; **YD**=You Decide; **PIAT**=Putting it All Together; **SQ**=Summary Questions; One **Time Block**=Approximately 50 minutes

# SUPPLY LIST

## Expendable Items and Unusual Materials

| Quantity | Unit | Item | Sections Used |
|---|---|---|---|
| 3 | mL | acetic acid, conc | B.8 |
| 25 | g | ammonium chloride | C.12 |
| 1.2 | g | barium chloride dihydrate | B.8 |
| 8 | g | calcium chloride dihydrate | B.8,D.2 |
| 1 | box | Calgon | D.2 |
| 100 | g | cation exchange resin | D.2 |
| 150 | g | charcoal, decolorizing | A.2 |
| | | clay | B.3 |
| 28 | g | copper(II) sulfate pentahydrate | B.3, C.12 |
| | | cotton | B..3 |
| 2 | L | distilled water | C.12 |
| 100 | mL | ethanol | C.12 |
| 36 | pieces | filter paper, Whatman #1 | A.2, D.2 |
| 1/2 | tbs | garlic powder | A.2 |
| 2 | kg | gravel, small | A.2 |
| 600 | mL | hexane | C.12 |
| 25 | g | iodine crystals | C.12 |
| 1.35 | g | iron(III) chloride hexahydrate | B.3 |
| 8 | g | iron(III) nitrate monohydrate | B.8 |
| 1.4 | g | iron(III) sulfate heptahydrate | B.8 |
| 250 | mL | liquid soap solution | D.2 |
| 50 | mL | milk/water mixture | B.3 |
| 25 | g | naphthalene | C.12 |
| 12 | | paper clips | A.2 |
| 12 | | paper cups | A.2 |
| 2.4 | g | potassium thiocyanate | B.8 |
| 2 | kg | sand | A.2 |
| 0.85 | g | silver nitrate | B.8 |
| 1.24 | g | sodium carbonate monohydrate | D.2 |
| 25 | g | sodium chloride | C.12 |
| 0.67 | g | sodium oxalate | B.8 |
| 1/2 | tbs | table salt | A.2 |
| 12 | | toothpicks | B.3 |
| 25 | g | urea | C.12 |
| 1/2 | cup | vegetable oil | A.2 |

This first *ChemCom* unit focuses on the fictional town of Riverwood and problems with its water supply. To solve a fish-kill mystery, students explore the properties of water and aqueous solutions and examine procedures for obtaining adequate supplies of pure water. Fundamental concepts related to formulas, nomenclature, and solubility are introduced and developed along with the Riverwood story.

The unit demonstrates the *ChemCom* approach: create a need-to-know in the minds of the students via a science-technology-society (STS) problem or issue, and then develop the chemical concepts that help to resolve the issue. This is a departure from traditional chemistry courses, where students encounter the concepts first, and then use societal problems and issues to illustrate the use of the chemistry. It is important for you to resist teaching more about a particular concept than is necessary to understand the issue or problem. For example, acids, bases, salts, and pH are introduced in this unit. It is unnecessary to teach them in greater depth at this time, as all of these topics will be revisited in subsequent units.

# LEARNING GOALS

Upon completion of this unit, the student will be able to

1. Use metric units in measurements of length and volume. [A.1]
2. Purify a water sample using separation and filtration techniques. [A.2]
3. Collect data on family water usage. [A.3]
4. Relate the purity of water supplies to community health. [A.4]
5. Discuss direct and indirect water use and their importance for water conservation. [A.5, A.8, A.9]
6. Describe the hydrologic cycle, and indicate the primary storage reservoirs of Earth's water supply. [A.6, A.7]
7. Discuss the impact of the unusual properties of water on Earth's life forms. [B.1]
8. Define the terms solution, solvent, and solute, and apply them in an example. [B.2]
9. Classify matter in terms of elements, compounds, and mixtures; and distinguish among different types of mixtures (solutions, colloids, and suspensions) in a laboratory setting. [B.2–B.4]
10. Interpret the symbols and formulas in a balanced chemical equation in terms of atoms and molecules. [B.5]
11. Describe the relationship of charged particles to the concept of polarity. [B.6]
12. Define the terms insoluble, unsaturated, saturated, and supersaturated, and calculate solution concentration as a percentage. [C.1, C.2]
13. Use solubility curves to describe the effect of temperature on solubility, and calculate percent saturation. [C.1, C.2, C.4]
14. Demonstrate the ability to organize and interpret data in graphs or tables. [C.1, C.5, C.10]
15. Given the pH of a solution, classify it as acidic, basic, or neutral. [C.6]
16. Determine the formula and name of simple ionic compounds when provided with the charge and name of the anion and cation. [C.7]
17. Describe, on a molecular level, what happens when a solid dissolves in water. [C.8]
18. Discuss the effects on human health when heavy-metal ions are present in a water supply. [C.9]
19. Evaluate the risks of contaminants in a community's water supply, with particular attention to heavy-metal ions of lead, mercury, and cadmium. [C.10]
20. Compare and contrast natural and artificial water purification systems, and assess the risks and benefits of water softening and chlorination. [D.1, D.3–D.5]
21. Present opinions and supporting facts, both orally and in written form, concerning the responsibility for the fish kill in Riverwood. [E.1, E.2]

# A. THE QUALITY OF OUR WATER

Two newspaper articles introduce students to a water-quality problem in Riverwood and set the stage for the major "story line" of the unit. Students follow the progress of Riverwood citizens in dealing with this issue. Encourage your students to note the difference between fact and opinion and to identify the kind of scientific information useful in solving such issues.

It is also worth noting that a large problem can often be solved more easily if it is first reduced to a series of smaller sub-problems. Students can identify some of the sub-problems involved in the Snake River water-quality crisis.

*ChemCom* units provide the framework for student involvement in various local STS issues. Students can collect relevant newspaper articles and carry out individual projects related to local issues. (The Extending Your Knowledge sections at the end of the unit provide suggestions.) Additional teacher background and student activities are found in the Enrichment Materials section at the end of this unit.

## A.1 MEASUREMENT AND THE METRIC SYSTEM *(pages 5–7)*

Emphasize that the modernized metric system (SI) is used worldwide, and thus serves as an important international "language." Encourage students to consider items that are measured in metric units (photographic equipment, skis, ammunition, 2-L plastic soft drink containers) to get an idea of the approximate size of metric units. Both the text and the student activity focus on "thinking metric" rather than on doing English/metric or metric/English conversions.

## YOUR TURN *Meters and Liters* *(pages 6–7)*

Providing drill and practice in metric manipulation and use of prefixes, this activity encourages students to use metric measurements in their everyday lives. It also illustrates why scientists prefer metric over the English units commonly used in the United States.

### Answers to Questions

1. a. Penny 18 mm, nickel 21 mm, dime 17 mm, quarter 23 mm
   b. Penny 1.8 cm, nickel 2.1 cm, dime 1.7 cm, quarter 2.3 cm
2. a. $10 \text{ cm} \times 10 \text{ cm} \times 10 \text{ cm} = 1{,}000 \text{ cm}^3$
   b. $1{,}000 \text{ cm}^3 \times \frac{1 \text{ mL}}{1 \text{ cm}^3} = 1{,}000 \text{ mL}$
   c. $1{,}000 \text{ mL} \times \frac{1 \text{ L}}{1{,}000 \text{ mL}} = 1 \text{ L}$
   d. $1{,}000 \text{ cm}^3 \times \frac{1 \text{ sugar cube}}{1 \text{ cm}^3} = 1{,}000 \text{ sugar cubes}$
3. Most beverage containers report volumes in both U.S. customary units and metric units. One liter is slightly larger than a quart (1 qt = 0.947 L or 0.947 $dm^3$). One fluid ounce is approximately 30 times larger than one milliliter.
4. a. The conversion of the magnitude of metric units, based on the number 10, is accomplished by decimal point changes. The units of volume are related to the length by 1 L = 1 $dm^3$, 1 mL = 1 $cm^3$. For conversion in U.S. customary units, one must memorize many different factors such as 5,280 feet = 1 mile, 32 fluid ounces = 1 quart. Because metric units are used by most of the rest of the world, commerce and trade are more easily accomplished.
   b. In the United States we sometimes need to translate metric units into U.S. customary units. Such conversions often have the effect of making metric–based units seem awkward. SI units are also less familiar to many U.S. citizens. Exclusive use of metric units would require retooling for some manufacturers.

## CHEMQUANDARY *Water, Water Everywhere* *(page 7)*

Consider closing a class session by assigning this *ChemQuandary* for homework without discussing the explanation. Point out that the numbers given for one egg and a can of juice mean that eating two eggs and drinking 1.3 L of juice require enough water for taking 10 showers, washing 10 loads of clothes, or 25 loads of dishes, or flushing a toilet 100 times.

Begin the discussion of this *ChemQuandary* by asking students to list processes involved in obtaining eggs or cans of juice. These include using water for raising food for the chickens, for drinking water, for trees providing wood to build chicken houses, for washing the eggs, and for producing the paper cartons in which eggs are sold. To produce canned juice, obviously the fruit trees must be watered and the fruit washed. But also much water is used indirectly to produce the can that contains the juice. In addition, digging the iron ore from the ground and moving it to the steel mill requires 4 L of water per can; the steel mill uses about 76 L of water to produce each can; and the rolling mill that fabricated the metal for the can consumed another 46 L of water. Thus, the total indirect volume of water used to make the juice can is about 126 L. Canned juice may help solve the short-term problem of water shortage in Riverwood, but it does not contribute to long-term water conservation.

## A.2 LABORATORY ACTIVITY: FOUL WATER *(pages 7–10)*

This laboratory activity is designed to stimulate student interest in some of the practical problems associated with water quality. The goal of the activity is for students to purify a sample of highly impure "foul" water to the point that it could be used to wash their hands. Do not extend the challenge to making the water suitable for drinking, since no laboratory-based materials should be swallowed. The activity introduces students to the techniques of filtration, absorption, and separation.

### Time

50 minutes for student work
50 minutes for distillation demonstration

### Materials (for a class of 24 working in pairs)

2–3 L of foul water (100 mL/pair) (see recipe below)
12 100-mL graduated cylinders
24 150-mL beakers
12 funnels (73-mm inside diameter)
12 ring stands
12 ring clamps
12 clay triangles
12 cups (paper preferred)
12 paper clips (for punching holes in cups)
12 filter papers (Whatman #1, 11 cm diameter)
12 125-mL Erlenmeyer flasks (for mixing charcoal and water)
150 g decolorizing charcoal
2 kg sand, cleaned
2 kg gravel, cleaned (small grind; roughly 5 mm diameter)
1 distillation apparatus
1 conductivity meter
rubber tubing

### Foul Water Recipe

Add the following ingredients to an empty two-liter soft-drink bottle (all specified quantities can be estimated):

1/2 tablespoon table salt (NaCl)
1/2 teaspoon garlic powder
1/2 cup used coffee grounds
1/3 cup vegetable oil

Fill the bottle with tap water, cap tightly, and shake well. Save the excess for section B.2 and the discussion of mixtures.

## Safety

Students should wear safety goggles and aprons throughout the laboratory activity. During the post-lab distillation demonstration, the liquid should not become boiled to dryness, as the residue will be etched into the container, which makes cleaning difficult. More important, an overheated flask may break. Leave about 1/4 of the original liquid in the flask at the end of the distillation. Do not permit students to drink any of the water samples used in this activity.

## Advance Preparation

Encourage lab teams to compete with others to obtain the largest volume of "pure" water. Set up an apparatus to demonstrate distillation of water at the close of the activity. An illustration of a distillation apparatus is shown in Figure 8 (page 168) in the Petroleum unit. Directions for assembling a conductivity apparatus are given in the reference below. A commercial version is also available from Flinn Scientific

Katz, D. and Willis, C., *Journal of Chemical Education,* 1994, p.330.

## Pre-Lab Discussion

Because this is the first laboratory activity in the course, emphasize safety procedures, identification and use of safety equipment supplied in your classroom (such as the eye wash, shower, or fire blanket), and the general rules of safe laboratory conduct. Be sure to discuss the Safety in the Laboratory section in the preliminary pages of the student text, as well as general safety precautions. To encourage careful reading of the safety section and compliance with the rules, give a short quiz and have each of your students sign a statement saying that they have read the safety rules and agree to obey them.

Explain the meaning of the protective goggles symbol found at the beginning of each laboratory activity. The goggles sign indicates that students should wear goggles for the activity and heed any other safety precautions. Highlight all of your own school's laboratory safety policies.

Discuss the potential hazards of wearing contact lenses in the laboratory. Students should not wear contact lenses in the laboratory. Any reagent splashed into the eye is held directly onto the eye's surface by the contact lens; even an eye-wash fountain cannot rinse away this trapped reagent.

Display the foul water mixture by pouring it into a large beaker. Review the methods of purification used in the activity. Provide containers into which students can pour used sand, gravel, and charcoal.

Point out that the data table should be larger than the sample version shown on page 8 in the text—the tables are purposely reduced in size in the unit to discourage students from writing in their books.

## Lab Tips

Dispense about 100 mL of the foul water to each pair, shaking the mixture before dispensing to ensure homogeneity among samples.

Some forms of decolorizing charcoal are quite messy and can be irritating if inhaled; it is helpful to prepare a thick slurry of the charcoal with water beforehand. Have your students use a spoonful of the slurry for this step. Granular charcoal, used in aquaria, is a cheaper alternative than powdered charcoal.

Do not reveal the composition of the original foul water sample until the Post-Lab Discussion.

Filtration does not remove the dissolved table salt; therefore, the final sample will conduct electricity. Compare the conductivity of distilled water with the student-purified sample.

## Expected Results

| | Color | Clarity | Odor | Presence of oil | Presence of solids | Volume, mL |
|---|---|---|---|---|---|---|
| Before treatment | dark | cloudy | strong | yes | yes | 80–100 |
| Oil-water separation | dark | cloudy | strong | less | yes | 75–95 |
| Sand filtration | dark | clearer | strong | none | none | 70–90 |
| Charcoal filtration | lighter | clearer | less | none | none | 65–85 |

## Calculations

1. Percent of water purified = $\frac{\text{Volume of water purified}}{\text{Volume of foul water sample}} \times 100$
2. Volume of water before treatment – volume of water after charcoal filtration
3. Percent of water lost = $\frac{\text{Volume before} - \text{Volume after}}{\text{Volume before}} \times 100$

## Answers to Questions

1. Color, clarity, odor, presence of immiscible liquids, presence of ions as shown by the conductivity tests
2. Distillation is too expensive for large volumes of water because of the time, energy, and equipment required.

## Post-Lab Activities

Check the electrical conductivity of the student samples and distilled water. Compare the results. Distilled water acts as a control for this test. Using the previously assembled distillation apparatus, distill a student team's sample of purified water. Discard the initial 20% of the distillate because it may contain dissolved gases, and allow about 20% of the sample to remain in the distillation flask. Test the collected distillate with the conductivity apparatus. Compare the results with previous tests, and relate the results to the usefulness of the purification procedures.

Many students think liquid samples are pure if they are clear and have no odor, but the presence of some contaminants cannot be detected with the unaided senses. The positive conductivity test on the sample before distillation should lead them to question their assumption.

## A.3 YOU DECIDE: INFORMATION GATHERING *(pages 10–11)*

Throughout the *ChemCom* course, students will have many opportunities to explore the role of chemistry in their own lives. This activity, which requires participation of family members, generally receives a positive response. The activity explores water use in the home and promotes awareness of the large quantities of water used routinely in a household. The activity also emphasizes information gathering as an important first step in making decisions. The data collected will be analyzed later in *You Decide* A.8.

The time needed for data collection is three days. If possible, include one weekend day, in order to obtain data from a typical sample period.

## A.4 WATER AND HEALTH *(pages 11–12)*

This section describes how important clean water is to human beings.

## A.5 WATER USES *(pages 12–14 )*

This section discusses regional differences throughout the United States in sources and uses of water. It also introduces indirect use of water. Although students may have the impression that direct home use represents the major use of water, the fact is that growing, preparing, and manufacturing foods and consumer goods far exceed the more visible (direct) use of water.

## YOUR TURN U.S. *Water Use* *(page 12)*

This *Your Turn* provides practice in interpreting a figure to answer questions about regional water use.

### Answers to Questions:

1. a. manufacturing
   b. irrigation
   c. irrigation
   d. irrigation
   e. mining
   f. irrigation
2. Possible reasons include land forms (mountains, etc.), available precipitation, location of rivers and lakes, population density, climate for growing, and location of factories.
3. Since the East is more densely populated, 32% of the total water used in the East is for homes and offices, compared with 5% in the West. Since the East contains more manufacturing plants, 40% of the total water used in the East is for manufacturing as compared with 2% in the West. In the East, 13% of the water use is for steam–electrical plant generating, while the West, with extensive hydroelectric resources, uses only 1% for generating electricity.

## A.6 BACK THROUGH THE WATER PIPES *(pages 14–15)*

This section traces tap water "back through the pipes" to its source, supporting the fact that Earth has a finite (though recyclable) supply of water. Students can investigate the source of water for their own community: reservoirs, water towers, pumping stations, wells, rivers, or other site-specific locations. A field trip to a municipal pumping station can be highly motivational and informative.

## A.7 WHERE IS THE EARTH'S WATER? *(pages 15–17)*

Students are generally not surprised to learn that most of Earth's water supply (97%) is in the oceans. However, they may be puzzled to find that the next largest store of Earth's water is locked in glaciers and ice caps. Neither of these sources of water is readily available—it would be very costly to change either to a useful form. It is fortunate that many areas of the United States have an abundant water supply. Some cities, particularly in the arid parts of the West, already find it necessary to regulate water usage at times. Remind students that even though this country's total water-supply profile is favorable, there are local and regional problems. The fish-kill problem in Riverwood is a typical example. *You Decides* A.8 and A.9 challenge students to make thoughtful decisions about how water should be used during a shortage.

## CHEMISTRY AT WORK *(pages 16–17)*

This new feature in the 3rd edition of *ChemCom* continues the theme of demonstrating how chemistry is used in the real world. The two segments in each unit show chemistry in the workplace. Each one focuses on an individual who is either a chemist using their knowledge in an unusual way or an individual who is not a chemist using chemistry knowledge. Many emphasize new methods and use of technology as well as an interdisciplinary approach to problem solving. The variety of people involved communicate to students that anyone can be successful in these careers. The vignettes suggest follow-up student activities.

## Purifying Water Means More than Going with the Flow

The Chief Plant Operator discusses the source of the water supply and how the water is purified for use.

## A.8 YOU DECIDE: WATER-USE ANALYSIS *(pages 18–19)*

In this activity, students analyze the data they collected earlier about daily family use of water. Students gain useful experience transforming raw data into more interpretable forms. The data processing includes finding the total volume of water used by the family during the survey, and then calculating the total water use per person, as well as the daily water use. Students compare their results with individual water-use information provided in the unit.

The activity takes about 25–35 minutes. A data sheet for reporting the results may be helpful, and group work on calculations may help some students to complete them successfully. These calculations provide a good opportunity to stress the importance of keeping track of units associated with any quantities being computed.

## A.9 YOU DECIDE: RIVERWOOD WATER USE *(page 19)*

This activity simulates typical water-use decisions individuals might be required to make during a water shortage. Students may have experienced some actual short-term community emergency, such as a water main break, that would increase the realism of this exercise. Students are challenged to plan how to use a limited supply (375 L) of water for maximum benefit. This activity is appropriate for group work and takes 15–20 minutes to complete. An additional ChemQuandary is found in the Enrichment Materials section at the end of this unit.

### Answers to Questions

1. a. Eliminate washing cars, windows, pets, clothes, watering indoor and outdoor plants, watering lawn. Some may answer bathing, showers, washing hair!
   b. Dirty cars, dirty pets, plants may die, lawns may turn brown, odor levels increase
2. a. Reduce water for bathing, showering, washing clothes and dishes, flushing toilet
   b. Shorter showers, lower bath water level, wash dishes by hand, do not flush toilet every time, low-flow toilets and shower heads
3. a. Use impure water for washing car, flushing toilet, possibly watering plants and lawn
   b. Use water from washing hands, preparing food

It has been estimated that 30–50% of the water used in the United States is wasted. In the discussion, lead students to see how individual acts of conservation can lead to significant savings at the household level and more at the community or national level, where the impact is multiplied by 260 million.

## Part A: Summary Questions *(page 20)*

1. a. dime = 1 mm
   b. glass of water = 250 $cm^3$
   c. pencil diameter = 7 mm
   d. gallon = 3.8 L
2. a. 735 $cm^3$ = 735 mL = 0.735 L
   b. 10.7 mm = 1.07 cm
3. a. The family would have to adopt conservation measures as discussed in sections A.8 and A.9.
   b. There would be restrictions on watering plants, washing cars, and other nonessential uses.
   c. Agriculture and manufacturing would suffer. Eventually, the population would decrease.

d. Mandatory conservation measures would be adopted. The economy would suffer greatly as manufacturers would be forced to cut back. People would have to do without many kinds of consumer goods.

4. a. Indirect water use refers to water used in the production or processing of foods and goods.
   b. Water for growing grain, water for mixing dough, growing yeast, washing pans, packaging materials, grinding and transporting grain and flour
5. Water is very stable, and Earth's gravity prevents losses to outer space. Our planet is nearly a closed system, keeping the amount of water on Earth constant.
6. Cholera, typhoid fever, dysentery (possibly others)
7. a. There is concern over the availability of pure water, since more than 97% of the total water is in the oceans and is unfit for human use. Owing to increasing population, we are using more and contaminating more.
   b. We could use more seawater by distillation (using solar power) or purification (by using reverse osmosis). Plans for desalinating ocean or brackish water may become economically competitive with the cost of transporting fresh water over long distances.
   c. Ocean water could be used in coastal cities for tasks that do not require salt-free water, such as flushing toilets.

## EXTENDING YOUR KNOWLEDGE *(page 20)*

- The hydrologic cycle works because liquid water evaporates to form a lower-density gas, which rises into the atmosphere to cool and form clouds. Evaporation also leaves behind the dissolved particles present in impure water. Water in clouds may freeze, producing higher-density liquid and solid states that fall back to Earth as precipitation.

- The maximum and minimum temperatures on Earth are 45 °C and -40 °C, respectively. For the Moon the values are 130 °C and -173 °C. On Venus the values are 480 °C and 90 °C. The large amount of water on Earth's surface helps to stabilize the surface temperature during Earth's rotation and revolution around the sun. Compared to other common substances, large quantities of heat energy are involved when the temperature of water rises or goes down. Also, a large quantity of energy is used or released when water changes from one physical state to another.

*Heat of fusion* is defined as the thermal energy needed to melt (convert a solid to a liquid) a specified quantity of a solid at its melting point.

*Heat of vaporization* is defined as the thermal energy needed to vaporize (convert to gas) a specified quantity of liquid at its boiling point.

*Heat capacity* refers to the quantity of thermal energy needed to change the temperature of a given sample of matter by one degree Celsius.

- Compared with other common substances, water has an unusually high heat of fusion, heat of vaporization, and heat capacity. The next best candidate for a liquid that could support some type of life is ammonia.

| Formula | Melting point, °C | Boiling point, °C | Heat of fusion, J/g | Heat of vaporization, J/g |
|---|---|---|---|---|
| $H_2O$ | 0 | 100 | 333 | 2,260 |
| $NH_3$ | -78 | -33 | 341 | 1,370 |

- Ammonia has a much lower freezing point and boiling point than water. Thus, even if ammonia were as good a solvent, an ammonia-based organism would have to function within a much narrower and lower temperature range (-78 °C to -33 °C).

# B. A LOOK AT WATER AND ITS CONTAMINANTS

Part B returns to Riverwood's problem with another news report about the fish kill. To understand the reported issues, students need to know more about the chemistry of water and how substances interact with water. Some properties of water are given, and students learn to classify mixtures as suspensions, colloids, or solutions in a laboratory activity. To explain these observations, a simple particulate model of matter is proposed. The universal symbolic language of chemistry is also introduced. Students then learn the difference between pure water and clean water, which leads them to laboratory testing for dissolved ions in water samples.

Many new chemical terms and concepts are introduced in Part B. It is important to remember that because these concepts will be revisited many times during the *ChemCom* course, it is not necessary to teach for mastery of each concept now. The Summary Questions at the end of the unit provide important clues concerning the depth of knowledge required at this time.

With more information, students go back to the newspaper articles to separate fact from interpretation. A knowledge of chemistry provides a framework for asking questions about societal problems that may help lead to solutions.

## MEETING RAISES FISH-KILL CONCERNS *(page 21)*

Chemists and other authorities report their progress in the investigation of the Snake River fish kill, using such terms as *dissolved oxygen, suspended*, and *solubility*. In order to understand the fish-kill issue, students need to understand basic chemical concepts regarding the behavior of water.

## B.1 PHYSICAL PROPERTIES OF WATER *(pages 22–24)*

This section focuses on the physical properties of water—density, physical state, surface tension, boiling and freezing points, and solvent characteristics. Consequences of some of water's physical properties are discussed. The important property of density is carefully developed. To emphasize the unusual properties of water, begin the class session with some of the demonstrations described below. Ask students to interpret and explain the results of each demonstration.

### Demonstration Ideas

1. Surface tension: Float a needle on the surface of water as shown in the student text. Show that more water can be added to a full glass of water.
2. Density of water: Find the mass of 10.0 mL water in a 100-mL graduated cylinder. Have the students calculate the density of liquid water. Then place the cylinder in the freezing compartment of a refrigerator. Find the new volume and have students calculate the density of ice.
3. A density column: Carefully place the following materials in a large graduated cylinder (to show density in terms of g/mL) in this order: lead sinker (density 11.3), Karo syrup (density 1.36), glycerin (density 1.26), rubber stopper (density 1.2), ethylene glycol (density 1.11), water (density 1.0), plastic (density 0.9), oil (cooking or mineral, density 0.9), oak (density 0.9), alcohol (either ethanol or 2-propanol, density 0.79), and cork (density 0.2). To obtain discrete layers, pour the materials slowly down the side of the tilted graduated cylinder, or add each liquid layer using a long-stemmed funnel.
4. Place an ice cube in a beaker of alcohol. The ice cube will sink rather than float because it is more dense than the alcohol.

## B.2 MIXTURES AND SOLUTIONS *(pages 25–26)*

This section introduces water mixtures and their classification. A sample of the foul water sample used in Laboratory Activity A.2 is an effective visual aid. Dissolve table salt in water to demonstrate the formation of a solution.

## B.3 LABORATORY ACTIVITY: MIXTURES *(pages 26–28)*

This laboratory activity will provide students with the experience of applying knowledge of colloids, suspensions, and solutions to the classification of mixtures. Suspensions contain particles that settle out or can be filtered; neither colloids nor solutions settle or can be filtered. Colloids are differentiated from solutions by using the Tyndall effect described in the student text. The macroscale version of this Laboratory Activity is found in the Enrichment Materials section at the end of this unit.

### Time

50 minutes

### Materials (for a class of 24 working in pairs)

12 24-well wellplates
48 Beral pipets
48 Pasteur pipets
cotton
light sources—overhead projector, penlight, flashlight
unknown #1: 50 mL milk/water mixture (instant creamer can be used to prepare the mixture)
unknown #2: 50 mL 0.1 M $CuSO_4$ solution (1.24 g $CuSO_4 \cdot 5H_2O$ in 50 mL solution)
unknown #3: 50 mL of a small amount of clay suspended in 50 mL water
unknown #4: 50 mL 0.1 M $FeCl_3$ (1.35 g $FeCl_3 \cdot 6H_2O$ in 50 mL solution)
12 toothpicks, for packing cotton in the filter

### Safety

Students should wear safety goggles and aprons throughout the laboratory activity.

### Advance Preparation

1. The water-milk colloidal mixture should be quite dilute— add milk dropwise to the water (with stirring) until a very faint cloudiness is evident. Test the colloid for an observable Tyndall effect before dispensing the mixture to students.
2. The actual amount of clay used to prepare the clay suspension is not important. Adding a small amount of powdered charcoal may increase the visibility of the clay in the filtering experiment.
3. The $FeCl_3$ and $CuSO_4$ solutions should be freshly prepared, as they may form colloidal dispersions (because of the formation of hydroxides under alkaline conditions) upon standing. If the solutions appear cloudy, add a drop of hydrochloric acid or sulfuric acid, depending on the anion present, to clarify them.
4. Movement around the classroom during the activity is minimized if you label the pipets, fill them, and distribute them upside down in the wellplate. Ask for these supplies to be returned in the same manner. The same pipets and solutions can be used for several classes.

### Lab Tips

A piece of cardboard having a pencil-size hole, placed over a filmstrip projector, slide projector, or overhead projector, can be used by students. Turn off the light sources between uses to avoid overheating. The clay mixture may be classified as both a suspension and a colloid, if the filtrate shows the Tyndall effect. Caution students to shake each mixture before filling their Beral pipets and before dispensing liquid into the microplates or when filtering. Wellplates should be thoroughly rinsed immediately after use.

## Pre-Lab Discussion

The pre-lab discussion should review the text material on solutions, suspensions, and colloids. Note that the color of a solution does not necessarily indicate the presence of a colloid. A laser borrowed from the physics classroom can be used to demonstrate the Tyndall effect.

## Expected Results

**Original Sample**

| Unknown | Color | Clarity | Settle? | Tyndall effect? | Classification |
|---|---|---|---|---|---|
| 1 | white | very slight | no | yes | colloid |
| 2 | blue | clear | no | no | solution |
| 3 | light tan | cloudy | yes | yes | suspension |
| 4 | yellow | clear | no | no | solution |

**Filtrate**

| Unknown | Color | Clarity | Settle? | Tyndall effect? | Filter paper | Classification |
|---|---|---|---|---|---|---|
| 1 | white | slightly cloudy | no | yes | nothing | colloid |
| 2 | blue | clear | no | no | nothing | solution |
| 3 | light tan | cloudy | no | yes | residue | colloid |
| 4 | yellow | clear | no | no | nothing | solution |

## Post-Lab Discussion

After the students have classified the mixtures, reveal the contents of each beaker. Milk was discussed in the previous section, and copper(II) sulfate and iron(III) chloride will be discussed as ionic compounds in section B.6. Ask students to apply their new knowledge of colloids, suspensions, and mixtures to analyze the results of the foul water lab.

## B.4 MOLECULAR VIEW OF WATER *(pages 28–29)*

Moving beyond directly observable properties of water, this section begins to build a model based on the idea of atoms and molecules.

## YOUR TURN *Matter at the Microlevel* *(page 29)*

In this activity, students use the theory of molecules to explain observations. Additional details on the atomic theory and the kinetic-molecular theory are found in subsequent units. At this point it is enough for students to account for certain aspects of the macroscopic world in terms of submicroscopic particles. The three observations can be demonstrated in the classroom and discussed by student groups.

## Answers to Questions

1. a. Molecules of gas in a balloon can escape slowly through small holes in the balloon wall. The helium-filled balloon shrinks more rapidly than other gas-filled balloons because the helium gas particles are smaller and faster-moving than those of most other gases.
   b. Metal-foil (Mylar) balloons do not shrink as quickly as plastic balloons because of smaller-size holes in the metal foil wall. (This can be set up as a two-day demonstration. Use two identical balloons filled equally, one with helium and the other with human breath; then use two same-size balloons—one rubber, one Mylar—filled with the same gas.) The point is that all matter is porous.
2. Solids and liquids are more dense than gases because the particles are much closer together in solids and liquids.
3. When the perfume bottle is open some perfume molecules escape from the surface of the liquid and travel through the air.

## B.5 SYMBOLS, FORMULAS, AND EQUATIONS *(pages 30–31)*

Students are introduced to symbols, formulas, and equations and to the significance of subscripts and coefficients. Emphasize the importance of a common, worldwide chemical language. These topics are further developed in the Resources unit. It is not necessary for students to memorize symbols or learn to balance equations now. Instead, show students where to find a listing of symbols and elements. Emphasize that an equation represents a chemical reaction that happens or that someone predicts might happen.

## YOUR TURN *Working with Symbols, Formulas, and Equations* *(pages 30–31)*

This activity reinforces the information just read and discussed by giving students practice recognizing and interpreting symbols, formulas, and equations.

### Answers to Questions

1. a. phosphorus; b. nickel; c. nitrogen; d. cobalt; e. bromine; f. potassium; g. sodium; h. iron (note the Latin-name symbols for the last three elements)
2. a. 2 hydrogens, 2 oxygens
   b. 1 calcium, 2 chlorines
   c. 1 sodium, 1 hydrogen, 1 carbon, 3 oxygens
   d. 2 hydrogens, 1 sulfur, 4 oxygens
3. a. One molecule of nitrogen reacts with one molecule of oxygen to form two molecules of nitrogen oxide.
   b. Two atoms of nitrogen react with two atoms of oxygen to form two molecules of nitrogen oxide, each containing one atom of nitrogen and one of oxygen. Note that there are two atoms of nitrogen and two atoms of oxygen on each side of the equation.

**Extra items for practice**

ethylene glycol (car antifreeze): $C_2H_6O_2$ — 2 carbons, 6 hydrogens, 2 oxygens; ammonium phosphate (a fertilizer): $(NH_4)_3PO_4$— 3 nitrogens, 12 hydrogens, 1 phosphorus, 4 oxygens; iron(III) oxide (common rust): $Fe_2O_3$ — 2 irons, 3 oxygens

## B.6 THE ELECTRICAL NATURE OF MATTER *(pages 31–33)*

The electrical nature of matter is discussed in terms of the interaction among charges, and in terms of neutrons, protons, and electrons. Water is described as a polar molecule whose structure helps explain certain observations, such as water's relatively high boiling point. To help students understand ions, ask them to interpret the final paragraph of B.6 in the student text by drawing a picture.

## B.7 PURE AND IMPURE WATER *(pages 33–34)*

This section distinguishes between pure water and clean water. It also discusses some unwanted substances that enter our water supplies.

## B.8 LABORATORY ACTIVITY: WATER TESTING *(pages 34–37)*

Students test water samples for the presence of dissolved ions— iron(III), calcium, chloride, and sulfate. The qualitative analysis procedures in this activity are based on a "confirming test" approach. Emphasize that a negative test does not necessarily mean that the ion was absent, only that it was not present in detectable quantities. The macroscale version of this Laboratory Activity is found in the Enrichment Materials section at the end of this unit.

### Time

50 minutes

## Materials (for a class of 24 working in pairs)

12 24-well wellplates
120 Beral pipets

*Test solutions*

50 mL 0.5 M KSCN (potassium thiocyanate, 2.4 g/50 mL) [$SCN^-$ ions]
50 mL 0.1 M $Na_2C_2O_4$ (sodium oxalate, 0.67 g /50 mL) [$C_2O_4^{2-}$ ions]
50 mL 0.1 M $HC_2H_3O_2$ (acetic acid, 0.3 mL conc. acetic acid/50 mL)
50 mL 0.1 M $AgNO_3$ (silver nitrate, 0.85 g/50 mL) [$Ag^+$ ions]
50 mL 0.1 M $BaCl_2$ (barium chloride, 1.2 g $BaCl_2 \cdot 2H_2O$/50 mL) [$Ba^{2+}$ ions]

*Reference ion solutions*

50 mL 0.1 M $Fe(NO_3)_3$ ( 8.0 g $Fe(NO_3)_3 \cdot 9\ H_2O$/50 mL) [$Fe^{3+}$ ions]
50 mL 0.1 M $CaCl_2$ ( 5.9 g $CaCl_2 \cdot 2H_2O$/50 mL) [$Ca^{2+}$ ions and $Cl^-$ ions]
50 mL 0.1 M $FeSO_4$ ( 1.4 g $FeSO_4 \cdot 7H_2O$/50 mL) [$SO_4^{2-}$ ions]

## Safety

Students should wear safety goggles and aprons throughout the laboratory activity. Caution students that sodium oxalate is a poison and silver nitrate stains skin.

## Advance Preparation

Movement around the classroom is minimized during the activity if the pipets are labeled and filled with the solutions and distributed upside down in the wellplate. The same pipets and solutions can be used for several classes.

## Pre-Lab Discussion

Repeat the philosophy behind this confirming test approach. Also emphasize the use of distilled water as a control. Remind students that although small quantities are used, it is necessary to exercise care in handling all materials in the laboratory. If your tap water is particularly pure, you may wish to "manufacture" your own tap water sample, which will contain at least one of the ions being tested.

Remind students to clean the wellplates immediately after use.

## Lab Tips

If a test color is too dark, students may add small amounts of distilled water in order to see the color better.

## Expected Results

| Reference solution | Color | Precipitate |
|---|---|---|
| $Fe^{3+}$ | red | no |
| $Ca^{2+}$ | white | yes |
| $Cl^-$ | white | yes |
| $SO_4^{2-}$ | white | yes |

## Answers to Questions

1. a. The control represents the absence of the ion the chemist is attempting to detect and therefore is useful in interpreting the experimental results.
   b. Distilled water is presumed to be essentially free of the ions detected by the tests.
2. The concentration of some ions may be too low to detect. Or if more than one ion is present, the test for one ion may interfere with or mask the behavior of another ion.
3. These tests can confirm the presence of an ion but cannot prove the absence of ions.
4. Improper cleaning may leave stray ions behind that can contaminate and interfere with tests of solutions in the wellplate.

## Post-Lab Discussion

Focus on student answers to questions. These questions develop analytical skills in interpreting laboratory data, which will be useful throughout the course.

## B.9 YOU DECIDE: THE RIVERWOOD MYSTERY *(pages 37–39)*

This activity introduces students to an important aspect of a scientific approach to solving problems—distinguishing facts from opinions.

Facts relevant to help figure out the cause

After listing facts and factual questions, students attempt to eliminate causes for the fish kill. Students should recognize that if the cause of the fish kill is water-related, it must be due to something suspended or dissolved in the water. This at least narrows the range of possible fish-kill causes.

Assign students to groups of four or five to work on this activity. The entire class compares and discusses the group answers. Assign the rereading of the newspaper articles as homework. Direct student groups to prepare answers to the three questions to share with the class. Suggest that groups might review the chemical knowledge from the previous sections to help in forming questions.

1. Possible facts or data
   Dead trout and other fish were found floating at the base of the Snake River Dam, five miles upstream from the Riverwood water intake. Preliminary tests revealed no evidence of toxic substances. The town council voted to shut down the water plant for three days. Analysis of the fish revealed signs of biological trauma, including hemorrhaging and small bubbles under the skin along the lateral line. Suspended particles were ruled out as a cause of the kill.
2. Factual questions that might be posed
   What types of substances might be dissolved in water?
   What are possible emission sources upstream from the dam?
   What happens to water as it goes over the dam? Do any additional substances dissolve in the water as it goes over the dam?
   What is the nature of the bubbles found in the fish?
   What substances might harm fish but not humans?
   Could the fish have died from a deficiency or overdose of a naturally occurring and necessary substance?
3. There are many possible answers, which depend on the student questions. At this point, closure or a correct answer is not possible—there is more information coming to help clarify the situation.

## CHEMISTRY AT WORK/Environmental Cleanup: It's a Dirty Job...But That's the Point *(page 38)*

This vignette illustrates how the work of an environmental contaminants specialist involves investigation of problems. One question that must be asked is if the cleanup will do more harm than good.

## B.10 WHAT ARE THE POSSIBILITIES? *(page 39)*

This section identifies the process just used by students as typical of the work of scientists. It is a bridge to Part C, where students will investigate the types of substances that can be dissolved or suspended in water, and determine which ones are potentially harmful to aquatic life.

Knowledge of scientific facts, theories, laws, and principles helps the working scientist to form questions or hypotheses just as chemical knowledge helps students to reject some possible causes as improbable.

## PART B: SUMMARY QUESTIONS *(page 40)*

1. a. Unusual properties of water include its lower density as ice [solid]; unusually high boiling point [liquid]; high surface tension [liquid]; excellent solvent [liquid].
   b. Ice floats on a lake, insulating life below the surface. Water remains a liquid at higher temperatures than do other substances of comparable size and mass. Surface tension allows water to bead and form raindrops and be transported in plants. Water

transports materials essential for human life to all cells in an organism.

2. Gasoline
3. a. liquid water; b. lead; c. AgCl precipitate; d. cold air
4. a. suspension; b. solution; c. suspension; d. colloid; e. solution; f. suspension; g. colloid
5. a. compound; b. element
6. a. "100% chemical-free water" is not possible because water itself is a chemical substance.
   b. "Pure" refers to water such as distilled water, without any dissolved ions or molecules. Clean and safe water samples, in contrast, may still contain dissolved gases and ions that give water a pleasing taste.
7. a. 3 hydrogens, 1 phosphorus, 4 oxygens
   b. 1 sodium, 1 oxygen, 1 hydrogen
   c. 1 sulfur, 2 oxygens
8. a. One molecule of methane reacts with two molecules of oxygen to form one molecule of carbon dioxide and two molecules of water.
   b. methane—one carbon atom, four hydrogen atoms; oxygen—four oxygen atoms; carbon dioxide—one carbon and two oxygen atoms; water—four hydrogen atoms, two oxygen atoms. Each side contains the same number of each type of atom; the equation is balanced.

## EXTENDING YOUR KNOWLEDGE *(page 41)*

- Because 100 Celsius degrees = 180 Fahrenheit degrees, 1.8 °F = 1 °C or 1 °F = 0.56 °C, the following expression can be written: °F = (1.8 × °C) + 32; if this expression is solved for °C, we obtain °C = (°F – 32)/1.8.
- Since the density of oil is less than the density of liquid water, the oil will float on the water. An oil spill floating on water can ignite and burn or block the sunlight from reaching the aquatic plants. It can slow the exchange of gases between the water and the atmosphere and can be blown by the winds and carried by the currents and tides to beaches. On the other hand, if oil were more dense than water, cleanup efforts would probably be more difficult.
- Symbols for some elements are derived from the Latin name for the element.

| Element | Symbol | Latin origin |
|---|---|---|
| Gold | Au | Aurum; "shining dawn" |
| Iron | Fe | Ferrum |
| Lead | Pb | Plumbum |
| Mercury | Hg | Hydrargentum; "liquid silver" |
| Potassium | K | Kalium; first isolated by electrolysis in 1807 |
| Silver | Ag | Argentum |
| Sodium | Na | Natrium; also first isolated by electrolysis in 1807 |
| Tin | Sn | Stannum |

# C. INVESTIGATING THE CAUSE OF THE FISH KILL

Part B established that a possible cause of the Snake River fish kill was something dissolved in the water. Part C provides students with deeper understanding of the concepts of solubility and solutions, as well as the chemistry of ionic and molecular substances.

Part C begins by introducing basic terms and concepts related to solubility. After gaining a working knowledge of these ideas, students study several kinds of substances that dissolve in water. They first consider the importance of dissolved oxygen (DO) for the survival of aquatic life. To examine the possibility that a shortage of DO caused the fish kill, students find out how much oxygen is available to aquatic life and how much DO fish need to survive. This leads to questions about the influence of temperature on solubility. After examining and graphing the DO data from the Snake River, students rule out the possibility of DO shortages induced by thermal pollution and move on to consider other possibilities.

Acid contamination is then considered as another possible cause of the fish kill. Students learn to recognize common acids and bases from their formulas. Acid contamination is ruled out as a fish-kill possibility. Students next consider heavy-metal ions. A decision-making activity leads them to rule out this as well. Part C ends by examining some molecular substances that dissolve in water.

## C.1 SOLUBILITY *(pages 42–44)*

Section C.1 presents vocabulary and concepts related to solubility: *solute, solvent, solution, saturation*, and *solubility curve*. The temperature dependence of the solubility of gases and solids and the effect of pressure on gaseous solubility are explained. A solubility curve is used to help clarify the terms *saturated, unsaturated*, and *supersaturated*. Additional teacher background is found in the Enrichment Materials section at the end of this unit.

### Demonstration Ideas

1. Supersaturation can be demonstrated by dissolving 160 g sodium acetate trihydrate, $NaC_2H_3O_2{\cdot}3H_2O$, in 30 mL distilled water. Heat the mixture until all solid is dissolved. Allow the solution to cool to room temperature. Add a single crystal of sodium acetate to the solution, and observe the crystallization. The solution can be reused by reheating to dissolve the solid.
2. The effect of temperature on gas solubility can be demonstrated as follows: chill a bottle of carbonated drink (colored beverages show the result better) to just above 0 °C. Prepare three beakers of water. Have one at 0 °C (ice-water bath), one at room temperature, and the third at about 40 °C. Open the bottle and pour soda gently into three test tubes until full. Close each tube with a one-hole stopper, and with your finger over the stopper hole, invert the tube and immerse it in one of the water baths. Do the same with the other two. Having two students assist you, so that all three tubes are inverted and immersed simultaneously, is more effective. In the beakers at room temperature and 40 °C, carbon dioxide is released from the solution and can be seen as bubbles in the test tube; the relative numbers of bubbles show that the solubility of $CO_2$ is less at higher temperatures.

## YOUR TURN *Solubility and Solubility Curves* *(pages 44–45)*

The purpose of this activity is for students to learn to interpret solubility curves such as those in Figures 18 and 19. The answers can be found by using proportions or by thinking through the steps. This activity could be completed in small groups and used to supplement a class discussion.

### Answers to Questions

1. a. 80 g
   b. 42 g
2. a. 17 g
   b. $\frac{42 \text{ g KNO}_3}{100 \text{ g H}_2\text{O}} = \frac{25 \text{ g KNO}_3}{x \text{ g H}_2\text{O}}$ $x = 60 \text{ g H}_2\text{O}$
3. a. 7.7 mg at 30 °C
   b. 9.2 mg at 20 °C
4. $\frac{9.2 \text{ mg O}_2}{1{,}000 \text{ g H}_2\text{O}} = \frac{x \text{ mg O}_2}{1{,}000 \text{ g H}_2\text{O}}$ $x = 0.92 \text{ mg O}_2$

## C.2 SOLUTION CONCENTRATION *(pages 45–46)*

This section introduces percentage as a way to express solution concentration and modifies that concept to include parts per hundred (pph), parts per thousand (ppt), and parts per million (ppm). Students may have read the term *ppm* in news articles about air and water pollution.

## YOUR TURN *Describing Solution Concentrations* *(pages 46–47)*

In this activity, students use the concept of percent to calculate concentrations of solutions.

### Answers to Questions

1. a. Sugar
   b. Water
2. a. $\frac{17 \text{ g sucrose}}{200 \text{ g solution}} \times 100 = 8.5\%$ (8.5 pph)
   b. $\frac{34 \text{ g sucrose}}{400 \text{ g solution}} \times 100 = 8.5\%$ (8.5 pph)
3. A solution is *saturated* when the solvent holds as much of a solute as it can at a given temperature.
4. a. Less oxygen would be dissolved in the warmer water.
   b. The solubility of gases in liquids decreases with increasing temperature, according to Figure 19.

## C.3 OXYGEN SUPPLY AND DEMAND *(pages 47–49)*

This section points out that dissolved molecular oxygen gas ($O_2$) is needed by aquatic animals. The oxygen atom in the water molecule is not available because of the high energy required to free it from the hydrogen. The oxygen balance in a body of water is also discussed. The concentration term *ppm*, introduced in the last section, is used.

## C.4 TEMPERATURE AND GAS SOLUBILITY *(pages 49–50)*

Increased water temperature increases the fishes' need for oxygen at the same time that it decreases the available dissolved oxygen. The chemical reactions involved in the metabolism of cold-blooded animals increase at higher temperatures. In addition, anaerobic bacteria (which also require more oxygen at higher temperatures) compete with fish for the available supply. This explains why fish kills are found after long periods of hot weather.

Analogies help students understand concentrations expressed as parts per million. If a goldfish is swimming inside a blue whale, the goldfish represents one part fish (by weight) per million parts of aquatic animals. An additional ChemQuandary is found in the Enrichment Materials section at the end of this unit.

## C.5 YOU DECIDE: TOO MUCH, TOO LITTLE? *(pages 50–52)*

In this activity, students plot graphs using data in Tables 6, 7, and 8. After the graphs are completed, they are used to search for patterns and draw conclusions about the likelihood that the amount of dissolved oxygen is related to the Riverwood fish kill.

This activity requires 90–120 minutes to complete. If necessary, review the meaning of parts per million (ppm) and the concept of saturation. Be sure that all students read and understand the graphing hints. A common graphing error made by beginning students is to make the equal divisions on the graph paper represent unequal values. Some students like to draw very small graphs. Direct their attention to the first hint. An average class requires continual monitoring and assistance to complete this first graphing exercise successfully.

## Answers to Questions *(page 52)*

1. As temperature increases, dissolved oxygen decreases.
2. Check student graphs for plotting techniques.
3. a. The average water temperature in June is 11 °C and in December is 7 °C. One would thus expect a lower oxygen ppm in June, which is verified by the data.
   b. Since March and November exhibit the same water temperature (7 °C), one would expect the oxygen concentrations to be similar. In fact, both months show 11.0 ppm of dissolved oxygen.
4. The data for August of this year show a higher concentration of dissolved oxygen than the corresponding data for last year. The average August temperature was lower than that of last year. The difference in the concentrations of oxygen in water is probably explained by the 2 °C difference in the average temperatures.

## Answers to Questions *(pages 52–54)*

1. a. On both graphs the levels rise during the day and fall in the evening.
   b. Although the changes do not exactly correspond, the oxygen levels are generally higher when the temperatures are higher.
   c. Oxygen levels are higher during the day, probably because of photosynthesis.
2. Average water temperature = 21.2 °C. Average oxygen concentration = 9.13 ppm.
3. It is unlikely that the one-day average chosen for this year would be representative of the average weather for last year—there may have been drastic changes in weather, cloud cover, temperature, etc.
4. For a given day, maximum and minimum values provide more useful information than does an average value. Events occurring during the course of the day, such as increased photosynthesis at midday, might be overlooked if only averages were examined.
5. Point out to students that because so many sets of values are identical, they do not need to do a division problem for each hour of the day.

**Data for One-Day, Hourly Samples**

| Time | Temp., °C | $O_2$, ppm | Sat. $O_2$, ppm | % of sat. |
|---|---|---|---|---|
| 8 a.m. | 21 | 9.1 | 9.0 | 101 |
| 9 | 21 | 9.1 | 9.0 | 101 |
| 10 | 21 | 9.1 | 9.0 | 101 |
| 11 | 21 | 9.1 | 9.0 | 101 |
| 12 noon | 22 | 9.2 | 8.8 | 105 |
| 1 p.m. | 23 | 9.3 | 8.7 | 107 |
| 2 | 23 | 9.3 | 8.7 | 107 |
| 3 | 23 | 9.2 | 8.7 | 106 |
| 4 | 23 | 9.2 | 8.7 | 106 |
| 5 | 23 | 9.2 | 8.7 | 106 |
| 6 | 23 | 9.2 | 8.7 | 106 |
| 7 | 23 | 9.2 | 8.7 | 106 |
| 8 | 22 | 9.2 | 8.8 | 105 |
| 9 | 22 | 9.2 | 8.8 | 105 |
| 10 | 22 | 9.2 | 8.8 | 105 |
| 11 | 21 | 9.1 | 9.0 | 101 |

| | | | | |
|---|---|---|---|---|
| 12 midnight | 21 | 9.1 | 9.0 | 101 |
| 1 a.m. | 21 | 9.1 | 9.0 | 101 |
| 2 | 19 | 9.0 | 9.3 | 97 |
| 3 | 19 | 9.0 | 9.3 | 97 |
| 4 | 19 | 9.0 | 9.3 | 97 |
| 5 | 19 | 9.0 | 9.3 | 97 |
| 6 | 19 | 9.0 | 9.3 | 97 |
| 7 | 19 | 9.0 | 9.3 | 97 |

6. Based on the above data, and the information in the text about dissolved oxygen levels needed to support river-water fish, the oxygen concentration in the Snake River appears adequate for most fish species to survive.
7. The amount of dissolved oxygen in the Snake River is not a likely cause of the Riverwood fish kill. The levels of dissolved oxygen, as calculated from the collected data, are well within the acceptable range for most fish species to survive.

## C.6 ACID CONTAMINATION *(pages 54–56)*

Acidity and basicity play a major role in water quality and the survival of aquatic life. Students are introduced to general properties of acids and bases, and learn about some common acids and bases. The concepts of neutralization and the pH scale are also discussed briefly. The textbook's treatment of pH as a scale for distinguishing acids and bases is generally easily understood by students, who have encountered the term in advertisements (pH-balanced shampoo or deodorant) or in hot-tub or swimming pool maintenance. The treatment of pH and acids and bases illustrates part of the *ChemCom* philosophy—the chemistry is introduced only as needed. Students will learn more about acids and bases later in the course.

Students learn that pH measurements taken in the Snake River near the site of the fish kill were within the normal range at the time of the fish kill. These data eliminate another possible cause of the fish kill.

## C.7 IONS AND IONIC COMPOUNDS *(pages 56–57 )*

The naming of ionic compounds and a simple method for writing formulas for ionic compounds are introduced and discussed very briefly. Be sure your students understand that ionic crystals are composed of regular arrangements of positive and negative ions (see Figure 25). The formula NaCl thus represents the 1:1 ratio of the number of cations and anions in the structure.

### YOUR TURN *Ionic Compounds* *(page 57)*

This activity provides practice in recognizing, naming, and writing the formulas of ionic compounds.

### Answers to Questions

| Cation | Anion | Formula | Name |
|---|---|---|---|
| 1. $K^+$ | $Cl^-$ | KCl | potassium chloride |
| 2. $Ca^{2+}$ | $SO_4^{2-}$ | $CaSO_4$ | calcium sulfate |
| 3. $Ca^{2+}$ | $PO_4^{3-}$ | $Ca_3(PO_4)_2$ | calcium phosphate |
| 4. $NH_4^+$ | $NO_3^-$ | $NH_4NO_3$ | ammonium nitrate |
| 5. $Fe^{3+}$ | $Cl^-$ | $FeCl_3$ | iron(III) chloride |
| 6. $Al^{3+}$ | $SO_4^{2-}$ | $Al_2(SO_4)_3$ | aluminum sulfate |
| 7. $Na^+$ | $HCO_3^{1-}$ | $NaHCO_3$ | sodium hydrogen carbonate |
| 8. $Mg^{2+}$ | $OH^-$ | $Mg(OH)_2$ | magnesium hydroxide |
| 9. $Fe^{3+}$ | $O^{2-}$ | $Fe_2O_3$ | iron(III) oxide |
| 10. $Ca^{2+}$ | $CO_3^{2-}$ | $CaCO_3$ | calcium carbonate |

Note: Some students may use the common names for some compounds (baking soda or rust). Point out that such names give less information than does the more systematic naming of both cation and anion.

## C.8 DISSOLVING IONIC COMPOUNDS *(pages 58–59 )*

Emphasize that dissolving can be viewed as involving two steps—overcoming attractive forces among the particles of the solute (and solvent) and forming new attractive forces between solute and solvent particles.

Careful examination of Figure 26 shows how the solution might look if we could see the ions and water molecules.

## C.9 HEAVY-METAL ION CONTAMINATION *(pages 59–62)*

Another candidate for responsibility of the fish kill is heavy-metal ion contamination. The cations of Pb, Hg, and Cd are discussed in relation to their chemistry, commercial uses, and toxic effects on living organisms. The heavy metals are particularly dangerous because they accumulate in the body and have serious health consequences. Emphasize that metal ions, not metal atoms, are dissolved in water. Discuss the idea of allowable limits.

## C.10 YOU DECIDE: HEAVY-METAL IONS *(pages 62–63)*

This decision-making activity provides students with data collected by scientists, which they analyze to determine whether heavy-metal contamination could be related to the fish kill.

The estimated time for this activity is 45 minutes if done in small groups (three or four students), which reduces the time spent on repetitive calculations. Discuss the risk factor equation in Question 3 and review calculations involving parts per million (ppm). A full class discussion should follow completion of the questions. Emphasize how knowledge of ions in solution, concentration (ppm), and the concept of a risk factor enable citizens to seek appropriate information from experts.

### Answers to Questions

1. $Hg^{2+}$, $Cl^-$, $NO_3^-$
2. $Cd^{2+}$, $Pb^{2+}$, $Se^{2-}$, $SO_4^{2-}$
3. $Cd^{2+} = 0.67$, $Pb^{2+} = 0.27$, $Se^{2-} = 0.031$, $SO_4^{2-}$ = no risk limit
4. $Cd^{2+}$ is highest risk factor for aquatic life
5. $Cd^{2+} = 0.1$, $Pb^{2+} = 0.4$, $Se^{2-} = 0.8$, $SO_4^{2-} = 0.14$
6. Lead ions and selenium ions have the greatest risk factor for humans, but neither is over EPA limits. The suggestion may be made that the community continue to monitor these two ions, even though there is no immediate danger.

## C.11 MOLECULAR SUBSTANCES IN THE RIVER *(pages 63–64)*

Although several possible explanations for the fish kill have been eliminated, the role of molecular substances remains to be explored. This section gives a structural explanation for why some substances dissolve in water while others do not. The simple explanation offered for polarity is sufficient for now.

## C.12 LABORATORY ACTIVITY: SOLVENTS *(pages 64–65)*

In this activity, students will explore the concept of "like dissolves like" by testing the solubility of several substances in water and hexane, examples of polar and nonpolar solvents.

### Time

30 minutes

### Materials (for class of 24 working in pairs)

24 test tubes (assuming tubes are reused), 13 × 100 mm
12 test tube racks
12 10-mL graduated cylinders
12 marking pencils

12 spatulas
12 medicine droppers
25 g urea, $CO(NH_2)_2$
25 g iodine, $I_2$
25 g ammonium chloride, $NH_4Cl$
25 g sodium chloride, NaCl
25 g naphthalene flakes, $C_{10}H_8$
25 g copper(II) sulfate, $CuSO_4 \cdot 5H_2O$
100 mL ethanol (denatured alcohol), $C_2H_5OH$
1 L distilled water, $H_2O$
600 mL hexane, $C_6H_{14}$

## Safety

Students should wear safety goggles and aprons throughout the laboratory activity. Direct skin contact with iodine is to be avoided. Consider demonstrating this part of the activity. Hexane and naphthalene are flammable. **No flames are allowed in the laboratory.**

## Pre-Lab Discussion

It is important that students follow the directions and use small solute samples to judge the solubilities accurately. The gentle mixing is also important. Provide labeled containers for discarded hexane samples. After class, the hexane can be evaporated in the fume hood. (If you have no hood, carry out the evaporation outdoors under your careful supervision.)

## Lab Tips

Careful supervision is necessary during the laboratory activity because of the toxicity of iodine and naphthalene, and the flammability of hexane.

## Expected Results

| Solute | Water | Hexane |
|---|---|---|
| Urea, $CO(NH_2)_2$ | Soluble | Insoluble |
| Ammonium chloride, $NH_4Cl$ | Soluble | Insoluble |
| Naphthalene, $C_{10}H_8$ | Insoluble | Soluble |
| Iodine, $I_2$ | Insoluble | Soluble |
| Copper(II) sulfate, $CuSO_4$ | Soluble | Insoluble |
| Ethanol, $C_2H_5OH$ | Soluble | Soluble |
| Sodium chloride, NaCl | Soluble | Insoluble |

## Answers to Questions

1. $NH_4Cl$, $CuSO_4$, NaCl, and $CO(NH_2)_2$ are all more soluble in water than in hexane. Note: Urea is often used as a fertilizer owing to its high nitrogen content and solubility in water.
2. Naphthalene and iodine are more soluble in hexane than in water.
3. Ionic compounds tend to dissolve in water, which is polar, and molecular compounds tend to dissolve in hexane, which is nonpolar. This reinforces the concept of polar with polar, and nonpolar with nonpolar, the "like dissolves like" concept.
4. Ethanol dissolves in both water and hexane, suggesting its structure has both polar and nonpolar features. (Its structure—and structures of other common organic substances including hexane—will be considered further in the Petroleum unit.)

## Post-Lab Discussion

Center the post-lab discussion on the questions at the end of the activity. The newspaper clipping "Fish Kill Remains a Mystery" eliminates organic compounds from sources such as pesticides, fertilizers, and industrial waste as potential causes of the fish kill.

## PART C: SUMMARY QUESTIONS *(pages 65–66)*

1. Carbon dioxide gas is dissolved in the liquid of a bottle of capped soda. Because the solubility of a gas in a liquid decreases with increasing temperature, gas comes out of solution at warmer temperatures, producing more "fizz" than at colder temperatures.
2. The phrase "like dissolves like" means that polar solvents will tend to dissolve polar solutes and many ionic substances, while nonpolar solvents will tend to dissolve nonpolar solutes.
3. Sodium chloride is an ionic substance. The ions of sodium chloride are attracted by the polar water molecules. Table salt will not dissolve in nonpolar cooking oil because there are no strong attractive forces between the ions and the polar solvent molecules.
4. a. Rivers with rapids have a greater concentration of dissolved oxygen because of aeration.
   b. A lake in springtime will have a greater concentration of dissolved oxygen than in the summer owing to lower spring temperatures. However, the argument of more photosynthesis occurring in summer may also be suggested by students.
   c. A lake containing trout will have a higher concentration of dissolved oxygen. Trout need a higher dissolved oxygen level to survive than do catfish.
5. a. The density of liquid water is about 1 g/mL; therefore, 100 mL of water has a mass of 100 g. This mass of water dissolves about 80 g $KNO_3$, according to Figure 18.
   b. The solution is supersaturated. At 60 °C the same solution will be unsaturated.
6. The abbreviation ppm means that 9 parts of oxygen are dissolved in 1 million parts water.
7. Sea water must be basic because its pH is greater than 7.
8. a. sodium nitrate
   b. magnesium sulfate
   c. aluminum oxide
   d. barium sulfate
9. The total mass of the solution is 35 g + 115 g = 150 g.

$$\frac{x \text{ g ethanol}}{100 \text{ g solution}} = \frac{35 \text{ g ethanol}}{150 \text{ g solution}} \quad x = 23 \text{ g ethanol}$$

10. Heavy metals are metallic elements, such as lead, mercury, and cadmium, that are heavier than the essential elements, such as iron, calcium, and magnesium. Heavy metals dissolve in water as ions. Small amounts of heavy metals are toxic to humans and accumulate in the body. Heavy metals are also concentrated in the food chain. Heavy metals damage the nervous system, kidneys and liver, causing mental retardation and even death. Heavy metals are especially toxic to infants and children.

## EXTENDING YOUR KNOWLEDGE *(page 66)*

- Because DDT dissolves in the fatty tissue of animals (composed of nonpolar substances), it is probably nonpolar. Substances soluble in water generally have polar molecules, which have low solubility in nonpolar solutes such as fats and oils.
- Student reports can be shared with the class to illustrate application of the solubility principles developed in this unit to another science field.

# D. WATER PURIFICATION AND TREATMENT

In Part D the focus is on water quality in the local community. This part features information needed to make intelligent decisions about water quantity and quality. It addresses questions such as, How pure is our water? How can we gain the water quality required? How can waterborne wastes be treated?

The natural water purification system, the hydrologic cycle, is described. The problem of hard water and techniques of water softening are explored. A laboratory activity gives students the opportunity to find out more about water softening. Attention is then turned to wastewater treatment and providing safe water for a community. In a *You Decide* activity, student groups consider one of three alternatives to chlorination of water supplies.

## D.1 NATURAL WATER PURIFICATION *(pages 67–68)*

Three basic steps of the water cycle are defined so students can recognize later that human water purification and treatment methods resemble aspects of the hydrologic cycle. The problem of high concentration of dissolved minerals in some water supplies (hard water) is briefly described.

## D.2 LABORATORY ACTIVITY: WATER SOFTENING *(pages 69–71)*

Students explore three ways to soften water. Full discussion of the chemistry of these processes can be delayed until section D.3, where it is considered in greater detail.

### Time

50 minutes

### Materials (for 24 students working in pairs)

12 10-cm diameter funnels
12 ring stands
12 ring clamps
12 clay triangles
72 pieces of 11-cm filter paper (Whatman No. 1)
48 test tubes (18 × 150 mm)
24 test tube racks
12 20-cm glass stirring rods
12 10-mL graduated cylinders
12 washing bottles
6 dropper bottles for sodium carbonate solution
Several burets for dispensing hard water
1.5 L hard water, containing 0.01 M $Ca^{2+}$ (2.2 g $CaCl_2 \cdot 2H_2O$/1.5 L)
100 mL 0.1 M sodium carbonate in dropper bottles
(1.2 g $Na_2CO_3 \cdot H_2O$ / 100 mL)
Liquid soap solution, Ivory or equivalent
12 wash bottles
1 L distilled water in wash bottles
1 lb sand
1 box Calgon (sodium hexametaphosphate)
100 g cation-exchange resin, ionic form $Na^+$ Dowex 50W-X8 or equivalent, 50–100 mesh

### Safety

Students should wear safety goggles and aprons throughout the laboratory activity.

## Advance Preparation

Prepare the ion-exchange resin as a water slurry. Add 2 parts water to 1 part resin in a beaker. Decant excess water. Soak the resin overnight before it is used. Purchase Calgon or equivalent water softeners, sodium hexametaphosphate for bath use, not sodium sesquicarbonate for dishes.

## Pre-Lab Discussion

Emphasize that hard water is caused by excess concentrations of calcium, magnesium, and/or iron(III) ions in water. Point out that water softening is done in many homes and will be discussed later.

## Lab Tips

The exchange resin contains 40–50% moisture and thus must be stored in sealed condition to avoid drying. The ion-exchange resin and sand can be reused from period to period. The resin can be regenerated for future use by washing it three times in a beaker with 1 M NaCl, and then three times with distilled water.

## Expected Results

| | Filter paper | Filter paper and sand | Filter paper and Calgon | Filter paper and ion-exchange resin |
|---|---|---|---|---|
| Reaction with sodium carbonate ($Na_2CO_3$) | cloudy | cloudy | no reaction | no reaction |
| Degree of cloudiness (turbidity) with Ivory soap | very clear | very clear | blue and turbid | very turbid |
| Height of suds | none | none | 1–2 mm | 1–2 mm |

## Answers to Questions

1. The ion–exchange resin should remove the hard-water ions best, but Calgon results may turn out better.
2. The higher the concentration of $Ca^{2+}$ remaining in the filtrate (that is, the harder the water), the less soluble the soap.
3. The cleaning action of a soap depends on how dispersed or dissolved it is in the water. The more soap molecules that are bound to hard-water ions such as $Ca^{2+}$, the less there are available for cleaning.
4. Soaps react with the hard-water ions to form a precipitate in the water. When the water drains, this precipitate sticks to the sides of the tub, leaving a characteristic "ring." Softeners, such as Calgon, contain compounds that remove some of the ions responsible for hard water. The label on the box suggests that Calgon be used for laundry, bathing, hand washing, and general cleaning. Calgon will bind the troublesome positive ions found in hard water, but it does not form a precipitate with those ions.

## Post-Lab Discussion

Organize your post-lab discussion around the questions provided. Students may be asked to interpret their observations. Consider mentioning that the ion-exchange process is used in most home water-softener systems. Such information provides a good transition to section D.3.

## D.3 HARD WATER AND WATER SOFTENING *(pages 71–73)*

This section identifies sources of hard-water ions. Explanations are given for common problems in the home caused by hard water. Section D.3 ends with a discussion of ion-exchange resins sometimes used to soften hard water. Additional teacher background is found in the Enrichment Materials section at the end of this unit.

## D.4 MUNICIPAL WATER PURIFICATION *(pages 73–77)*

The major steps in the purification of a municipal water supply are shown in Figure 31 and 32. Each step is also described in the text. Direct students' attention to the figures and the text.

## D.5 CHLORINE IN OUR WATER *(page 77)*

Chlorination of water offers both benefits and burdens. Students may have personal experience with chlorinated water in swimming pools or fish tanks. Encourage such contributions from students, since they relate the chemistry to everyday experience.

## D.6 YOU DECIDE: CHLORINATION AND THMs *(pages 77–78)*

This *You Decide* points out that technological solutions to many problems create new problems. The formation of trihalomethanes (THMs), such as chloroform ($CHCl_3$) from the chlorination of water, offers one example. Assign three or four students to a group, and assign each group to evaluate one of the three alternative methods of water purification. Groups can present their reasoning to the class in oral reports.

### Answers to Questions

1. Activated charcoal filters are expensive to install and operate on a large scale. Ozone or ultraviolet radiation disinfects water but does not provide added protection once the water leaves the plant. Eliminating the prechlorination step results in formation of smaller quantities of THMs. One disadvantage of this idea is that the decreased chlorine concentration makes the water more susceptible to bacterial growth.
2. Other alternatives to chlorination include the use of chlorine dioxide, ozone, chloramines, UV irradiation, iodination, or some combination of these.

## PART D: SUMMARY QUESTIONS *(page 78)*

1. Thermal energy from the sun causes surface water from the ocean and other sources to evaporate, leaving behind dissolved minerals. The water vapor rises and condenses into tiny droplets (clouds). The water eventually falls as precipitation. It then joins the surface water or seeps into the ground. Some of the groundwater eventually becomes surface water, and the cycle begins again.
2. a. Evaporation and condensation, bacterial action, and filtration take place in nature and in municipal water purification processes.
   b. The chlorination and flocculation steps in municipal water treatment procedures do not have direct counterparts in the hydrologic cycle.
3. a. The chlorination of water provides some assurance that the water is bacteria-free.
   b. Conversely, chlorine in water presents a health risk. Chlorine may react with organic compounds to produce trihalomethanes (THMs). Chloroform ($CHCl_3$), a suspected carcinogen (cancer-causing agent), is a representative trihalomethane.

## EXTENDING YOUR KNOWLEDGE *(page 78)*

• Increased population growth has raised the demand for goods and services. The increased direct and indirect demand for "clean" water can no longer be met by nature's recycling system. New knowledge of the adverse health effects of poor-quality water has increased the trend toward regulation of water quality by government agencies. In addition, new chemical and instrumental tests can determine smaller concentrations than was previously possible.

• Tertiary water treatment refers to a series of specialized chemical and physical processes that reduce the quantities of pollutants after primary and secondary treatment. The removal of organic wastes is typically the goal of tertiary water treatment. Sample procedures include activated carbon beds, chemical precipitation, ion exchange, electrodialysis, reverse osmosis, and distillation. The cost of tertiary water treatment is high owing to the large quantity of energy required.

- $Fe^{3+}(aq) + 3\ NaR \rightarrow FeR_3 + 3\ Na^+(aq)$. In this equation, R represents a cation-exchange resin site containing fixed negative charges, which interacts with mobile cations. Ion-exchange resins are macromolecules or insoluble polyelectrolytes having fixed charges distributed uniformly throughout the structure. During the ion-exchange reaction, the resins release sodium ions and bind the undesirable ion. Note that the charges are balanced as well as the numbers of atoms.
- About 50% of all U.S. residents (and more than 95% of all rural U.S. residents) rely on groundwater, to supply their water needs. Twenty large U.S. cities rely exclusively on groundwater and 12 states use it for more than half of their public supplies. The topic was not addressed in the student text, but it makes an excellent topic for optional student reports. Refer students to the supplemental references.

# E. PUTTING IT ALL TOGETHER: FISH KILL— WHO PAYS?

## FISH KILL CAUSE FOUND *(page 79)*

These newspaper articles contain the resolution of the Riverwood fish-kill mystery as well as Riverwood reactions to the final answer. Gas bubble disease, caused by an excess of air dissolved in the river water, has been identified as the cause of the fish kill. The most likely source of this supersaturated solution of gases in water is the power company's release of water from a dam upstream from Riverwood. Gas bubble disease is well documented in the technical literature.

## E.1 DIRECTIONS FOR THE TOWN COUNCIL MEETING *(pages 81–85)*

The simulated town council meeting addresses two questions: Who is responsible for the fish kill? Who should pay for the costs incurred because of the fish kill? The purpose of this culminating activity is to encourage students to use the chemistry they have learned to think through real-world problems resembling those they will be called upon to consider as voting adults. The preparation for the town meeting and the actual meeting should not require more than three 50-minute class periods.

Student groups prepare briefs for presentation and develop presentations that express the viewpoint of the group. Individually, students are assigned to write an editorial or conduct an interview about the Riverwood problem. Both group and individual responsibility are built into this activity.

This activity helps students recognize that science does not provide answers to social problems but scientific principles can often be used to help formulate possible solutions to such problems. Different people can reach different conclusions from the same facts, based upon their own perspectives or values. Science enables us to pose productive questions. Although scientific knowledge provides information on many of these problems, the decisions to take action must always involve additional considerations from social, economic, political, moral, and related perspectives. The activity is intentionally open-ended. There is no "right answer" for the class to reach.

### Pre-Activity Preparation and Discussion

Many of the detailed procedures for this activity are summarized for students in the text. Students should be assigned their roles in the town council meeting at least one class period before the activity begins. The two town council members are needed to organize and monitor the debate. Consider the personalities of the students you choose for these roles.

If you wish to have smaller groups or more speakers, add other "special interest" groups to the meeting. For example, there could be two groups of engineers: one as a source of information, and the other as original designers of the dam. The Riverwood Corporation, with some fictitious product, could be represented. The only limit to this is time and your imagination!

Encourage students to bring appropriate props such as hard hats or lab coats to increase the authenticity of group identities. Each interest group could design a logo to be displayed on a badge or name tag symbolizing their identity. Encourage students to use talents in other areas such as art, graphics, computer science, business, drama, journalism, to create the town meeting atmosphere. If possible, rearrange the classroom tables and chairs to resemble a meeting room. A microphone adds realism to the setting. Videotaping the meeting lends an element of the theater to the meeting, and students thoroughly enjoy seeing themselves.

### Post-Activity Discussion

Following completion of this activity, conduct a class discussion regarding students' experiences in this Riverwood simulation. At this point students should abandon their assigned roles and critically analyze the entire decision-making process. This discussion is central to the success of the activity. Consider focusing on points such as these:

1. Informed decisions concerning any problem should take into account opposing views rather than jumping to conclusions.
2. Ask students to identify points in the debate where chemistry knowledge was essential.
3. Point out the benefits and consequences of seeking consensus solutions to complex issues. Discourage students from viewing the debate in terms of winning or losing.
4. Highlight, with students' help, examples of factual, opinion-based, emotional, and logical arguments presented during the debate.
5. Point out that each group had worthwhile viewpoints to contribute to the debate. The final resolution often becomes a compromise among the different group positions.

The evaluation of student performance in this activity should include both individual and group grades. Group grades can be based upon the oral presentation or the simulated interview. It is helpful to have a written list of items to evaluate during the town council meeting. The list can be shared with the students before the debate begins. Items to consider include active participation in group work, effective use of facts, use of persuasive arguments, adherence to time requirements, evidence of research beyond the text. Teachers will want to generate their own lists. Written letters to the editor permit you to evaluate individual performance.

## E.2 LOOKING BACK AND LOOKING AHEAD *(page 85)*

This section provides a transition from this initial *ChemCom* unit to the next unit, Conserving Chemical Resources. Several important points are made in this section regarding the story of the Riverwood water crisis. Discuss these ideas briefly or assign the reading as homework.

# ENRICHMENT MATERIALS

Optional activities are presented to enrich the learning environment for students. Some classes and individuals learn very quickly; optional activities may be assigned to such students. Some of the ChemQuandaries may be used for evaluation. Be sure not to assign so many optional activities that the classroom pace through the unit is adversely affected. Students should not be held responsible for recalling particular figures and statistics found here.

## TEACHER BACKGROUND

### Part A

On a worldwide basis enough rain and snow fall each year to cover Earth's total land area to a depth of 83 centimeters. And yet, uneven distribution and inadequate technology make high-quality fresh water a major constraint on economic growth, food production, and general health in many nations. This is especially true of Asia and Africa, where periodic droughts lead to the deaths of millions through famines. Even when water is available in adequate quantities, it may be of poor quality. It has been estimated that 80% of all the world's sickness is caused by contaminated water, and that persons with water-borne diseases occupy half of the world's hospital beds and die at a rate of 25,000 per day.

### Indirect Water Use

Like energy, water is an essential "behind-the-scenes" ingredient in nearly every human endeavor. In the United States, where overall water use is over 7,000 L/person/day, industrial processes, such as electrical energy production, primary metals production, chemical products, petroleum refining, pulp and paper manufacturing, and food processing, account for about 51% of the use, agriculture for about 40%, and the municipal sector for about 9%. The point to make with students is that not only do they use large quantities of water directly (325 L/person/day), but even larger quantities are used on their behalf behind the scenes.

### Section A.8 and A.9

Although household and other municipal water use accounts for about only 9% of the total water use in the United States, delivering adequate quantities of water of sufficient quality for this purpose is increasingly expensive for individuals and communities. Families can purchase conservation devices, install the devices themselves to decrease water consumption, and save money in the process. Investments in household water-saving devices tend to pay off through lower water, sewer, and energy costs within a few months to a year. Encourage students to visit hardware or plumbing stores to examine such devices.

**Potential Water Savings with More Efficient Household Fixtures**

| Fixture | Water use (L) | Water saved (%) |
|---|---|---|
| Toilets (per flush) | | |
| Conventional | 19 | |
| Low flush | 13 | 32 |
| Air-assisted | 2 | 89 |
| Clothes washers (per load) | | |
| Conventional | 140 | |
| Wash recycle | 100 | 29 |
| Front-loading | 80 | 43 |
| Shower heads (per min) | | |
| Conventional | 19 | |
| Flow-limiting | 7 | 63 |

| | | |
|---|---|---|
| Air-assisted | 2 | 89 |
| Faucets (per min) | | |
| Conventional | 12 | |
| Flow-limiting | 6 | 50 |

Source: *Chemistry in Context,* Table 5.2, p. 126, American Chemical Society, Brown (1994).

**Section C.1**

The solubility of a solute in a solvent (water in the present discussion) depends on the relative strengths of solute-solute and solute-solvent attractions. When a substance dissolves, bonds between solute molecules and between solvent molecules are broken. New bonds form between solute and solvent molecules. Breaking bonds always requires energy, and making bonds always produces energy.

For many solids, the overall dissolving reaction is endothermic because more energy is needed to overcome the solute-solute attractions than is released when the solute-solvent bonds are formed. In this case, adding thermal energy by raising the temperature will permit more solid to dissolve. This is reflected in the solubility curves in Figure 18. Pressure has no significant effect on the solubility of solids, Grinding and stirring change the rate of dissolving but do not affect how much solid can dissolve.

In gases there are virtually no forces between solute molecules, so the process is to make new bonds between solute and solvent molecules. The overall reaction produces energy. At higher temperatures less gas dissolves in a given quantity of water. The solubility of gases is also directly proportional to the pressure. This is demonstrated by observing canned or bottled soft drinks, in which the carbon dioxide is dissolved under as much as 10 atm of pressure. When the pressure is released by opening the container, bubbles of gas escape.

**Section D.3**

The following equation summarizes the reaction that occurs in water softeners. R represents an ion-binding site in the resin:

$$2\ NaR + Ca^{2+} \rightarrow CaR_2 + 2\ Na^+$$

The resin is prepared by rinsing thoroughly with solutions containing sodium ions, which bind to the anion sites. When hard water containing calcium or magnesium ions is filtered through the resin, the more highly charged ions are attracted to the anion sites more strongly, replacing the sodium ions, which are discharged with the flow of the liquid through the resin. The resin is regenerated by washing with a concentrated sodium ion solution (sodium chloride solution). Hard water is prepared with these types of resins. In some parts of the United States, students may have water softeners in their homes. Encourage them to report on the operation of these devices.

Students will need to understand the concept of charge balance illustrated in this equation if they are asked to answer the Extending Your Knowledge questions. The notion is simply that the total electrical charge must remain constant on the resin and in the water— here we see that two $Na^+$ ions are freed for each $Ca^{2+}$ ion bound to the resin.

## CHEMQUANDARIES

### Section A.9

Explain the relevance of the Benjamin Franklin quotation (1746), "When the well is dry, we learn the worth of water."

**Answer:** It is in the nature of economic systems to place a higher value on scarce commodities. Even though water is a renewable resource, it is possible to deplete and pollute it at a rate that exceeds natural chemical recycling processes.

### Section C.4

One 40-L aquarium is kept at 25 °C, while a second identical aquarium is kept at 20 °C. Which aquarium can support a larger population of fish? Why?

**Answer:** Fish require dissolved oxygen to survive. The fish tank that is kept at the lower temperature (20 °C) will support a larger fish population.

## Section D.3

Ion–exchange water-softening units substitute sodium ions for hard water ions such as $Ca^{2+}$. Does this create any new problems?

**Answer:** Yes, since the water now contains higher concentrations of sodium ions. People with high blood pressure needing low-sodium intake must find alternate sources of drinking water.

## Other Activities

**Water Unit**

Students can maintain a "Chemistry in the News" bulletin board where local, regional, and national newspaper or magazine articles, cartoons, etc., are assembled related to the main idea of the unit being studied. Many variations are possible.

Such activities help emphasize that Riverwood-type scenarios are not just stories but happen regularly. Chemistry-related issues consistently appear in both national and local news, and *ChemCom* topics focus on a wide variety of community settings and events.

## Part A

1. Investigate the differences in water usage in various parts of the United States: Industrial East, Agricultural Midwest, Desert Southwest.
2. Investigate the uses of cooling water in electrical plants, and the relative numbers and types in the West and East. Explore the difference in water use between coal–fired, oil–fired, and nuclear power plants.

# MACROSCALE LABORATORY ACTIVITIES

For teachers preferring the macroscale activity, student instructions are printed on the following pages, which are ready for duplication. Larger quantities of substances and solutions are often needed for the macroscale activity. Some macroscale laboratory activities also require significantly more time than do the small-scale versions. Questions for students to answer are found in the student text following the small-scale version of the activity. Additional instructions for the teacher are found in the corresponding sections of this guide.

## B.3 LABORATORY ACTIVITY: MIXTURES

### Teachers' Directions

Directions for preparing the unknown solutions are found in the Advance Preparation part of the instructions for the small-scale version of the activity. The student version is found on separate pages following the Teacher's Directions.

### Materials

24 beakers (100-or 150-mL)
48 pieces of filter paper
12 funnels
12 ring stands
12 ring clamps
12 clay triangles
12 glass rods
Several light beam sources (flashlight, slide projector, overhead projector, laser)
Unknown #1: 1.0 L milk-water colloidal mixture
Unknown #2: 1.0 L 0.1 M $CuSO_4$ (25 g $CuSO_4 \cdot 5H_2O$/1.0 L)
Unknown #3: 1.0 L clay-water suspension mixture
Unknown #4: 1.0 L 0.1 M $FeCl_3$ solution (27 g $FeCl_3 \cdot 6H_2O$/1.0 L)

## B.8 LABORATORY ACTIVITY: WATER TESTING

### Teacher's Directions

The student directions are found on separate pages, ready for duplication following the Teacher's Directions.

### Materials

12 wash bottles
36 test tubes
12 10-mL graduated cylinders
12 test tube racks
12 marking pencils
48 dropping bottles
4–5 burets for dispensing reference solutions
4 L distilled water

## B.3 LABORATORY ACTIVITY: MIXTURES

### Student Directions

### Getting Ready

In this laboratory activity, you will examine four different water-containing mixtures and classify each as a suspension, a colloid, a solution or a combination of these. You will filter each sample and look for the Tyndall effect in both the filtered and unfiltered samples. Particles in a suspension can be separated by filtration, while those in a colloid or a solution are too small to be retained by the filter paper. The Tyndall effect will reveal the presence of colloidal particles.

In your laboratory notebook, prepare a data table for entering your observations on the original samples and filtrates, and your conclusions about each mixture.

**Data Table** **ORIGINAL SAMPLE**

| | | | | | Mixture Classification | | |
|---|---|---|---|---|---|---|---|
| Beaker | Color | Clarity | Settle out? | Tyndall effect? | Suspension | Colloid | Solution |
| 1. | | | | | | | |
| | | | | | | | |
| 2. | | | | | | | |
| | | | | | | | |
| 3. | | | | | | | |
| | | | | | | | |
| 4. | | | | | | | |

**Data Table** **FILTRATE**

| | | | | | Mixture Classification | | |
|---|---|---|---|---|---|---|---|
| Beaker | Color | Clarity | Settle out? | Tyndall effect? | Suspension | Colloid | Solution |
| 1. | | | | | | | |
| | | | | | | | |
| 2. | | | | | | | |
| | | | | | | | |
| 3. | | | | | | | |
| | | | | | | | |
| 4. | | | | | | | |

### Safety

Students should wear safety goggles and aprons throughout the laboratory activity.

### Procedure

1. Obtain two 100- or 150-mL beakers. Fill one beaker half-full of one of the unknown mixtures.
2. Carefully examine the unknown mixture. Record observations on its appearance.
   a. Is the sample colored?
   b. Is the sample clear or cloudy?
   c. Do any particles settle out after you stir the mixture?
3. Perform the Tyndall effect test on the mixture as demonstrated by your teacher. If the liquid is relatively clear, shine a light beam through the side of the beaker and observe the light from the top. (Suggestion: If you are uncertain how to interpret a "faint" test result, compare it with a Tyndall effect test on a reference beaker half-filled with pure water.) If the liquid is very cloudy, shine the beam across the liquid surface and observe the beaker from the side. Can you see the light beam as it passes through the sample? Record your observations.

4. Filter the unknown sample, using a filter paper in a funnel. (If you have forgotten how to fold a filter paper, look back at Figure 5 in your text.) Use a fresh filter paper for each sample. Collect enough filtrate in a clean, numbered beaker to repeat steps 2 and 3.
5. Examine the filter paper. Were any particles removed from the liquid? Record your observations.
6. Repeat step 2 with the filtrate; record your observations. Compare properties of the filtrate with those of the original liquid sample. Note and record any differences.
7. Repeat step 3 with the filtrate; record your observations.
8. Based on your observations, decide whether the unknown mixture was a solution, a colloid, a suspension, or some combination. Record your classifications.
9. Clean and rinse your two beakers.
10. Repeat steps 1 through 9 with another unknown. Continue until four unknown mixtures have been tested.
11. Compare your observations and conclusions with those of other laboratory teams.
12. Wash your hands thoroughly before leaving the laboratory.

## B.8 LABORATORY ACTIVITY: WATER TESTING

## STUDENT DIRECTIONS

### Getting Ready

Chemists can detect and identify ions in water solution in several different ways. In this activity you will use some chemical tests to check for the presence of certain ions in aqueous solution. Positively charged ions have a deficiency of electrons and are called cations; negatively charged ions have an excess of electrons and are called anions. You will investigate two cations and two anions.

The tests you will perform are **confirming tests**. That is, if the test is positive, it confirms that the ion in question is present. In each confirming test, you will look for a change in solution color, or the appearance of an insoluble material called a **precipitate**. A negative test (no color or precipitate) does not necessarily mean the ion in question is not present. The ion may simply be present in such a small amount that the color or precipitate cannot be seen.

You will test for the presence of the cations iron(III), $Fe^{3+}$, and calcium, $Ca^{2+}$, as well as the anions chloride, $Cl^-$, and sulfate, $SO_4^{2-}$. (Notice that when the positive or negative charge of an ion is more than one, the amount of the charge is shown with the plus or minus sign for the ion.) You will perform each confirming test on three different samples:

- a **reference solution** (known to contain the ion of interest)
- **tap water** (which may or may not contain the ion)
- a **control** (distilled water, known not to contain the ion)

### Data Table

| Solutions | Color | Precipitate | Is ion present? |
|---|---|---|---|
| $Fe^{3+}$ reference | | | |
| Tap water | | | |
| Control | | | |
| $Ca^{2+}$ reference | | | |
| Tap water | | | |
| Control | | | |
| $Cl^-$ reference | | | |
| Tap water | | | |
| Control | | | |
| $SO_4^{2-}$ reference | | | |
| Tap water | | | |
| Control | | | |

### Safety

Students should wear safety goggles and aprons throughout the laboratory activity.

### Lab Tips

Prepare a data table similar to the one shown above in your laboratory notebook. Here are some suggestions to guide your ion analysis:

1. If the ion is in tap water, it will probably be present in a lower amount than in the same volume of reference solution. Thus the color or quantity of precipitate produced in the tap water sample will be less than in the reference solution.
2. When completing an ion test, mix the test tube contents thoroughly, unless the instructions give other directions.
3. In a confirming test based on color change, so few color-producing ions may be present that you remain doubtful the reaction actually occurred. Here are several ways to decide whether the expected color is actually present:

- Look through the side of the tube. Placing a sheet of white paper behind or below the test tube may make any color more visible.
- Compare the color of the control (distilled water) test to that of tap water. Distilled water does not contain any of the ions tested. Thus even a faint color in the tap water confirms that the ion is present.

4. In a confirming test involving a precipitate, you may remain uncertain as to whether a precipitate is present even after thoroughly mixing the solutions. In this case, shine a light beam through the tube to see whether the beam's path can be clearly seen in the liquid (the Tyndall effect test). Such light scattering confirms the presence of a precipitate in the form of colloidal particles.

## Procedure

1. Wash three test tubes thoroughly with tap water and rinse with distilled water.
2. Measure 2 mL of tap water in a graduated cylinder; pour the water into one test tube. Use a grease pencil to mark the 2-mL level on the outside of the tube. Mark the other two tubes with a line at the same level. Label the three test tubes Reference (R), Tap Water (TW), and Control (C).
3. Complete each of the following four tests on the respective reference solution provided by your teacher, on tap water, and on a control sample of distilled water.

**Iron(III) Ion Test ($Fe^{3+}$)**

1. Pour 2 mL of iron(III) reference solution into the clean Reference test tube.
2. Add three drops of potassium thiocyanate (KSCN) solution.
3. Mix the test tube contents thoroughly. Record your observations. The confirming test you observed for $Fe^{3+}$ can be represented as follows:

| Iron(III) ion (reference solution) | | Thiocyanate ion (test solution) | | Iron(III) thiocyanate ion (red color) |
|---|---|---|---|---|
| $Fe^{3+}(aq)$ | + | $SCN^{-}(aq)$ | → | $Fe(SCN)^{2+}(aq)$ |

4. Repeat the iron(III) ion test on a 2-mL sample of tap water and on a 2-mL sample of distilled water, placing each sample in its properly labeled test tube. Record your observations and conclusions.
5. Discard the test tube contents as instructed by your teacher. Wash the tubes thoroughly with tap water and rinse with distilled water before continuing with the next test.

**Calcium Ion Test ($Ca^{2+}$)**

1. Pour 2 mL of calcium ion reference solution into the clean Reference test tube.
2. Add three drops of dilute acetic acid ($HC_2H_3O_2$).
3. Add three drops of sodium oxalate ($Na_2C_2O_4$) solution to the tube.
4. Mix the test tube contents thoroughly. Record your observations. The confirming test you observed for $Ca^{2+}$ can be represented as follows:

| Calcium ion (reference solution) | | Oxalate ion (test solution) | | Calcium oxalate (precipitate) |
|---|---|---|---|---|
| $Ca^{2+}(aq)$ | + | $C_2O_4^{2-}(aq)$ | → | $CaC_2O_4$ (s) |

5. Repeat the calcium ion test on a 2-mL sample of tap water and on a 2-mL sample of distilled water, placing each sample in its properly labeled test tube. Record your observations and conclusions.
6. Discard the test tube contents as instructed by your teacher. Wash the tubes thoroughly with tap water and rinse with distilled water before continuing with the next test.

**Chloride Ion Test ($Cl^-$)**

1. Pour 2 mL of chloride ion reference solution into the clean Reference test tube.
2. Add three drops of silver nitrate ($AgNO_3$) test solution. Avoid contact with skin.
3. Mix the test tube contents thoroughly. Record your observations. The confirming test you observed for $Cl^-$ can be represented as follows:

| Silver ion (test solution) | | Chloride ion (reference solution) | | Silver chloride (precipitate) |
|---|---|---|---|---|
| $Ag^+(aq)$ | + | $Cl^-(aq)$ | → | $AgCl(s)$ |

4. Repeat the chloride ion test on a 2-mL sample of tap water and on a 2-mL sample of distilled water, placing each sample in its properly labeled test tube. Record your observations and conclusions.
5. Discard the test tube contents as instructed by your teacher. Wash the tubes thoroughly with tap water and rinse with distilled water before continuing with the next test.

**Sulfate Ion Test ($SO_4^{2-}$)**

1. Pour 2 mL of sulfate ion reference solution into the clean Reference test tube.
2. Add three drops of barium chloride ($BaCl_2$) test solution.
3. Mix the test tube contents thoroughly. Record your observations. The confirming test you observed for $SO_4^{2-}$ can be represented as follows:

| Barium ion (test solution) | | Sulfate ion (reference solution) | | Barium sulfate (precipitate) |
|---|---|---|---|---|
| $Ba^{2+}(aq)$ | + | $SO_4^{2-}(aq)$ | → | $BaSO_4(s)$ |

4. Repeat the sulfate ion test on a 2-mL sample of tap water and on a 2-mL sample of distilled water, placing each in its properly labeled test tube. Record your observations and conclusions.
5. Discard the test tube contents as instructed by your teacher. Wash the tubes thoroughly with tap water and rinse with distilled water.
6. Wash your hands thoroughly before leaving the laboratory.

# SUPPLEMENTAL READINGS

The articles and publications in the following units may be used for several purposes: (1) background support material for the teacher--given the strong applications focus of *ChemCom* and the more theoretical nature of many college-level chemistry courses, you may wish to undertake a selected updating of your knowledge base with respect to science-technology-society (STS) issues; (2) background reading material for students involved in simulations and role playing activities; and (3) optional reading/reporting assignments. Of course, any such listing can become dated with respect to specific topics and should be updated and revised with newer materials periodically.

It is especially important that students realize that (1) no textbook contains all the "right" answers (or even all the right questions), (2) as our knowledge base expands, textbooks are constantly becoming somewhat out of date, and (3) some journals/magazines can be generally relied upon for valid scientific information (such as those published by scientific associations), while others (such as many of the more popular newspapers and magazines found in grocery check-out lanes) cannot. Encourage your students to follow national and local news on STS issues and to bring relevant items to the attention of the class. "Chemistry in the News" bulletin boards, cartoon contests, etc., can be organized by students to help make the *Chemistry in the Community* text a relevant, living, ever-growing course of study.

"Chemistry Words." *CHEM MATTERS,* February 1984.
"How Soon Will We Measure in Metric?" *National Geographic,* August 1977, 287-294.
"Inching Toward the Metric System." *American Biology Teacher,* April 1989, 213-218.
"Metric Mania: Metric-ulate Your Students." *The Science Teacher,* October 1992, 32-35.
"Mad for Metic Measure." *Science and Children,* January 1994, 23-25+.
"Our Most Precious Resource." *National Geographic,* August 1980, 144-179.
"Megameanings." *CHEM MATTERS,* December 1984.
"Memory Metals." *CHEM MATTERS,* October 1993.
"The Chemical Detective." *CHEM MATTERS,* October 1983.
"The Groundwater Terrarium: Geology on a Small Scale." *The Science Teacher,* May 1982, 44-46. [Instructions for constructing a model and demonstration to examine the water cycle.]
"Deadly Cholera." *CHEM MATTERS,* October 1994.
"Groundwater: A Critical Resource." *Science Scope,* November/December 1993, 32-34.
"Saint's Blood." *CHEM MATTERS,* February 1993.
"An Atomic Tour." *CHEM MATTERS,* October 1983.
"Polywater." *CHEM MATTERS,* December 1987.
"Water in the Geosystem: Phase Relationships." *The Science Teacher,* May 1974, 39-43. [Discusses the role of phase changes in the hydrologic cycle.]
"Microcosms for Lake Adidification Studies." *American Biology Teacher,* October 1994, 433-437.
"Bizarre New Molecules Reported!" *CHEM MATTERS,* October 1992.
"What to Do about Water." *The Science Teacher,* January 1985, 34-37. [Lists the causes of freshwater supply depletion and highlights several successful reclamation projects. Advocates a national water policy.]
"Protection of the Water: Who Is Accountable?" *American Biology Teacher,* November/December 1993, 478-479.
"Swimming Pools." *CHEM MATTERS,* April 1994.
"Swimming Pool Chemistry." *CHEM MATTERS,* April 1983. [Issue focuses on acids and bases in everyday life.]
"Acids and Bases: Chemical Principles Revisited." *Journal of Chemical Education,* July 1978, 459-464.
"Difficult Decisions: Acid Rain." *The Science Teacher,* March 1989, 33-35.
"The Challenge of Acid Rain." *Scientific American,* August 1988, 30-38.
"The Use of Oxygen in the Treatment of Sewage (The Real World of Industrial Chemistry)." *Journal of Chemical Education,* February 1980, 137-138. [Discusses the advantages, problems, and costs of using oxygen rather than air in the second-stage treatment of municipal wastewater.]

"Crystal Growing." *CHEM MATTERS,* October 1983.

"Breakfast of Crystals." *CHEM MATTERS,* October 1983.

*Groundwater.* Chemical Manufactures Association, Washington, DC. [Free booklet explains in laymen's terms where groundwater comes from, its uses, and some principles of proper management and national policy.]

"Ground Water in a Fish Tank." *Science Scope,* September 1992, 50-52.

"The Creeping Poison Underground." *Discover,* February 1985, 74-78. [Hazardous chemical and metals affect on groundwater.]

"The Transformation of a River--From 'Sewer' to Suburbs in 20 Years." *Smithsonian,* April 1990, 71-75+.

*Cleaning Our Environment: A Chemical Perspective.* American Chemical Society, Washington, DC, 1978, 457 pages. [Includes chapters on chemical analysis and monitoring, toxicology, air, water, solid wastes, pesticides, and radiation.]

"Fox River Fish Kill." *CHEM MATTERS,* October 1990.

"The Invisible Menace." *Trout,* Summer 1994, 18-25+.

"Survival at Sea." *CHEM MATTERS,* October 1992.

"Threats to the World's Water." *Scientific American,* September 1989, 80+.

"Joseph Priestly and the All-American Lunch." *CHEM MATTERS,* February 1983.

"At Water's Edge: Students Study Their Rivers." *Educational Leadership,* September 1993, 80-83.

"Acid-Base Indicators." *CHEM MATTERS,* April 1983.

"Acrostic." *CHEM MATTERS,* April 1983.

"Ion Search." *CHEM MATTERS,* December 1988.

"Mercurial Risks from Acid's Reign." *Science News,* March 9, 1991, 152-156.

"Buried in Ice." *CHEM MATTERS,* April 1994.

"Lead Poisoning." *CHEM MATTERS,* December 1983.

"Macroinvertebrates as Indicators of Stream Health." *American Biology Teacher,* November/December 1991, 462-466.

"The Pollution Detectives: Part II. Lead and Zinc Mining." *School Science Review,* June 1988, 721-728.

"Deer Kill." *CHEM MATTERS,* October 1992.

"Ape Antibiotic." *CHEM MATTERS,* February 1987.

"Magic Sand." *CHEM MATTERS,* April 1994.

"Car Cooler." *CHEM MATTERS,* February 1993.

"Wastewater." *CHEM MATTERS,* April 1992.

"Great Lakes Environmentalist Push for Zero Chemical Pollution." *Chemical & Engineering News,* March 18, 1991, 12-13.

"Get Your Feet Wet–Scientifically: A Guide to Water Testing as a School Science Project." *Outdoor Communicator,* Spring/Summer 1989, 26-32.

# TEACHING SCHEDULE

| | TIME BLOCK 1 | TIME BLOCK 2 | TIME BLOCK 3 | TIME BLOCK 4 | TIME BLOCK 5 |
|---|---|---|---|---|---|
| Class Work | Read Introduction<br>Discuss Intro.<br>CQ Part A<br>Prelab LA A.1<br>Getting Ready for LA A.1 | LA A.1 | Discuss A.2-A.3<br>YT A.3 | LA A.4 | Discuss A.6<br>CQ A.6<br>YD A.7 |
| Homework | | Read A.4 | | Read A.5-A.6 | SQ Part A<br>Read B.1-B.2 |
| | **TIME BLOCK 6** | **TIME BLOCK 7** | **TIME BLOCK 8** | **TIME BLOCK 9** | **TIME BLOCK 10** |
| Class Work | Review Part A<br>Discuss B.1<br>YT B.1 | Quiz Part A<br>Discuss B.2<br>YT B.2<br>Prelab LA B.3<br>Getting Ready for LA B.3 | LA B.3 | YD B.5 | Discuss B.6<br>YTs B.6 (2)<br>Prelab LA B.8<br>Getting Ready for LA B.8 |
| Homework | | | Read B.4 | Read B.6-B.7 | |
| | **TIME BLOCK 11** | **TIME BLOCK 12** | **TIME BLOCK 13** | **TIME BLOCK 14** | **TIME BLOCK 15** |
| Class Work | Discuss B.7<br>LA B.8 | Discuss B.9-B.10<br>YD B.11 | Review Part B<br>Quiz Part B | Discuss C.1<br>Discuss C.2<br>YT C.2 | Discuss C.3<br>YT C.3 |
| Homework | Read B.9-B.10 | SQ Part B<br>Read C.1 | Read C.2 | Read C.3 | Read C.4-C.5 |
| | **TIME BLOCK 16** | **TIME BLOCK 17** | **TIME BLOCK 18** | **TIME BLOCK 19** | **TIME BLOCK 20** |
| Class Work | YD C.5 | Review Part C<br>Quiz Part C<br>YT D.1 | Discuss D.2<br>YT D.2 | Discuss D.3-D.4<br>CQ D.3<br>YT D.3<br>Prelab LA D.5 | LA D.5 |
| Homework | SQ Part C<br>Read D.1-D.2 | | Read D.3-D.6 | | |
| | **TIME BLOCK 21** | **TIME BLOCK 22** | **TIME BLOCK 23** | **TIME BLOCK 24** | **TIME BLOCK 25** |
| Class Work | Discuss D.6<br>YT D.6<br>SQ Part D | PIAT Part E | Review Resources Unit | Resources Unit Exam | |
| Homework | Read E.1-E.2 | | | | |

**LA**=Laboratory Activity; **CQ**=ChemQuandary; **YT**= Your Turn; **YD**=You Decide; **PIAT**=Putting It All Together; **SQ**=Summary Questions; One **Time Block**=Approximately 50 minutes

# SUPPLY LIST

## Expendable Items and Unusual Materials

| Quantity | Unit | Item | Sections Used |
|---|---|---|---|
| 12 | 9-V | batteries | D.5 |
| 12 | 9-V | battery connectors | D.5 |
| 5 | g | carbon (pencil lead) | B.3, D.5 |
| 24 | | carbon rods | D.5 |
| 36 | | copper pennies | A.1 |
| 12 | cm | copper wire | A.4, B.8 |
| 4 | g | copper(II) chloride | B.3 |
| 3.4 | g | copper(II) chloride dihydrate | D.5 |
| 4.8 | g | copper(II) nitrate trihydrate | B.8 |
| several | small | hammers | B.3 |
| 1 | mL | hydrochloric acid | B.3 |
| 5 | g | iron, thin wire | B.3 |
| 5.1 | g | magnesium nitrate hexahydrate | B.8 |
| 12 | cm | magnesium ribbon | B.3, B.8 |
| 25 | mL | nitric acid | A.4 |
| 5 | g | silicon | B.3 |
| 3.4 | g | silver nitrate | B.8 |
| 100 | g | steel wool | A.1 |
| 5 | g | sulfur | B.3 |
| 5 | g | tin, mossy or foil | B.3 |
| 136 | g | zinc chloride | A.1 |
| 36 | g | zinc, mossy or granulated | A.1 |
| 15 | g | zinc, mossy or wire | B.3 |
| 6 | g | zinc nitrate hexahydrate | B.8 |
| 12 | cm | zinc wire | B.8 |

The Resources unit gives students some insight into the worldwide challenge of limited chemical resources. As Earth's human population has grown, the demand for chemical resources such as metal ores and fossil fuels has increased. Although our developing technology provides us with a comfortable way of life, we face decreasing reserves of resources and increasing disposal problems associated with large quantities of "waste" materials. These trends generate a host of economic, environmental, and political issues.

Conservation strategies—including reuse, recycling, and substitution—are considered. For many resources a combination of these conservation strategies is desirable. All three options require knowledge of the properties and behavior of matter. In that context, the chemical concepts covered in this unit include equation-balancing, properties and reactivities of metals, and periodicity.

While studying this unit, students should become familiar with source books such as the *CRC Handbook of Chemistry and Physics, Lange's Handbook of Chemistry*, the *Merck Index*, and the *McGraw-Hill Encyclopedia of Science and Technology.*

# LEARNING GOALS

Upon completion of this unit, the student will be able to:

1. Compare and contrast science and technology. [Introduction]
2. State the law of conservation of matter, and apply the law by determining whether a given chemical equation is balanced. [A.2, A.3]
3. Explain why it is not possible to "throw something away." [A.4, A.5, A.7]
4. List common types and sources of municipal waste, and describe attempts to reuse and recycle waste. [A.5, A.6]
5. Define and give examples of renewable and nonrenewable resources. [A.5, A.7]
6. Distinguish between chemical and physical changes and/or properties when given specific examples of each. [B.1]
7. Classify selected elements as metals, nonmetals, or metalloids based on observations of their chemical and physical properties. [B.2, B.3]
8. Use the periodic table to (a) predict physical and chemical properties of an element; (b) write formulas for various compounds; (c) identify elements by their atomic masses and atomic numbers; and (d) locate periods and groups (families) of elements. [B.4, B.6]
9. Construct a workable periodic table and explain its organization, given chemical and physical properties of a set of elements. [B.5]
10. Compare the reactivities of selected elements, and explain the differences in terms of the structure of their atoms. [B.7–B.9]
11. Discuss the development of new materials as substitutes for dwindling resources. [B.10]
12. Explain from a chemical viewpoint the problems and solutions involved in restoring the Statue of Liberty. [B.11]
13. List the three primary layers of our planet and some resources that are "mined" from each region. [C.1]
14. Write balanced chemical equations and relate them to the law of conservation of matter. [C.2]
15. Define the term "mole," and calculate the molar mass of a compound when provided with its formula and the atomic masses of its elements. [C.3]
16. Explain the economic and environmental reasons for recycling. [C.4, C.5]
17. Outline the production of a metal from its ore (using copper as an example) and list four factors that determine the profitability of mining. [D.1]
18. Calculate the percent by mass of a specified element in a given compound, given the formula of the compound. [D.2]
19. Define oxidation and reduction, and compare the three most common redox–reaction methods for separating metals from their ores. [D.4]
20. Design a new coin, applying the main idea of this unit—the properties of a material dictate its uses. [E.1]

# A. USE OF RESOURCES

Part A examines (from a chemical perspective) what it means to "use" resources and how we "dispose" of waste products. The Law of Conservation of Mass is explained in terms of atoms, with the aid of common examples. Balancing equations is introduced as a way of keeping track of atoms. In the first laboratory exercise, students observe how the appearance of a penny can be changed through chemical reactions. They also have an opportunity to consider what happens when copper dissolves in nitric acid. Finally, the use of resources, along with generation and disposal of waste, are considered.

## CHEMQUANDARY A *Penny for Your Thoughts* *(pages 90–91)*

Hobbyists and coin collectors see some pennies as having special worth, depending on their condition and the number of pennies in circulation. For example, a 1909 uncirculated penny is worth \$3,200 today. An uncirculated 1950 penny is worth \$80.

## A.1 LABORATORY ACTIVITY: STRIKING IT RICH *(pages 91–93)*

This laboratory activity is intended to generate student interest in metallic resources and the chemistry involved in their transformation into useful materials. Students produce brass, an alloy of zinc and copper. The shiny brass color may remind students of gold, and an intermediate step may suggest the appearance of silver on the coin. Do not give away the anticipated results—let students discover the changing appearance of the penny for themselves.

### Time

30 minutes

### Materials (for a class of 24 working in pairs)

24 protective goggles
36 250-mL beakers
12 25-mL graduated cylinders
12 crucible tongs
12 ring stands
24 ring clamps
12 watch glasses
12 wire gauze squares
12 Bunsen burners or hot plates
12 matches or strikers
36 copper pennies
36 g granulated zinc (Zn) or zinc foil
350 mL of 1 M $ZnCl_2$ (136 g $ZnCl_2$/1 L solution)
2 L of distilled water
100 g steel wool

### Safety

Students should wear safety goggles and aprons throughout the laboratory activity. Students must wear protective goggles for this and all other laboratory activities. Hot plates are safer than burners.

### Advance Preparation

Either pre- or post-1982 pennies can be used. However, since post-1982 pennies contain a much higher percent of zinc than pre-1982 pennies (97.6% vs. 5%), they are more likely to warp if overheated (melting point of copper: 1,083 °C, vs. zinc: 420 °C).

## Pre-Lab Discussion

During a short pre-lab discussion, demonstrate how to operate a Bunsen burner, as presented on page 91 of the student text.

## Lab Tips

Remind students to refer to Figure 3 for the correct way to hold the coin in the flame.

## Teacher Background

The net reaction is $Zn_{Zn} + Zn^{2+}(aq) \rightarrow Zn^{2}+(aq) + Zn_{Cu}$. The reduction potential for zinc ions being reduced to zinc on copper ($Zn_{Cu}$) is more favorable than zinc ions being reduced to zinc on zinc ($Zn_{Zn}$). The silvery product, formed first, is a type of brass with a relatively high concentration of zinc. The golden surface, formed after heating, is α-brass with less than 35% zinc. The deposition process can be repeated by returning the golden penny to the zinc chloride solution, where it will again become silvery. For a more complete discussion of the theory involved in this experiment, consult the *Journal of Chemical Education 72*: 387 (May 1995).

Today's students will not remember when pennies that were 95% copper were used regularly. Before 1982, a penny contained 2.95 g of copper and 0.15 g zinc; currently, pennies contain only 0.06 g copper and 2.46 g of zinc. Current prices of copper and zinc are found in the financial pages of most newspapers. If time permits, have students find the current prices of copper and zinc and calculate the value of a pre- and post-1982 penny. This information can be referenced for the final activity of this unit.

## Answers to Questions

1. a. Untreated—copper color; heated in Zn and $ZnCl_2$ solution—shiny "silver" color; heated in Zn and $ZnCl_2$ solution and by flame—"gold" color
   b. Treated pennies resemble silver and gold.
2. By comparing physical properties like density (Zn: 7.1, Cu: 9.0, Ag: 10.5, Au: 19.3 g/mL), melting point (pure Zn: 420 °C, Cu: 1,083 °C, Ag: 962 °C, Au: 1,064 °C), and chemical reactivity (for example, zinc, copper, and silver all react with nitric acid, while gold does not)
3. Inexpensive substitute for silver or gold jewelry, cooking pan with different heat-conducting properties (Student answers will vary.)
4. a. A solid solution of brass was produced; the copper atoms were merely mixed with zinc.
   b. Yes, allowing the zinc to react with acid would remove the zinc and leave the copper layer.

## Post-Lab Discussion

Focus the discussion on the change in properties that occurs in the activity. Each substance has a unique set of properties, just as humans have unique sets of fingerprints. The change in properties in this activity allows us to predict that a chemical change has occurred. Referring to the properties of a substance also permits identification of that substance. This idea of chemists as "matter detectives" occurs repeatedly throughout the *ChemCom* course.

The concept of physical and chemical properties will be further developed in Part B, where students will consider the practical significance of physical and chemical properties. The goal of this activity is to motivate the students to want to learn more about the changes that take place in matter. They are not ready to consider chemical equations for the reactions.

## A.2 USING THINGS UP *(pages 93–94)*

Students can easily learn to repeat the Law of Conservation of Matter by rote without being convinced that it is impossible to throw something away. Emphasize that scientific laws are simply generalizations or summaries of observations. To help them clarify their thinking, ask students to identify where things came from and where they went in changes such as those described in the text. In chemical reactions, atoms do not disappear; they are just rearranged.

## A.3 TRACKING ATOMS *(pages 94–95)*

The Law of Conservation of Matter is illustrated each time a chemical equation is balanced. Counting atoms of each element is the first step in balancing an equation. Later, coefficients will be used to balance the number of atoms of each element on each side of the reaction symbol.

## YOUR TURN *Balanced Equations* *(pages 95–97)*

This activity provides practice in balancing equations and reinforces the idea that it is necessary to count atoms of each type on both sides of the arrow.

### Answers to Questions

1. (No answer required.)
2. Cellulose reacts with oxygen gas to form carbon dioxide gas and liquid water.

| **Reactants:** | **Products:** |
|---|---|
| 6 C atoms | 6 C atoms |
| 10 H atoms | 10 H atoms |
| 17 O atoms | 17 O atoms |

This is a balanced equation.

3. Liquid nitroglycerin decomposes to form nitrogen gas, one oxygen gas, carbon dioxide gas, and water vapor.

| **Reactants:** | **Products:** |
|---|---|
| 12 C atoms | 12 C atoms |
| 20 H atoms | 20 H atoms |
| 12 N atoms | 12 N atoms |
| 36 O atoms | 36 O atoms |

This is a balanced equation.

4. Solid silver reacts with hydrogen sulfide gas and oxygen gas to form solid silver sulfide and liquid water. (This is the reaction behind the tarnishing of silver and explains why silver vessels should not be used to heat sulfur-containing food such as eggs,)

| **Reactants:** | **Products:** |
|---|---|
| 4 Ag atoms | 4 Ag atoms |
| 8 H atoms | 8 H atoms |
| 4 S atoms | 2 S atoms |
| 2 O atoms | 4 O atoms |

This is not a balanced chemical equation.

### Demonstration Idea

To assess understanding of the Law of Conservation of Matter do the following demonstration. Place a candle on a balance pan. Have a student read the mass. Ask students to predict what will happen to the mass if the candle is lighted. Light the candle and read the mass at intervals. Ask students to explain why the mass changed as it did. (The mass decreases because the candle wax combined with oxygen from the air to form carbon dioxide gas and water vapor.)

## A.4 LABORATORY ACTIVITY: USING UP A METAL *(pages 97–98)*

This activity demonstrates copper being "used up" by the action of nitric acid. Post-Lab questions, demonstration, and discussion emphasize the fallacy of the phrase "used up." The macroscale version and a Demonstration Idea are found in the Enrichment Materials section at the end of the unit.

Either before or after the activity, students will enjoy hearing Ira Remsen's (an American chemist) story of his first encounter with nitric acid when he was high school age. The following excerpt from Geuman, F. H., *The Life of Ira Remsen*, is on pp. 9–10 of the *Journal of Chemical Education*, (1940).

"While reading a textbook of chemistry," said he, "I came upon the statement, 'nitric acid acts upon copper.' I was getting tired of reading such absurd stuff and I determined to see what this meant. Copper was more or less familiar to me, for copper cents were then in use. I had seen a bottle marked 'nitric acid' on a table in a doctor's office where I was then 'doing time!' I did not know of its peculiarities, but I was getting on and likely to learn. The spirit of adventure was upon me. Having nitric acid and copper, I had only to learn what the words 'act upon' meant. Then the statement 'nitric acid acts upon copper,' should be something more than mere words. All was still. In the interest of knowledge I was even willing to sacrifice one of the few copper cents then in my possession. I put one of them on the table; opened the bottle marked 'nitric acid'; poured some of the liquid on the copper; and prepared to make an observation. But what was this wonderful thing which I beheld? The cent was already changed, and it was no small change either. A greenish blue liquid foamed and fumed over the cent and over the table. The air in the neighborhood of the performance became colored dark red. A great colored cloud arose. This was disagreeable and suffocating—how should I stop this? I tried to get rid of the objectionable mess by picking it up and throwing it out the window, which I had meanwhile opened. I learned another fact—nitric acid acts not only upon copper but it acts upon fingers. The pain lead to another unpremeditated experiment. I drew my fingers across my trousers and another fact was discovered. Nitric acid acts upon trousers. Taking everything into consideration, that was the most impressive experiment, and relatively, probably the most costly experiment I have ever performed. I tell of it even now with interest. It was a revelation to me. It resulted in a desire on my part to learn more about that remarkable kind of action. Plainly the only way to learn about it was to see its results, to experiment, to work in a laboratory."

## Time

15–20 minutes

## Materials (for a class of 24 working in pairs)

12 24-well wellplates
12 Beral pipets
50 mL of 6 M nitric acid (19 mL conc. $HNO_3$ diluted to 50 mL)
12 cm copper wire cut in 1-cm lengths

## Safety

Student should wear safety goggles and aprons throughout the laboratory activity. Nitric acid is corrosive and a strong oxidant. Caution students to use only the amount of nitric acid and copper specified in the procedure.

## Advance Preparation

Prepare the nitric acid solution by slowly adding (with stirring) the concentrated acid to water. **Caution: Nitric acid is corrosive and a strong oxidant.** Fill the Beral pipets and carefully wipe the ends to remove any excess nitric acid. Store the pipets with the points up to prevent spillage. Cut the copper wire into 1-cm lengths.

## Pre-Lab Discussion

Students should be cautioned regarding the hazards of nitric acid and reminded to keep their faces away from the reaction plate. Even the use of small-scale procedures does not completely eliminate the danger of toxic gases. If this is the first time your students have used pipets, caution them that using them as squirt guns is dangerous and strictly forbidden.

## Lab Tips

Be sure your laboratory room is well ventilated. Hoods must be used. Extra supplies of nitric acid and copper should not be available to students. The reaction is interesting, and students may be tempted to repeat it with larger quantities.

## Answers to Questions

1. The copper metal reacted with the nitric acid and seemed to disappear in the liquid.
2. The blue color results from the product formed when copper reacts with nitric acid.
3. (No answer required.)
4. a. Copper—reddish brown solid; nitric acid—colorless liquid; copper nitrate solution—blue liquid; nitrogen dioxide—brown gas; water—clear liquid
   b. To the copper nitrate solution
   c. Nitrogen and oxygen
   d. From the nitric acid
   e. From the nitric acid

## Post-Lab Discussion

Focus on the Law of Conservation of Matter. Emphasize that although the copper seems to have been "used up", each and every copper atom is still present, although now in the form of $Cu^{2+}$ ions. You can evaporate a small sample of copper(II) nitrate solution to demonstrate the presence of the bright blue solid copper(II) nitrate. *Do not use a copper(II) nitrate sample made by allowing copper to react with nitric acid.* As nitric acid is likely to be present, noxious fumes will be produced during evaporation.

Discuss the physical and chemical changes that occurred during the procedure, and the use of symbols and equations to represent the chemical reaction. Point out that the brown gas produced, one of the pollutants formed by any high-temperature combustion process (such as in an automobile engine), contributes to the problem of acid rain.

## A.5 RESOURCES AND WASTE *(pages 98–100)*

The terms *renewable resources* and *nonrenewable resources* are explained in this section. Point out that both using and producing resources create waste and large quantities of discarded material. Recycling is introduced as a partial solution to part of this problem. An additional ChemQuandary is found in the Enrichment Materials section at the end of the unit.

## A.6 DISPOSING OF THINGS *(pages 100–101)*

One very concrete way to introduce this section is to dump the contents of a garbage bag on the desk or floor. It is recommended that you bring "selected" garbage from home to avoid any surprises from embarrassing or hazardous materials. The section points out that we are all responsible for the waste we generate and are therefore jointly responsible for reducing it. New uses of waste products are also discussed.

## CHEMQUANDARY *Wondering about Resources and Waste* *(pages 101–102)*

This ChemQuandary could serve as the basis of a group discussion, with each group proposing and evaluating the reasons for their choices. Students may need to consult printed resources to find facts to support their opinions.

## A.7 YOU DECIDE: CONSUMING RESOURCES *(pages 102–103)*

In this activity, students focus on their own use of resources and contributions to the problem of waste. Help students determine what resources are used in the manufacture of items. Focus on common materials such as TV sets, pencils, etc., encountered in everyday life, so that students are sensitized to the variety of things they use and the resources used to make them.

This activity should be completed in one class period. Each student will first answer Questions 1 through 3. Instruct them to list the first items that come to mind. It will be difficult to construct complete and accurate lists for the resources used to manufacture the items.

Each group should prepare a short report on the answers for items 4–7. A full class discussion can highlight major points developed by the groups.

## Answers to Questions:

1-5. (Answers will vary depending on the students. Check their understanding of renewable and nonrenewable resources in Question 3.)

6. Sampling of possible answers:

| Item | Nonrenewable Component | Possible Substitute |
|---|---|---|
| Aluminum beverage can | Aluminum metal | Paper/cardboard |
| Aluminum foil | Aluminum metal | Clear plastic wrap |
| Automobile tire | Synthetic rubber | Natural rubber |
| Batteries | Zinc or other metal | Rechargeable batteries, solar cells |
| Clothing | Synthetic fibers | Cotton, wool, silk |
| Fast food container | Petroleum | Paper |
| Gasoline | Petroleum | Grain alcohol |
| Soap, perfume | Petroleum products | Natural fats, oils, plant and animal products |
| Styrofoam cup | Petroleum products | Paper, washable glasses |
| Tungsten light bulb wire | Tungsten metal | Electroluminescent panels |

7. (Answers will depend on items identified in previous questions. Help students focus on the problems and benefits of each method of disposal. )

## PART A: SUMMARY QUESTIONS *(pages 103–104)*

1. a. In a chemical reaction, matter is neither created nor destroyed. There is no known instance in which matter has not been conserved in a chemical reaction.
   b. Governmental laws, being created by people, are subject to social and economic pressures; scientific laws are part of a natural system immune to human intervention. Well-established scientific laws are not generally broken; there is no observable instance where the Law of Conservation of Matter has not been followed in a chemical reaction. Governmental laws, in contrast, are regularly broken by people—observe the many speeders caught by state troopers on the highways!
2. a. $Sn + HF \rightarrow SnF_2 + H_2$

   | Reactants | Products |
   |---|---|
   | 1 Sn, 1 H, 1 F | 1 Sn, 2 H, 2 F |

   This equation is not balanced.

   b. $SiO_2 + 3\ C \rightarrow SiC + 2\ CO$

   | Reactants | Products |
   |---|---|
   | 1 Si, 2 O, 3 C | 1 Si, 2 O, 3 C |

   This is a balanced equation.

   c. $Al(OH)_3 + 3\ HCl \rightarrow AlCl_3 + 3\ H_2O$

   | Reactants | Products |
   |---|---|
   | 1 Al, 6 H, 3 O, 3 Cl | 1 Al, 6 H, 3 O, 3 Cl |

   This a balanced equation.
3. According to the Law of Conservation of Matter, atoms cannot be created or destroyed. Materials are changed in form, but the atoms are never destroyed. It is impossible to throw anything away because Earth is self-contained and does not lose materials to space. "Throwing it in the trash can" is a more proper statement than "throwing it away."
4. a. The Earth may be viewed as a spaceship, since Earth and its atmosphere represent (for matter) a nearly closed system. (A small amount of matter does enter Earth via meteorites.) The amount of resources is limited because matter can neither be lost

nor gained. The analogy is useful in helping students realize that there is a finite amount of resources—water, fuel, raw materials—on board, and no place to go for more (although students love to employ science fiction scenarios about space explorations and mining).

b. The analogy is a little misleading, since Earth continually gains radiant energy from the sun (energy that is partially returned to space) while a spaceship more closely resembles an isolated (self-contained) system. A truly isolated system exchanges neither mass nor energy with its surroundings.

5. Discarding less waste decreases the waste-storage problem; and recycling waste may lower the demand for resources, resulting in a decrease in the price. Other benefits include pollution reduction, energy savings, creation of new jobs, and political stability.
6. Reuse is more advantageous than recycling, owing to lower costs and energy use.
7. a. Renewable: water, air (nitrogen, oxygen, etc.), paper (trees), biomass.
   b. Nonrenewable: metals (aluminum, iron, copper, etc.), fossil fuels (coal, petroleum, natural gas).

# B. WHY WE USE WHAT WE DO

Part B introduces the students to the chemical and physical properties of the elements. Before they are introduced to the modern periodic table based on atomic numbers, students discover patterns in the properties of the elements by constructing their own tables. Laboratory activities involve investigating properties of elements and comparing chemical reactivities of metals.

Part B also demonstrates how properties of resources are matched to specific uses. Using the periodic table, chemists can modify structures or physically blend substances to produce materials with the desired properties. In the final activity of Part B, students apply their chemical knowledge to the restoration of the Statue of Liberty.

## B.1 PROPERTIES MAKE THE DIFFERENCE *(pages 105–106)*

This section introduces chemical and physical properties and changes by describing some of the materials used for building the Statue of Liberty. Remind students of the changing properties observed in the Striking It Rich Laboratory Activity in section A.1, To further clarify the differences between chemical and physical change, perform some of the teacher-led demonstrations found in the Enrichment Materials at the end of this unit.

### YOUR TURN *Properties: Physical or Chemical?* *(page 106)*

In this activity, students practice recognizing the difference between physical and chemical properties. Chemical properties become evident when one substance is transformed into a new substance, whereas physical properties (such as melting point, boiling point, thermal and electrical conductivity, density, color, hardness) do not depend on the presence of another substance. In a physical change, the identity of the substance is not altered.

#### Answers to Questions

1. Physical—no change in material, depends on nothing else
2. Chemical—the metal reacts with the oxygen from the air
3. Chemical—new substances are formed
4. Chemical—material changes (taste), depends on bacteria in milk
5. Physical—no change
6. Physical—no change in identity, even though shape is changed
7. Chemical—bread rising is due to the production of carbon dioxide by yeast: this involves a change of material
8. Physical—argon does not react with the filament
9. Physical—no change in the metals

## B.2 THE CHEMICAL ELEMENTS *(pages 107–108)*

All the substances that make up our world are composed of a limited number of chemical elements. Some general properties of elements are given in Table 2 (page 108), and the properties and symbols of common elements (those listed in Table 1) are reviewed through the following *Your Turn* activity. Other activities are described in the Enrichment Materials section at the end of the unit.

### YOUR TURN *Chemical Elements Crossword Puzzle* *(page 108)*

This activity acquaints students with some properties and uses of selected elements. A blank copy of the crossword puzzle matrix is included in this Teacher's Guide (see Black Line Masters at end of book) so that you can copy it for your students.

Key

| Down | | Across | |
|---|---|---|---|
| 1. Helium | 22. Lead | 3. Fluorine | 21. Silver |
| 2. Barium | 25. Tungsten | 5. Calcium | 23. Oxygen |
| 4. Lithium | 27. Manganese | 7. Platinum | 24. Silicon |
| 6. Phosphorus | 28. Zinc | 8. Iodine | 26. Cadmium |
| 9. Krypton | 29. Chromium | 10. Boron | 30. Bromine |
| 10. Beryllium | 30. Bismuth | 11. Nickel | 32. Argon |
| 12. Magnesium | 31. Cesium | 13. Aluminum | 35. Antimony |
| 14. Carbon | 33. Copper | 14. Cobalt | 36. Sodium |
| 18. Mercury | 34. Iron | 15. Uranium | 37. Nitrogen |
| 19. Hydrogen | 38. Gold | 16. Neon | 40. Chlorine |
| 20. Potassium | 39. Tin | 17. Sulfur | |

To extend the activity, ask students, or groups of students, to propose uses based on the listed properties.

## B.3 LABORATORY ACTIVITY: METAL, NONMETAL? *(pages 108–110)*

In this laboratory activity, students will learn how to distinguish metallic from nonmetallic elements after examining their properties. Seven elements are tested for appearance, malleability, reaction with hydrochloric acid, reaction with copper(II) chloride solution, and electrical conductivity. The macroscale version of this activity is found in the Enrichment Materials section at the end of the unit.

### Time

50 minutes

### Materials (for 24 students working in pairs)

several small hammers
24 pipets
12 24-well wellplates
7 small vials for samples of elements
25 mL 0.5 M hydrochloric acid solution (1 mL conc. HCl/25 mL)
25 mL 0.1 M copper(II) chloride solution (0.43 g $CuCl_2 \cdot 2H_2O$/25 mL)
The seven elements to be tested (codes for student samples shown in parentheses):
(a) 5 g carbon (C), pencil lead
(b) 12 cm magnesium (Mg), ribbon, cut to 1-cm lengths
(c) 5 g silicon (Si), 12 small lumps
(d) 5 g tin (Sn), 12 small pieces mossy tin or tin foil
(e) 5 g sulfur (S), 12 small lumps
(f) 5 g iron (Fe), 1-cm pieces of thin wire (better for conductivity)
(g) 15 g zinc (Zn), 12 small pieces mossy zinc or 12 1-cm pieces wire (better for conductivity)

### Safety

Students should wear safety goggles and aprons throughout the laboratory activity. Remind students to always wear goggles when using hammers.

### Advance Preparation

Fill the pipets with 0.5 M HCl and 0.1 M $CuCl_2$. Wipe the stem of each pipet with paper towels. Cut the pencil lead, magnesium ribbon, tin foil, iron wire, and zinc wire to the appropriate lengths. Select small lumps of silicon and sulfur. Place the samples in small vials, labeled with codes as suggested above.

## Pre-Lab Discussion

Review the differences between physical and chemical properties. Caution students not to handle any element samples with their hands. Review the procedures for using acids safely.

## Lab Tips

The electrical conductivity portion of this activity is optional. If the 110-V apparatus is used, the teacher should demonstrate the activity. The microconductivity setups, using 9-V batteries and light-emitting diodes (LED), could also be used. (See description in Enrichment Materials.)

If there are fewer hammers than students, groups could demonstrate this part of the activity to the class.

## Expected Results

| Code | Element | Appearance | Result of crushing | Conductivity (optional) | Reaction with acid | Reaction with $CuCl_2(aq)$ |
|---|---|---|---|---|---|---|
| a | C | gray, dull | brittle | yes | no | no |
| b | Mg | shiny, silver | malleable | yes | yes, bubbles | yes, darkens |
| c | Si | gray, shiny | brittle | yes | no | no |
| d | Sn | shiny, silver | malleable | yes | yes, bubbles | yes, darkens |
| e | S | yellow, dull | brittle | no | no | no |
| f | Fe | gray, shiny | malleable | yes | yes, bubbles | yes, darkens |
| g | Zn | shiny, silver | malleable | yes | yes, bubbles | yes, darkens |

## Answers to Questions

1. Group 1: magnesium, zinc, iron, and tin; Group 2: carbon, silicon, and sulfur
2. Silicon
3. Metals—magnesium, zinc, iron, and tin; nonmetals—carbon and sulfur; metalloids—silicon. Students may also classify carbon as a metalloid based on its conductivity. Ask students to consider the other properties of metals. Because carbon lacks the characteristic luster, it is not classified as a metal.

## Post-Lab Discussion

Provide time for the student groups to compare results and subsequent grouping of the elements. Discuss the criteria for assigning elements to groups. Allow students to resolve disagreements, but be ready to give suggestions if requested.

## Optional Information

Some everyday uses of the elements:

Carbon as graphite—steelmaking, lubricants, brake linings, pencils

Carbon as diamond—gemstone, abrasive

Carbon as activated carbon—decolorizing and purifying agent

Magnesium—high-strength/low-weight alloys, flares, fireworks

Silicon— transistors, semiconductors, solar cells

Tin—coating ("tin" cans), alloys (solder, bronze, pewter)

Sulfur—manufacturing sulfuric acid (no.1 industrial chemical), matches, gunpowder, vulcanized rubber

Iron—structural metal (steel)

Zinc—coating (galvanizing), alloys (brass), dry cell batteries

## B.4 THE PERIODIC TABLE *(pages 110–111)*

This section describes the development of the periodic table as a search for patterns and regularities. Demonstration Ideas can be found in the Enrichment Materials section at the end of the unit.

## B.5 YOU DECIDE: GROUPING THE ELEMENTS *(pages 111–112)*

Data cards containing selected chemical and physical properties of 20 elements are used to classify the elements according to similarities in their properties, an approach used by Mendeleev (who worked with 63 elements and similar data). The cards are found on page 59 and can be duplicated to make a copy for each student. This works well as a group project in class. Groups can display their periodic tables and be ready to describe how patterns they used are useful in organizing and predicting.

## B.6 THE PATTERN OF ATOMIC NUMBERS *(page 112)*

After introducing the concept of atomic number, this section expands on the earlier idea of grouping and leads to graphs relating atomic numbers and chemical formulas.

## YOUR TURN *Periodic Variation in Properties* *(pages 112–113)*

This activity demonstrates using graphing to discover regularities among observations. Students produce graphs to illustrate the periodic relationship of the chloride or oxide formulas and the relationship of melting or boiling points. From the graphs, students discover the cyclic (periodic) trends among the data and relate their interpretations to the modern periodic table. Such an approach was used in 1864 by Lothar Meyer, the second developer of the periodic table, when he plotted atomic volumes vs. atomic weights. Unfortunately, Meyer was reluctant to use his scheme to predict undiscovered elements and did not publish his findings until 1870—one year after Mendeleev had published the pattern evident in the formulas of compounds.

Some students will need assistance in dividing the y axis into appropriately sized units. Provide students with the following information, which will permit them to assign atomic numbers and symbols to the 20 elements on their cards. Make sure students distinguish between the two elements with mass 40; argon has no chlorine compound.

| Element | Symbol | Approximate Atomic Mass | Atomic Number |
|---|---|---|---|
| Hydrogen | H | 1 | 1 |
| Helium | He | 4 | 2 |
| Lithium | Li | 7 | 3 |
| Beryllium | Be | 9 | 4 |
| Boron | B | 10 | 5 |
| Carbon | C | 12 | 6 |
| Nitrogen | N | 14 | 7 |
| Oxygen | O | 16 | 8 |
| Fluorine | F | 19 | 9 |
| Neon | Ne | 20 | 10 |
| Sodium | Na | 23 | 11 |
| Magnesium | Mg | 24 | 12 |
| Aluminum | Al | 27 | 13 |
| Silicon | Si | 28 | 14 |
| Phosphorus | P | 31 | 15 |
| Sulfur | S | 32 | 16 |
| Chlorine | Cl | 36 | 17 |
| Argon | Ar | 40 | 18 |
| Potassium | K | 39 | 19 |
| Calcium | Ca | 40 | 20 |

### Questions

1. a. The melting point and boiling point graphs show repeating patterns. On both graphs, the valleys appear at hydrogen, helium, nitrogen, oxygen, fluorine, neon, and argon; the peaks are at carbon, silicon, and calcium. Of course, the melting

point trend is lower on the temperature scale, but maintains the same overall shape as the boiling point trend. The chemical properties of the oxides and chlorides show similar peaks and valleys.

b. The student grouping in the *You Decide* on page 111 may have produced the family groupings that are part of the periodic table. Students may have matched this grouping if they relied on chemical properties for their organizing of the elements, rather than physical properties.

2. The elements between two successive valleys correspond to a period. A table that orders the elements according to such repeating patterns is called a "periodic table."
3. The relative positions of the elements should be the same on the periodic table and on the tables made by the students.

## DATA SET OF CARDS FOR TWENTY ELEMENTS AND THEIR PROPERTIES

| | | | | | |
|---|---|---|---|---|---|
| Atomic mass | 1 | Atomic mass | 19 | Atomic mass | 36 |
| Melting point (°C) | -259 | Melting point (°C) | -218 | Melting point (°C) | -101 |
| Boiling point (°C) | -252 | Boiling point (°C) | -188 | Boiling point (°C) | -34 |
| O in oxide | 0.5 | O in oxide | 0.5 | O in oxide | 0.5 |
| Cl in chloride | 1 | Cl in chloride | 1 | Cl in chloride | 1 |

| | | | | | |
|---|---|---|---|---|---|
| Atomic mass | 9 | Atomic mass | 20 | Atomic mass | 27 |
| Melting point (°C) | 1,287 | Melting point (°C) | -248 | Melting point (°C) | 659 |
| Boiling point (°C) | 2,507 | Boiling point (°C) | -246 | Boiling point (°C) | 2,327 |
| O in oxide | 1 | O in oxide | - | O in oxide | 1.5 |
| Cl in chloride | 2 | Cl in chloride | - | Cl in chloride | 3 |

| | | | |
|---|---|---|---|
| Atomic mass | 16 | Atomic mass | 40 |
| Melting point (°C) | -219 | Melting point (°C) | -189 |
| Boiling point (°C) | -183 | Boiling point (°C) | -186 |
| O in oxide | - | O in oxide | -5 |
| Cl in chloride | 2 | Cl in chloride | - |

| | | | | | |
|---|---|---|---|---|---|
| Atomic mass | 7 | Atomic mass | 32 | Atomic mass | 23 |
| Melting point (°C) | 179 | Melting point (°C) | 119 | Melting point (°C) | 97 |
| Boiling point (°C) | 1,327 | Boiling point (°C) | 445 | Boiling point (°C) | 889 |
| O in oxide | 0.5 | O in oxide | 3 | O in oxide | 0.5 |
| Cl in chloride | 1 | Cl in chloride | 2 | Cl in chloride | 1 |

| | | | | | |
|---|---|---|---|---|---|
| Atomic mass | 39 | Atomic mass | 12 | Atomic mass | 4 |
| Melting point (°C) | 64 | Melting point (°C) | 3,470 | Melting point (°C) | -272 |
| Boiling point (°C) | 757 | Boiling point (°C) | 4,347 | Boiling point (°C) | -269 |
| O in oxide | 0.5 | O in oxide | 2 | O in oxide | - |
| Cl in chloride | 1 | Cl in chloride | 4 | Cl in chloride | - |

| | | | | | |
|---|---|---|---|---|---|
| Atomic mass | 40 | Atomic mass | 31 | Atomic mass | 14 |
| Melting point (°C) | 851 | Melting point (°C) | 44 | Melting point (°C) | -210 |
| Boiling point (°C) | 1,487 | Boiling point (°C) | 280 | Boiling point (°C) | -196 |
| O in oxide | 1 | O in oxide | 2.5 | O in oxide | 2.5 |
| Cl in chloride | 2 | Cl in chloride | 3 | Cl in chloride | 3 |

| | | | | | |
|---|---|---|---|---|---|
| Atomic mass | 10 | Atomic mass | 28 | Atomic mass | 24 |
| Melting point (°C) | 2,037 | Melting point (°C) | 1,407 | Melting point (°C) | 650 |
| Boiling point (°C) | 2,527 | Boiling point (°C) | 2,677 | Boiling point (°C) | 1,117 |
| O in oxide | 1.5 | O in oxide | 2 | O in oxide | 1 |
| Cl in chloride | 3 | Cl in chloride | 4 | Cl in chloride | 2 |

## YOUR TURN *Mendeleev's Periodic Table* *(pages 113–116)*

This activity demonstrates the usefulness of the periodic table in estimating properties of substances. Knowing major properties of a chemical family allows one to predict (via interpolation and extrapolation) the likely behavior of any element in that family.

### Answers to Questions

1. Germanium has a predicted melting point (between 1,410 °C and 232 °C) of 821 °C (actual value is 937 °C).
2. a. Rubidium has a predicted melting point (between 64 °C and 29 °C) of 46.5 °C (actual melting point is 38.9 °C).
   b. Sodium has a higher melting point than rubidium and potassium, at 97.8 °C.
3. a. $CF_4$
   b. $Al_2S_3$
   c. KCl
   d. $CaBr_2$
   e. BaO
4. The noble ("inert") gases helium, neon, argon, krypton, xenon, and radon.

## B.7 CHEMICAL REACTIVITY *(pages 116–117)*

We can compare the reactivities of elements by observing the relative ease with which different metals combine with oxygen. Gold, iron, and magnesium are discussed as examples of how a metal's reactivity helps determine its value for a specific purpose.

## B.8 LABORATORY ACTIVITY: METAL REACTIVITIES *(pages 117–118)*

This activity underscores the fact that different metals have varying degrees of reactivity. Students are to rank copper, magnesium, zinc, and silver according to their observed chemical reactivities. The macroscale version of this activity is found in the Enrichment Materials section at the end of the unit.

### Time

50 minutes

### Materials (for 24 students working in pairs)

12 24-well wellplates
48 pipets
12 cm magnesium ribbon cut into 1-cm lengths
12 cm zinc wire cut into 1-cm lengths
12 cm copper wire cut into 1-cm lengths
25 mL 0.2 M $Mg(NO_3)_2$, solution (5.1 g $Mg(NO_3)_2 \cdot 6\,H_2O$/100 mL solution)
25 mL 0.2 M $Cu(NO_3)_2$, solution (4.8 g $Cu(NO_3)_2 \cdot 3\,H_2O$/100 mL solution)
25 mL 0.2 M $Zn(NO_3)_2$, solution (6.0 g $Zn(NO_3)_2 \cdot 6\,H_2O$/100 mL
25 mL 0.2 M $AgNO_3$ solution (3.4 g $AgNO_3$/100 mL solution)

### Safety

Students should wear safety goggles and aprons throughout the laboratory activity. Caution students that silver nitrate solution stains skin (and clothing)—students should avoid any direct contact with it.

### Pre-Lab Discussion

The reactivities of the metals can be compared by observing their reactions with solutions of ions of other metals. The more reactive the metal, the more likely it is to react with other metal ions. Evidence for reaction includes formation of a flaky solid and appearance or disappearance of the color of the solution. Writing equations for displacement reactions is emphasized in the next section.

## Lab Tips

Record keeping is easier if the well plate is set up like the data table. Remind students to record initial appearance of the metal samples (after step 2) in their data table. Encourage students to observe immediately after the metal is added to the solution and a few minutes later. The magnesium/zinc nitrate reaction is difficult to detect. Very careful observations are necessary. Steel wool may be used in place of sandpaper or emery cloth.

## Expected Results

| | | Solutions | | | |
|---|---|---|---|---|---|
| **Metal** | **Appearance** | **$Cu(NO_3)_2$** | **$Mg(NO_3)_2$** | **$Zn(NO_3)_2$** | **$AgNO_3$** |
| Cu | shiny, copper | — | NR | NR | becomes silvery |
| Mg | shiny, silver | darkens | — | flaky coating | becomes silvery |
| Zn | shiny, silver | darkens | NR | — | becomes silvery |
| Ag | shiny, silver | NR | NR | NR | — |

NR, no reaction

## Answers to Questions:

1. Magnesium
2. Silver
3. Magnesium, zinc, copper, silver
4. Copper was chosen because it is lower than zinc in the activity series. Zinc reacts more readily with oxygen in the presence of water to form oxides.
5. a. Silver would have been a better choice than copper because of its lower chemical reactivity.
   b. Silver is more expensive than copper.
6. a. Silver is most likely to be found in the uncombined state because of its low reactivity.
   b. Magnesium is least likely to be found in the uncombined state because of its high reactivity.

## Post-Lab Discussion

In the post-lab discussion, resolve differences in results among student pairs. When students disagree, ask them for evidence supporting their claims.

This laboratory activity leads directly to the next section. Students can practice writing net ionic equations, such as the one illustrated on page 119, for the reactions they observed. Of course, they should not write equations for reactions that did not occur.

## B.9 WHAT DETERMINES PROPERTIES? *(pages 118–119)*

The differing reactivities of metals are explained by their differing numbers and arrangements of electrons. The more reactive a metal, the greater its tendency to give up electrons. Students do not yet have the tools to understand much about electron structures, but they should begin to build the understanding that there is a relationship between structure and properties.

## B.10 MODIFYING PROPERTIES *(pages 119–120)*

This section provides a variety of concrete examples of modifying materials to change their properties for meeting specific needs and desires.

## B.11 YOU DECIDE: RESTORING MS. LIBERTY *(pages 120–121)*

In this activity, students apply their knowledge of properties of materials to the problem of restoring the Statue of Liberty. They are asked to propose practical solutions to address the three major causes of the statue's deterioration. Proposed solutions should be based on the chemical and physical properties of the materials that compose the statue, as well as the costs of the proposed changes. All proposals should be tempered with common sense.

Have students work in groups for 20–25 minutes on the assignment. Each group should select a spokesperson to give a one-minute report to the class. Each report should contain:

1. reasons for restoring the statue;
2. proposed solutions to the deterioration problems; and
3. economic, chemical, and practical reasons supporting the solutions.

In the Enrichment Materials section at the end of the unit, you will find a table showing the chemical products, their quantities, costs, and uses that were used to restore Ms. Liberty.

## CHEMISTRY AT WORK/Preserving the Past...for the Future

*(page 122)*

The effects of acid rain on historic buildings and outdoor works of art are part of the work of a research associate employed by the National Park Service. Such individuals work in the laboratory as well as outdoors.

## PART B: SUMMARY QUESTIONS *(page 121)*

1. a. Physical properties: color, density, melting point. Chemical properties: gold much less reactive, won't tarnish
   b. Major difference is the nonreactivity of helium; contrast the Hindenburg dirigible and the Goodyear blimp
   c. Physical properties: color, density, melting point (tungsten's melting point is much higher); chemical properties: iron is more reactive
   d. Physical: elements in the lithium family have lower melting points and are much softer than silver. Chemical: silver is much less reactive than lithium, sodium, potassium, rubidium, cesium, or francium.
2. Many nonmetals are gases or liquids at normal temperatures and pressure. Nonmetals that are solid at normal conditions tend to be brittle and are nonconductors of electricity.
3. a. $MgF_2$.
   b. $B_2S_3$
4. a. KCl
   b. Potassium chloride has properties similar to sodium chloride because sodium and potassium are members of the same chemical family.
5. a. Magnesium
   b. Same family
   c. Arsenic
   d. Carbon
6. Any material chosen specifically to fit an intended use has disadvantages. The high reactivity of some elements is a drawback in construction, but using silver and gold would be prohibitively expensive, and the two elements are quite dense.
7. If two dissimilar metals are placed in contact, there is a good chance they will react through an electron–transfer reaction (oxidation–reduction). To prevent corrosion of the metals, a nonconducting material may be installed at the metal–metal contact points. The original designers of the Statue of Liberty used shellacked asbestos to perform this function. However, this insulation became damp and allowed an electrical current to flow between the copper plates and iron support structures.

# C. CONSERVATION IN NATURE AND THE COMMUNITY

Part C begins with an exploration of the chemical supplies of Earth—where they are located, how society uses them, how nature conserves them, and what conservation means in human terms. Balancing equations and using molar masses to account for masses of substances involved in reactions are introduced as extensions of the conservation principle.

Finally, students are asked to consider the economic, political, and environmental effects of recycling or reusing three important chemical resources.

## C.1 SOURCES OF RESOURCES *(pages 123–125)*

We rely mainly on the thin crust of Earth as our source of chemical resources. Unequal distribution of resources requires us to become a part of the world market of imports and exports.

## C.2 CONSERVATION IS NATURE'S WAY: Balancing Chemical Equations *(pages 125–126)*

This section focuses on balancing equations as another aspect of the Law of Conservation of Matter. The synthesis of water from hydrogen and oxygen is used as an example. The following *Your Turn* gives practice in this skill. Additional equations for balancing practice are found in the Enrichment Materials section at the end of the unit.

## YOUR TURN *Writing Balanced Chemical Equations* *(pages 126–127)*

Students practice balancing equations. Emphasize that equations represent chemical reactions. These equations represent particular aspects of actual resource processing or chemical manufacturing.

### Answers to Questions

1. a. $2\ C(s) + O_2(g) \rightarrow 2\ CO(g)$

   b. $Fe_2O_3(s) + 3\ CO(g) \rightarrow 2\ Fe(l) + 3\ CO_2(g)$
2. a. $Cu_2S(s) + 2\ O_2(g) \rightarrow 2\ CuO(s) + SO_2(g)$

   b. $2\ CuO(s) + C(s) \rightarrow 2\ Cu(s) + CO_2(g)$
3. $4\ NH_3(g) + 7\ O_2(g) \rightarrow 4\ NO_2(g) + 6\ H_2O(l)$
4. $2\ O_3(g) \rightarrow 3\ O_2\ (g)$
5. $Cu(s) + 2\ AgNO_3(aq) \rightarrow Cu(NO_3)_2(aq) + 2\ Ag(s)$
6. $2\ C_8H_{18}(l) + 25\ O_2(g) \rightarrow 16\ CO_2(g) + 18\ H_2O(g)$

## C.3 ATOM, MOLECULE, AND ION INVENTORY *(pages 128–130)*

This section introduces the concepts of the mole and molar mass. Point out the relationship of moles to balanced equations. Emphasize that the mole is a counting unit. The concept of a mole is generally a difficult one for students, and concrete activities are usually necessary to develop a thorough understanding. Several useful analogies can be found in videotapes listed at the end of this unit. Emphasize the analogy of the "chemist's dozen." Point out that the reason for Avogadro's number being so large is that the units being counted (atoms and molecules) are so small. Analogies that help emphasize the enormousness of the number $6.02 \times 10^{23}$ include its approximate equality to the number of grains of sand on all the beaches of the world or the number of glasses of water contained in all the oceans.

After students have an understanding of the mole as the unit for counting atoms and molecules, build the connection between the mole and molar mass. Chemists find the mass of macroscopic samples of matter rather than actually counting the number of atoms, molecules, or ions. A display of one mole of a variety of common substances—sugar, salt, aluminum, copper, water—reinforces the idea that the mass of one mole varies with the substance. Additional *ChemQuandaries* are found in the Enrichment Materials section at the end of the unit.

### YOUR TURN *Molar Masses* *(page 130)*

When students have an understanding of the concept of a mole, it is time to do some simple calculations.

#### Answers to Questions

1. 14.01 g/mol
2. 14.01 g/mol × 2 = 28.02 g/mol
3. 22.99 g/mol + 35.45 g/mol = 58.44 g/mol
4. (12.01 g/mol × 12) + (1.008 g/mol × 22) + (16.00 g/mol × 11) = 342.30 g/mol
5. 63.55 g/mol + 55.85 g/mol + (32.01 g/mol × 2) = 183.42 g/mol
6. (63.55 g/mol × 2) + 12.01 + (16.00 g/mol × 5) + (1.008 g/mol × 2) = 221.13 g/mol

### C.4 CONSERVATION MUST BE OUR WAY *(pages 131–132)*

This section explains that although atoms are always conserved, molecules are not. "Running out" of a resource may mean that it is too spread out for economical extraction.

The term *conservation,* when used by the general public, often refers to management of resources. Gathering, processing, and using resources generate unwanted materials. Conservation options that reduce resource demand and/or waste production include replacing, reusing, recycling, and rethinking. This leads to the next decision-making activity, planning a recycling program.

### C.5 YOU DECIDE: RECYCLING DRIVE *(pages 132–134)*

This activity examines resources and their recycling. Students decide which of three resources (paper, aluminum, glass) is the best candidate for recycling efforts by applying their current knowledge of chemistry. This activity reinforces the idea that trash can often assume greater economic worth as a resource when it is gathered and organized. The idea is to avoid dispersion of atoms throughout the environment. Waste and dispersion of nonrenewable resources may eventually threaten our present lifestyle. Because reuse requires less expenditure of energy than recycling or substitution, reusing resources should probably be a long-term goal.

The unit's discussion of glass, paper, and aluminum should be regarded as reference material for students to use in answering assigned questions. Specific facts about these three materials are not intended to be memorized but used as needed. As homework, students can organize their thoughts for the group discussion, which follows. Additional *ChemQuandaries* are found in the Enrichment Materials section at the end of the unit.

#### Answers to Questions

1. a. Aluminum is probably the most important of these resources to recycle for economic reasons, given that nearly 90% of our needed aluminum is imported, and considering the high energy costs in extracting aluminum from bauxite ore. All three resources require energy (either electrical or thermal) for processing. However, the energy required to produce aluminum is by far the greatest.
   b. Because of the large quantity of energy required, recycling aluminum is also important for environmental reasons. Some students, concerned about trees and forests, may argue that paper is the most important environmentally. Encourage students to think of the "hidden" environmental costs of producing aluminum—

energy sources, whether nuclear or coal power plants or hydroelectric dams, also have an impact on the environment.

2. a. (Student answers will vary.)
   b. (Student answers will vary.)
   c. One way to enforce significant recycling of aluminum and glass would be laws requiring deposits on all cans and bottles, which are already required in many states. The manufacturers of glass containers could be required to produce only returnable or reusable bottles. The federal government and some states require the purchase of products made from recycled materials for use in government offices. Some states and municipalities require that newspapers contain a minimum amount of newsprint from recycled sources. There is a continuing debate on whether or not government should be involved in recycling efforts.
3. Individuals can bundle old newspapers, participate in curbside recycling programs, and return nonreturnable bottles and aluminum cans to community collection centers for reprocessing. Paper can be conserved by buying goods in bulk quantities and avoiding excess packaging of goods. Glass containers may be reused in the home for other liquid storage.
4. a. Newspapers, books, magazines, personal stationery, textbooks
   b. Printing jobs, delivery people, textbook publishers and sales people, bookstores, reduction in forest product collection, pulp and paper mills
   c. Computer manufacturers, repair personnel, communications services (wiring, installation, etc.)
   d. A television screen or a computer screen
   e. More flexible schedule if possible, less travel (reduced school busing programs), more leisure time, home study, and working in the home

## PART C: SUMMARY QUESTIONS *(page 134)*

1. a. The three regions of Earth: atmosphere (100 km thick, composed mainly of gaseous nitrogen and oxygen); hydrosphere (5 km thick, composed mainly of water and ionic salts such as sodium chloride); and lithosphere—crust, mantle, and core (6,400 km thick, composed of silicates, fossil fuels, carbonates, oxides, and sulfides)
   b. The crust of Earth is the main storehouse of chemical resources used for manufacturing consumer goods.
2. a. $Ca_3(PO_4)_2(s) + 3\ H_2SO_4(aq) \rightarrow 2\ H_3PO_4(aq) + 3\ CaSO_4(s)$
   b. $WO_3(s) + 3\ H_2(g) \rightarrow W(s) + 3\ H_2O(l)$
   c. $2\ PbS(s) + 3\ O_2(g) \rightarrow 2\ PbO(s) + 2\ SO_2(g)$
3. Possible answers include recycling cans and paper at school; recycling papers, bottles, and cans at home; buying products in recyclable/reusable containers; buying products with less packaging.
4. a. 16.00 g/mol
   b. $16.00\ \text{g/mol} \times 2 = 32.00\ \text{g/mol}$
   c. $16.00\ \text{g/mol} \times 3 = 48.00\ \text{g/mol}$
   d. $24.32\ \text{g/mol} + 32.01\ \text{g/mol} + (16.00\ \text{g/mol} \times 4) = 120.33\ \text{g/mol}$
   e. $22.99\ \text{g/mol} + 16.00\ \text{g/mol} + 1.008\ \text{g/mol} = 40.00\ \text{g/mol}$
   f. $(12.01\ \text{g/mol} \times 9) + (1.008\ \text{g/mol} \times 8) + (16.00\ \text{g/mol} \times 4) = 180.15\ \text{g/mol}$
5. a. 2 mol $Cr_2O_3$, 3 mol Si, 4 mol Cr, 3 mol $SiO_2$
   b. $$2\ \text{mol}\ Cr_2O_3 \times \frac{152\ \text{g}}{1\ \text{mol}\ Cr_2O_3} = 304\ \text{g}\ Cr_2O_3$$

   $$3\ \text{mol Si} \times \frac{28.09\ \text{g Si}}{1\ \text{mol Si}} = 84.27\ \text{g Si}$$

$$4 \text{ mol Cr} \times \frac{52.00 \text{ g Cr}}{1 \text{ mol Cr}} = 208 \text{ g Cr}$$

$$3 \text{ mol } SiO_2 \times \frac{60.09 \text{ g } SiO_2}{1 \text{ mol } SiO_2} = 180.27 \text{ g } SiO_2$$

c. mass of reactants = 388.27 g; mass of products = 388.27 g

1pt

# D. METALS: SOURCES AND REPLACEMENTS

This part begins by examining the sources and uses of copper, and then introduces percent composition as a tool for the evaluation of ore quality. Three important techniques for refining a metal from its ore are explained, along with oxidation-reduction terminology. Part D concludes with a look toward the future, and an examination of materials such as ceramics, plastics, optical fibers, and polymer composites.

## D.1 COPPER: SOURCES AND USES *(pages 135–136)*

The properties of copper make possible a wide variety of uses. The United States is a major world supplier of this essential metal.

## YOUR TURN *Uses of Copper* *(pages 136–137)*

Students connect copper's properties with its uses, and propose technological changes that would alter copper usage.

### Answers to Questions

1. Electrical uses—low reactivity, high electrical conductivity, malleability, and ductility
   Nonelectrical uses—attractive color (jewelry); malleability (anything requiring shaping); low reactivity and high thermal conductivity (pots and pans)
2. a. Increased recycling would decrease our need for imports (the United States imports 13% of its copper from Canada, Chile, Peru, and Zambia) and decrease the need for mining and refining lower-quality ores.
   b. Recycling has a limit.
   c. Not all copper can be recycled—house wiring usually must stay in place for the lifetime of the home, and the same is true for plumbing and roofing. The benefits of recycling usually reach a maximum in terms of economic profit before maximum recycling is reached, which reduces the impetus to improve recycling.
3. a. Fiber optics, satellites, and cellular phones decrease the need for copper in communications wiring.
   b. Removing pennies from circulation, decreasing the copper in other coins, and using credit cards would lower copper usage.
   c. More use of solar power could reduce power generation by turbines, reducing copper used in the wiring of the turbines.
   d. New conducting alloys and aluminum may decrease the demand for copper.

## D.2 EVALUATING AN ORE *(page 137)*

In considering an ore as a source for a metal, its percent composition is an important factor to be considered. The *Your Turn* introduces calculations of percent composition for compounds and extends these calculations to finding the percent of copper in an ore.

## YOUR TURN *Percent Composition* *(pages 138–139)*

### Answers to Questions

1. a. $\frac{63.55 \text{ g Cu}}{183.54 \text{ g CuFeS}_2} \times 100 = 34.6\% \text{ Cu}$

   $\frac{127.1 \text{ g Cu}}{221.12 \text{ g Cu}_2\text{CO}_3(\text{OH})_2} \times 100 = 57.5\% \text{ Cu}$

$$\frac{190.65 \text{ g Cu}}{344.69 \text{ g } Cu_2(CO_3)_2(OH)_2} \times 100 = 55.3\% \text{ Cu}$$

b. Chalcocite has the highest percent of copper and would be the most profitable.

2. Magnetite has the greater amount of iron.

$$\frac{1{,}117 \text{ g Fe}}{159.7 \text{ g } Fe_2O_3} \times \frac{1{,}000 \text{ g } Fe_2O_3}{1 \text{ kg } Fe_2O_3} = \frac{699 \text{ g Fe}}{1 \text{ kg } Fe_2O_3}$$

$$\frac{167.55 \text{ g Fe}}{231.55 \text{ g } Fe_3O_4} \times \frac{1{,}000 \text{ g } Fe_2O_4}{1 \text{ kg } Fe_3O_4} = \frac{724 \text{ g Fe}}{1 \text{ kg } Fe_3O_4}$$

The mass of Fe in 1 kg $Fe_2O_3$ is greater.

3. a. Site A: 20 g $Fe_2O_3 \times 0.699 = 14.0$ g Fe

   Site B: 15 g $Fe_3O_4 \times 0.724 = 10.9$ g Fe

   b. The ore at site A has a greater percent of iron.

After the ore has been assayed for percent composition and mined, the metal must be processed to remove pure metal from the mineral. This leads to a discussion of reactivities of metals and metallurgy in the next two sections.

## D.3 METAL REACTIVITY REVISITED *(page 139)*

This section relates the reactivity of metals to the difficulty experienced in extracting them from their ores. The concept of an activity series is introduced and related to processes used to obtain the metal from its ore.

## CHEMQUANDARY *Discovery of Metals* *(page 139)*

Because of their lack of reactivity, copper, silver, and gold are likely to be found uncombined. More active metals will be found in nature in compounds.

## YOUR TURN *Metal Reactivity* *(page 140)*

This activity familiarizes students with the relative reactivity of metals, which determines whether or not certain chemical reactions may occur. It also explores the energy needed to free metals from their ores.

### Answers to Questions

1. Metallic reactivity generally decreases from left to right across a period of the periodic table. The most reactive metals are located on the left side. The least reactive metals, such as copper, silver, and gold, are in the transition metal section.
2. a. Yes

   b. No

   c. Iron is more reactive than lead, according to the metal activity series, and will release lead from a solution. Platinum is less reactive than lead.
3. a. Electrolysis will require the most energy, owing to the fact that this process is reserved for the most reactive elements, which require the most energy.

   b. Least expensive is roasting in air because no added materials are needed for the decomposition.
4. a. The least reactive metals are not the cheapest, despite having the lowest production costs.

   b. Other factors include demand for the metal (low reactivity is one of the attractive features of gold, silver, and copper—all three are long-lasting) and overall availability (gold, silver, and copper are much less abundant than aluminum and iron on Earth).

## CHEMISTRY AT WORK/All That's Gold Doesn't Glitter... Or Does It? *(page 141)*

The career spotlighted in this vignette is that of a metallurgist with a mining company whose company extracts gold from fairly low grade ore.

### Answers to Questions:

1. 23 metric tons
2. $154
3. Gold is very unreactive.

## D.4 METALS FROM ORES *(pages 142–143)*

Oxidation and reduction are defined, and the main reduction techniques for separating metals from their compounds in ores are presented.

## D.5 LABORATORY ACTIVITY: PRODUCING COPPER *(pages 143–144)*

This activity introduces one technique used to obtain metals from ores. Emphasize the fact that this process is endothermic, contributing to the higher cost of electrolysis for producing metals.

### Time

30 minutes

### Materials (for 24 students working in pairs)

12 U-tubes
24 carbon rods (pencil lead)
12 buret clamps
12 ring stands
100 mL 0.2 M $CuCl_2$ (3.4 g $CuCl_2 \cdot 2H_2O$/100 mL)
12 9-V battery connectors with alligator clips attached to the leads
12 9-V batteries

### Safety

Students should wear safety goggles and aprons throughout the laboratory activity. Extreme caution should be exercised when sniffing the electrodes (Step 7). Caution students about inhaling gases formed at the electrodes. Although only small amounts of chlorine gas are produced in the microscale procedure, good ventilation in the lab is important.

### Advance Preparation

Pencils sharpened at both ends can serve as electrodes. U-tubes can be purchased, or 10-cm lengths of clear plastic tubing may be wrapped around the battery in the shape of a U. If the U-tubes are small, adjust the pencil lead accordingly. The batteries, connectors, and small alligator clips can all be purchased from local electronics supply stores.

### Pre-Lab Discussion

Emphasize that this is an oxidation-reduction reaction: the two (oxidation and reduction) always go together. Stress the connection between this activity and previous experiences, the table of metal reactivity, and the concepts of oxidation and reduction.

### Lab Tips

Use fresh batteries. Rechargeable batteries are economical if they are used infrequently. The copper(II) chloride solution can be reused in subsequent classes, as the concentration is not critical to the success of this activity.

### Answers to Questions

1. A copper–colored substance is formed at the negative electrode. The odor of chlorine is present at the positive electrode.
2. a. The negative electrode is the cathode.
   b. Copper is formed at the cathode.
3. a. The positive electrode is the anode.
   b. Chlorine was formed at the anode.

4. $CuCl_2(aq) \rightarrow Cu(s) + Cl_2(g)$

## Post-Lab Discussion

Review the terminology of oxidation-reduction, and clarify the charges of the cathode and anode in the questions. Remind students of the energy cost of electrolysis. Point out that the more active elements of Groups I and II on the periodic chart, as well as aluminum, are commercially produced by electrolysis. The most active nonmetals, the halogens (Group VII), are also produced by electrolysis. Chlorine can be extracted from brine, a concentrated sodium chloride solution, by electrolysis. Students should be able to explain the results of reversing the leads in Step 8 of the lab procedure.

## D.6 FUTURE MATERIALS *(pages 144–147)*

This section discusses the substitution of new materials for less available or expensive resources. Begin by asking students to describe changes in the materials used to make sports equipment (fiberglass poles, graphite, golf clubs), medicine (heart valves, replacement joints, artificial skin), and automobiles (ceramic engines, plastic bumpers).

## YOUR TURN *Alternatives to Metals* *(page 147)*

## Answers to Questions

1. Wiring, motors, power distribution—aluminum
   Communication equipment, satellite transmissions—fiber optics
   Plumbing—plastic pipe
   Roofing—aluminum, fiberglass
   Coins—other alloys, more credit and debit cards, paper money
   Jewelry—silver, gold, chromium plated nickel
   Pots and pans—aluminum, ceramics
   Shell casings—less copper, more zinc
   Food preparation machinery—stainless steel
   Auto radiators—aluminum, plastics, ceramics
2. Car bumpers, automobile engines, plumbing materials (glass in place of copper), baseball bats, aircraft parts
3. Silver could replace copper in wiring owing to its higher conductivity. This could also improve efficiency of computers, with lower resistance to electron flow increasing the speed of signal transmission.

## PART D: SUMMARY QUESTIONS *(pages 147–148)*

1. Estimates of future resource supplies depend on future use or demand. Technology might increase or decrease the need, and improvement in mining techniques might alter the supply.
2. a. Recycling is important because it reduces waste disposal problems, and extends the supply of the resource.
   b. Recycling does not guarantee a future supply of the resource, if population growth and rising usage overtake the availability of the resource.
3.

| Use of copper | Property of copper utilized |
|---|---|
| wiring | high electrical conductivity, malleable |
| plumbing | malleable yet strong, low reactivity |
| cooking utensils | high heat conductivity |
| roofing | low reactivity, malleable, and strong |
| jewelry | low reactivity, pleasing appearance |
| car radiators | high heat conductivity, malleable |
| coins | appearance, low reactivity |

4. a. Silver sulfide. $\frac{215.8 \text{ g Ag}}{247.9 \text{ g } Ag_2S} \times 100 = 87.1\% \text{ Ag}$

b. Aluminum oxide. $\frac{53.96 \text{ g Al}}{101.96 \text{ g } Al_2O_3} \times 100 = 52.9\% \text{ Al}$

c. Calcium carbonate. $\frac{40.08 \text{ g Ca}}{100.09 \text{ g } CaCO_3} \times 100 = 40.0\% \text{ Ca}$

5. a. $\frac{10 \text{ g } PbSO_4}{100 \text{ g ore}} \times 100 = 10.0\% \text{ } PbSO_4$

b. $\frac{207.2 \text{ g Pb}}{303 \text{ g } PbSO_4} \times 100 = 68.3\% \text{ Pb}$

c. 10 g $PbSO_4 \times 0.683 = 6.83$ g Pb in 100 g ore = 6.83%

6. Reaction (b) would occur because zinc is more reactive than silver, and will displace silver from its compound.

7. a. No, it is not a good idea to stir lead(II) nitrate solution with an iron spoon. Iron is more active than lead and will displace it from its compound.

b. $Fe(s) + Pb(NO_3)_2(aq) \rightarrow Pb(s) + Fe(NO_3)_2(aq)$

8. Group I, the lithium family, also known as the alkali metals; and Group II, the beryllium family, also known as the alkaline earths

9. The three reduction processes all require some sort of electron source, either electricity or a reducing agent that will lose electrons easily. All of the processes are also endothermic reactions, requiring energy.

10. a. Zinc, as it has lost electrons

b. Nickel, as it has gained electrons

c. Zinc

# E. PUTTING IT ALL TOGETHER: MAKING SENSE OF CENTS

This section continues to develop the relationship between properties of metals and their uses, through consideration of metals used for coins.

## E.1 WHAT IS MONEY, ANYWAY? *(pages 149–150)*

Students are asked to work in small groups to design a metal currency for a new nation. Using the concepts developed in this unit, each group designs a currency and shares the design with the class. The answers to a series of questions given in the text can be the basis for your evaluation of each group.

## E.2 LOOKING BACK AND LOOKING AHEAD *(page 150)*

Students have encountered chemical language (symbols, formulas, and equations), laboratory techniques, and basic chemical concepts that relate to societal problems and decisions associated with use of resources. In these decision-making tasks, chemistry remains an important ingredient. Chemical insight can assist in the definition and rational resolution of many complex problems facing this nation and the world today, but most social problems are too complex to have simple remedies.

This unit has gone beyond the Water unit in considering the issues associated with general chemical resource use. In contrast to water, which is definitely a renewable resource, metallic resources can be used up or scattered so widely that it is economically impractical to retrieve them. In the next unit, students will examine a nonrenewable resource (petroleum) in detail, exploring the problems involved in its wise use—now and in the future. Since it is impossible to recycle petroleum after it has been burned, students will explore alternatives to petroleum, consistent with the approach introduced in this Resources unit.

# ENRICHMENT MATERIALS

## TEACHER BACKGROUND

### Recycling

Although participation is largely voluntary, recycling has become increasingly important to the citizens of many states and cities. Recyclables include plastic, glass, paper, metals, and yard waste. Municipal waste accounts for only about 2% of the solid waste generated in the United States, but recycling household waste is an area where students can make a positive contribution toward conserving our resources. One of the major problems in a recycling program, often not considered by the public, is finding a market for the recycled material.

It should be recognized that recycling is only a part of a waste management program and that recycling has limitations. The waste management system should include source reduction, recycling, waste combustion with energy recovery, and landfilling.

In 1990, 37% of the nation's solid waste was paper and paperboard; however, in 1992 the recycling rate for these materials was only 38.5%. Newspapers make up a large portion of the solid waste, but are also commonly recycled. Many paper products are made from 100% recycled fibers. Other recycled paper products are office paper and corrugated boxes. One major limitation to paper recycling is that fibers are gradually damaged as they go through the process repeatedly.

Only 15% of rigid plastic bottles and containers were recycled in 1992. One of the major problems in the process of recycling plastics is sorting, which must be done before it can be reused. Much of this sorting is done by hand, but recycling facilities are beginning to use automated systems. As automated systems become more common, the percentage of recycled plastic will become greater. Some of the items made from recycled plastics are drainage pipes, trash cans, grocery bags, traffic barrier cones, inner layers of new plastic bottles, fiberfill, carpeting, nonfood bottles, shower curtains, and paint brushes.

The 1992 rate for recycling glass bottles and jars was 33%. Color sorting is again a problem and must be done by hand. Most recycled glass is used to make new containers. Other products include glass fiber insulation, construction aggregate, water-main insulation, and road surfacing applications. About 60% of Baltimore's streets are composed of glassphalt, an aggregate of glass and asphalt. One factor that hinders recycling of glass is the presence of contaminants, which fouls the glass furnaces. Economically, the low cost of the raw materials, such as sand, to make new glass inhibits recycling.

Two metals most commonly recycled are aluminum and steel. The recycling of aluminum is a success story—in 1992, 68% of all cans were recycled. One of the major savings of using recycled aluminum is in energy costs. Most of the steel recycled from household wastes is also in the form of cans. Both steel and aluminum can be melted over and over without degradation. The major limitation to metal recycling is the presence of contaminants.

An often-forgotten type of recycling is the use of yard wastes for compost. Depending on the season and climate, as much as 35% of solid waste can be yard waste—grass, leaves, and brush. Composting, either individually or community-wide, can reduce the quantity of waste and create a useful product.

The information in this short essay on recycling was taken from *Recycling Information Pamphlet*, American Chemical Society, 1993. Copies of the 16-page pamphlet can be obtained by phoning 202/872-8725, or by writing the ACS Department of Government Relations and Science Policy, 1155 16th Street, NW, Washington, DC 20036.

## Constructing a Safe Conductivity Apparatus

Instructions for constructing a safe student conductivity apparatus from a 9-V battery, resistors, and LEDs available from electronic supply stores, are found in the reference below. A commercial version is also available from Flinn Scientific.

Katz, D. and Willis, C., *Journal of Chemical Education*, 1994, p. 330.

## CHEMQUANDARIES

**Sections A.5–A.7, C.4–C.5, and Part E**

Many homes are equipped with either home trash compactors or garbage disposals. While they may seem to reduce garbage, both pose their own set of ecological problems. Discuss this paradox.

**Answer** (a) Trash compactors reduce the volume of waste, not the total weight, and biodegradation of the compacted waste is slower. Compacted waste is harder to recycle.

(b) Disposals put organic, oxygen-demanding food waste into drains that enter waterways, increasing biological oxygen demand and creating a water pollution problem.

**Section C.3**

1. How large is a mole of rice grains?
   **Answers:** One mole of rice grains would cover all of the land area of planet Earth to a depth of 75 meters.
   One mole of rice grains is more grains than the number of grains of all grain grown since the beginning of time.
   A mole of rice would occupy a cube about 192 km (120 miles) on each edge.
2. Computers can count at the rate of about 80 million counts per second. At this rate, how long would take a computer to count to $6.02 \times 10^{23}$?
   **Answer:** 238 million years
3. If one mole of pennies were divided up and given to every person on Earth, how many pennies would each person receive?
   **Answer:** $1.55 \times 10^{14}$ pennies.
4. If each person spent $1 million a day, how long would the pennies last?
   **Answer:** 4,000 years

Source: David Tanis, *Chemunicator*, 6(11):12, 1988.

**Section C.5**

1. Explain the statement, Necessity is the mother of collection (and recycling).
   **Answer:** Materials are recycled when it is economically or strategically compelling to do so—witness the recycling efforts during World War II, the present-day recycling of aluminum, and the decision by the U.S. government to stockpile certain strategic metals. Recycling has reduced U.S. imports of chromium by 25%, and of cobalt 16%. (Source: United States Bureau of Mines Mineral Commodity Summary, 1989.)
2. Provide support for the statement, Waste is a human invention; in nature, nothing is wasted.
   **Answer:** Matter is continually recycled in nature—even the excretions and dead remains of organisms are broken down and reused.

## Demonstration Ideas

Use the usual precautions for demonstrations

**Section A.4**

The recovery of copper can be demonstrated if the reaction is conducted in a 250-mL flask in a fume hood. When the reaction with the acid is complete, place pieces of mossy zinc into the flask. The following reaction occurs.

$$\underset{\text{shiny silvery}}{Zn(s)} + \underset{\text{blue-green solution}}{Cu(NO_3)_2(aq)} \rightarrow \underset{\text{colorless solution}}{Zn(NO_3)_2(aq)} + \underset{\text{copper color}}{Cu(s)}$$

This reaction is especially dramatic if the copper nitrate solution is acidified with nitric acid first. In the absence of a fume hood, start with a solution of copper(II) nitrate. If enough zinc is used, the blue-green color will completely disappear and a mass of copper will be clearly evident. Point out that in recycling, "you can't get something for nothing." To get back a given resource requires use of another resource and/or energy. This demonstration is particularly effective as a transition to the next section.

**Section B.1**

To clarify the differences between physical and chemical properties and changes, some of the following demonstrations may be helpful. Ask students to observe the reactions carefully and record evidence for change before deciding if the change is chemical or physical. Choose demonstrations from the following list or other sources you may have in your classroom. Mix the order of the demonstrations.

**Physical changes**

1. Gently heat a few crystals of iodine in a 250-mL flask with a loosely fitting cork. Heat very carefully until the iodine vapors are just visible. Too much heating could cause the cork to become a projectile. **Do this demonstration only if a hood is available.** Iodine vapors are toxic.
2. Place a peanut-sized piece of Dry Ice ($CO_2$) in an inflated balloon, and tie the opening. As the Dry Ice warms, the balloon will expand.
3. Heat a bimetallic strip (20 cm or more in length) in a burner flame.
4. Shake and open a warm can of soft drink.
5. Dissolve any colored salt in water.

**Chemical changes**

1. There should be no open flames in the room during this demonstration. Place a few calcium turnings (calcium metal) in 200 mL of water in a 400-mL beaker. Hydrogen gas is evolved, which is made more noticeable if the beaker is placed on an overhead projector. Phenolphthalein added to the water causes the water to turn pink.
2. Burn a candle. This is especially effective if the candle is placed on a digital balance and the loss of mass is noted.
3. Fry or boil an egg.
4. Demonstrate spontaneous endothermic reactions such as that between 15 g $Ba(OH)_2 \cdot 8\,H_2O$ and 15 g $NH_4SCN$. Place the reactants in a 250-mL flask; stopper lightly, shake, wet the outside of the bottom of the flask with water, and set on a small piece of wood. $Ba(SCN)_2 + NH_3 + H_2O$ are formed with a temperature drop sufficient to freeze the flask to the block. Soluble barium salts are very toxic; however, barium sulfate is very insoluble. The solid formed in this demonstration should be dissolved in a minimum of water and reprecipitated with 1.0 M sulfuric acid. Collect the precipitate of barium sulfate, place it in a plastic bottle, and cap the bottle. Then it can be safely discarded in a landfill.

**Section B.4**

1. Alkali metal reactions can easily get out of hand. Do this demonstration only if you have the proper safety equipment. Sodium and potassium can be compared in terms of their rapid oxidation or "rusting" ($4\,M + O_2 \rightarrow 2\,M_2O$) when exposed to air, or their reactivity in water containing an indicator such as phenolphthalein ($2\,M + 2\,H_2O \rightarrow 2\,M\,OH + H_2$). Use only a very small, rice-grain-sized piece of metal and take proper safety precautions, such as placing a shield between the demonstration and the students. Contrast the behavior of these alkali metals to other metals such as iron, copper, or tin.
2. The reaction of magnesium or calcium with water containing phenolphthalein

provides a safer (though less dramatic) alternative to (1). Scrape the oxide coating from the magnesium with a razor blade. Calcium reacts with cold water, but magnesium requires boiling water. This reaction also points out the fact that metallic reactivity increases as one proceeds down a given family.

Point out that similar properties often lead to similar commercial uses, with abundance/cost often determining which element to use. For example, compare the uses of chlorine, bromine, and iodine to those of helium, neon, argon, and krypton.

## Other Activities

### Section B.2

A popular activity for making elements more familiar to students is selecting an "Element of the Week", in which each day has a small activity pertaining to that week's element. In a variation, each student "adopts" an element, prepares a report on it, and serves as the class expert on this element for the rest of the school year.

### Section B.11

**Some of the Chemical Products Used in the Restoration of the Statue of Liberty**

| Material | Quantity and cost | Uses |
|---|---|---|
| Aluminum oxide | 240 tons/$90,000 | Abrasive for blasting corroded layer of iron from frame |
| Liquid nitrogen (at 150 psi) | 4,000 gal/$75,000 | Removed seven layers of paint from the inside of the copper walls without damaging the penny-thin skin; works by temperature differential |
| Baking soda ($NaHCO_3$) | 40 tons/$12,000 | Used to strip/sandblast the dried-out, coal-tar weather seal from the interior copper—follow-up to liquid nitrogen step |
| Polytetrafluor-ethylene (powdered PTFE) | 400 sq. yards | Solid Teflon tape, replaced the asbestos as a barrier between the stainless steel saddles and the copper—thus preventing the recurrence of galvanic corrosion |
| Water-based potassium silicate primer containing zinc dust | 525 gal | Quick-drying paint with a high $SiO_2$/Zn ratio for hardness and good adhesion; generated no volatile organics |
| Waterborne epoxy polyamide | 1,000 gal | High gloss, good adhesion topcoat paint that resists graffiti, chemicals, and abrasion |
| Glass/polyvinyl butyral resin | 16,000 lb/300 sq. ft | Windows to two layers of PBR sandwiched between two layers of laminated glass |

### Section C.2

More practice in balancing equations

1. $Pb(s) + O_2(g) \rightarrow PbO(s) + SO_2(g)$
   coefficients: 2,3,2,2
2. $PbO(s) + C(s) \rightarrow Pb(l) + CO_2(g)$
   coefficients: 2,1,2,1
3. $NH_3(g) + O_2(g) \rightarrow NO(g) + H_2O(g)$
   coefficients: 4,5,4,6
4. $NO(g) + O_2(g) \rightarrow NO_2(g)$
   coefficients: 2,1,2

# MACROSCALE LABORATORY ACTIVITIES

Additional instructions for the teacher are found in the corresponding sections of this guide. For teachers preferring the macroscale activity, the student instructions are printed on separate pages, which are ready for duplication. Larger quantities of substances and solutions are often needed for the macroscale activity; in addition, some of the macroscale activities require significantly more time than the small-scale versions. Questions for students to answer are found in the student text following the small-scale version of the activity.

## A.4 LABORATORY ACTIVITY: USING UP A METAL Teachers' Directions

The larger amounts of copper and nitric acid used in this activity increase the hazard from the production of $NO_2$. In general, the quantities of solid copper compounds produced in the activities that follow can be discarded in the school trash. Small quantities of solutions of copper compounds can be discarded down the drain along with plenty of water. Because local regulations vary, it is important to check with school or municipal officials regarding required disposal methods.

### Materials (for a class of 24 working in pairs)

12 test tubes
6 dropping bottles for dispensing 8 M $HNO_3$
150 mL 8 M nitric acid solution (76 mL conc. $HNO_3$/150 mL solution)
30 cm copper wire (24 gauge, nonlacquered) cut in 2.5-cm lengths

## A.4 LABORATORY ACTIVITY: USING UP A METAL

### Getting Ready

In this activity, you will observe the chemical reaction between nitric acid, $HNO_3$, and copper metal, Cu. Many metals react with acids. As you will see, such a reaction causes the metal material to deteriorate or be worn away.

### Safety

*Caution:* Nitric acid also reacts with skin.

### Procedure

1. Obtain a piece of copper wire from the container at your table.
2. Place the wire in the test tube.
3. Carefully add just enough nitric acid ($HNO_3$) to the tube to cover the copper wire. *Caution:* Nitric acid is corrosive to skin. If any acid accidentally spills on you, wash the affected area with tap water for several minutes. Notify your teacher.
4. Observe the reaction. Record observations of changes in the wire, the solution, and the test tube. (Placing a white sheet of paper behind the test tube will make it easier to observe the changes.)
5. Discard the test tube contents according to your teacher's directions.
6. Wash your hands thoroughly before leaving the laboratory.

## B.3 LABORATORY ACTIVITY: METAL OR NONMETAL?

### Teachers' Directions

Additional instructions for the teacher are found in section B.3 of this guide.

### Materials

84 test tubes (13 × 100 mm)
12 test tube racks
12 test tube brushes
12 10-mL graduated cylinders
12 forceps or tweezers
84 vials for elements
12 marking pencils
12 dropper bottles for dispensing solutions
1 conductivity apparatus (optional)
several small hammers or mallets
100 mL 0.5 M hydrochloric acid solution (4.2 mL conc. HCl/100 mL solution)
100 mL 0.1 M copper(II) chloride solution (1.7 g $CuCl_2 \cdot 2\,H_2O$/100 mL solution)
(a) 25 g carbon (C), graphite sticks, pencil lead, or charcoal briquets
(b) 35 cm magnesium (Mg) ribbon, cut to 2-cm lengths
(c) 25 g silicon (Si), lump
(d) 25 g tin (Sn), mossy or foil
(e) 25 g sulfur (S), lump or roll
(f) 25 g iron (Fe), filings or thin wire
(g) 50 g zinc (Zn), mossy or wire

## B.3 LABORATORY ACTIVITY: METAL, NONMETAL?

### Getting Ready

In this activity, you will investigate properties of seven elements in order to classify them as metals, nonmetals, or metalloids. You will examine each for its physical properties of color, luster, and form (for example, is it crystalline, like table salt?). By attempting to crush each sample with a hammer, you will decide whether each element is malleable or brittle. You may also test for the physical property of electrical conductivity. (As an alternative, your teacher may demonstrate this test.)

Next, you will observe differences among these elements' chemical properties. You will find out whether each element reacts with hydrochloric acid, HCl(aq), and with a copper (II) chloride ($CuCl_{2)}$ solution.

Prepare a data table in your notebook, leaving plenty of space to record the properties of the seven element samples, which have been coded with letters a to g.

### Data Table

| Element | Appearance | Result of crushing | Conductivity | Reaction with acid | Reaction with $CuCl_2$(aq) |
|---|---|---|---|---|---|
| a. | | | | | |
| b. | | | | | |
| c. | | | | | |
| d. | | | | | |
| e. | | | | | |
| f. | | | | | |
| g. | | | | | |

### Procedure

**Physical Properties**

1. **Appearance:** Observe and record the appearance of each element. Include physical properties such as color and luster as well as the form.
2. **Crushing:** A material is malleable if it flattens without shattering when struck. A sample is brittle if it shatters into pieces. Gently rap each element sample with a hammer. Decide whether the samples are malleable or brittle.
3. **Conductivity (optional):** If a conductivity apparatus is available, use it to test each sample. *Caution:* Avoid touching the electrodes with your hands; you can get an electric shock. Touch both electrodes to the element sample. If the bulb lights, the sample has allowed electricity to flow through it. Such a material is called a conductor. If the bulb fails to light, the material is a nonconductor (see Figure 7).

**Chemical Properties**

1. Test each sample for reactivity with acid as described below. The formation of gas bubbles indicates that a chemical reaction has taken place.
   a. Prepare seven test tubes labeled a to g. Place 5 mL of water in one tube. Draw a grease pencil line on each test tube wall to indicate a 5-mL height. Pour out the water.
   b. Place a sample of each element in its appropriately labeled tube. The sample should be a 2-cm length of wire of ribbon, or 0.5–1.0 g of solid.
   c. Add 5 mL of hydrochloric acid (HCl) to each tube.
   d. Observe and record each result.
   e. Discard the test tube contents as instructed by your teacher.
2. Test each element sample for reactivity with copper(II) chloride ($CuCl_2$) solution as described below. Changes in the appearance of any element sample indicate a chemical reaction has taken place.
   a. Repeat Steps 1a and 1b.
   b. Add approximately 5 mL of $CuCl_2$ solution to each tube.
   c. Observe the test tubes for five minutes—changes may be slow.
   d. Discard the test tube contents as your teacher instructs.
3. Wash your hands thoroughly before leaving the laboratory.

## B.8 LABORATORY ACTIVITY: METAL REACTIVITIES Teachers' Directions

Additional instructions can be found in section B.8 of this guide.

## Materials

48 test tubes (13 × 100 mm)
12 10-mL graduated cylinders
12 test tube racks
12 test tube brushes
12 marking pencils
4 burets or 32 dropping bottles for dispensing
3 sheets of fine sandpaper
24 cm magnesium ribbon, cut to 0.5-cm lengths
144 cm copper wire, cut to 3-cm lengths
144 cm zinc wire, cut to 3-cm lengths
300 mL 0.2 M magnesium nitrate solution
  (15.4 g $Mg(NO_3)_2 \cdot 6\ H_2O$/300 mL solution)
300 mL 0.2 M copper(II) nitrate solution
  (14.0 g $Cu(NO_3)_2 \cdot 5\ H_2O$/300 solution mL)
300 mL 0.2 M zinc nitrate solution (17.9 g $Zn(NO_3)_2 \cdot 6\ H_2O$/300 mL solution)
300 mL 0.01 M silver nitrate solution
  (0.5 g $AgNO_3$/300 mL solution) (Store in brown bottle.)

## B.8 LABORATORY ACTIVITY: METAL REACTIVITIES

### Getting Ready

In this activity, you will observe and compare some chemical reactions of several metallic elements. Reactants include a metal and a solution containing ions of a different metal. You will investigate reactions of three metals (copper, magnesium, and zinc) with solutions of ionic compounds that contain metal ions (copper(II) nitrate, magnesium nitrate, zinc nitrate, and silver nitrate).

In your laboratory notebook, prepare a data table like the one below.

| | Solutions | | | |
|---|---|---|---|---|
| **Metal** | **$Cu(NO_3)_2$** | **$Mg(NO)_2$** | **$Zn(NO_3)_2$** | **$AgNO_3$** |
| Cu | ------- | | | |
| Mg | | ------- | | |
| Zn | | | ------- | |
| Ag | NR | NR | NR | ------- |

We have filled in results for reactions of silver metal, Ag (which is too expensive for this use). NR stands for "no visible reaction." The dashes indicate reaction combinations that are not needed.

**Procedure**

1. Obtain four strips (approx. 3 cm × 0.5 cm) of each of the three metals to be tested.
2. Clean the surface on each metal strip by rubbing it with sandpaper or emery paper. Record observations on the metals' appearance.
3. Label four test tubes for each of the four solutions to be used. Pour 2 mL of water in one tube. Mark all tubes with a grease pencil to show this 2-mL level. Discard the water from the tube.
4. Place 2 mL of test solutions in appropriately labeled tubes. *Caution:* Avoid touching skin or clothing with the $AgNO_3$ solution; it causes dark, nonwashable stains.
5. Add a copper strip to each solution to be tested (see Figure 9). Observe the results for 3–5 minutes. If no reaction is observed, write NR in your data table. If you observe a reaction, record the observed changes.
6. Remove the copper strip from the tubes. Dry the strip with a paper towel and clean it with sandpaper so it will be ready for the next group of students.
7. Discard the used solutions as directed by your teacher. Wash the test tubes.
8. Repeat Steps 4 through 7 with each of the other two metals. Complete all entries in the data table.
9. Wash your hands thoroughly before leaving the laboratory.

# SUPPLEMENTAL READINGS


"Chemical Matters." *CHEM MATTERS,* December 1984, 16.
"Archimedes." *CHEM MATTERS,* October 1987, 17.
"Camping Stoves." *CHEM MATTERS,* April 1982, 7.
"Chemists on Mt. Olympus." *CHEM MATTERS,* February 1987, 13.
"Dark Science." *CHEM MATTERS,* October 1985, 10.
"The Fascinating World of Trash." *National Geographic,* April 1983, 424-457. [Waste disposal and recycling.]
"Biodegradable Bags." *CHEM MATTERS,* October 1991, 4.
"BOHR." *CHEM MATTERS,* April 1984, 15.
"Eavesdropping on the Elements." *CHEM MATTERS,* December 1994.
"Element Soup!" *CHEM MATTERS,* October 1995.
"Element X." *CHEM MATTERS,* December 1987.
"Family Resemblance." *CHEM MATTERS,* December 1987.
"The Wild World of Compost." *National Geographic,* August 1980, 273-284. [Recycling.]
"Statue of Liberty." *CHEM MATTERS,* April 1985, 8-12.
"Chemicals Help Rejuvenate Miss Liberty." *Chemical Week,* July 2, 1986, v.139 (1). [A chemical trade magazine provides a brief rationale on the actual chemicals used in cleaning the statue.]
"Liberty Lifts Her Head Once More." *National Geographic,* July 1986, 2-20. [Historical perspective.]
"Brooklyn Bridge, the Structure of Metals." *CHEM MATTERS,* October 1983.
"The Disposal of Wastes in the Ocean." *Scientific American,* August 1974, 10-25.
"Cross-coin Puzzle." *CHEM MATTERS,* October 1987.
"It's Elementary." *CHEM MATTERS,* February 1989.
"Periodic Spiral." *CHEM MATTERS,* December 1990.
"Super-student Conductors." *CHEM MATTERS,* October 1987.
"Superconductivity." *CHEM MATTERS,* October 1987.
"Storing Up Trouble: Hazardous Wastes." *National Geographic,* March 1985, 318-351.
"Elements and Their Organization." *CHEM MATTERS,* April 1984.
"Aluminum, the Magic Metal." *National Geographic,* August 1978, 186-211.
"Going Against the Flow." *CHEM MATTERS,* December 1986.
"The Miracle Metal--Platinum." *National Geographic,* November 1983, 686-706.
"Element Search." *CHEM MATTERS,* October 1984.
"Gold, the Eternal Treasure." *National Geographic,* January 1974, 1-51.
"Love Always, Francium." *CHEM MATTERS,* December 1985.
"Periodically Puzzling." *CHEM MATTERS,* December 1985.
"The Wrong Knife." *CHEM MATTERS,* December 1985.
"Silver–A Mineral of Excellent Nature." *National Geographic,* September 1981, 280-313.
"Comic." *CHEM MATTERS,* February 1986.
"Beyond the Era of Materials." *Scientific American,* June 1986, 34-41.
"Tainted Water." *CHEM MATTERS,* February 1988.
"High-Tech Ceramics: a *C&EN* Special Report." *Chemical & Engineering News,* July 9, 1984.
"Old News, New Paper." *CHEM MATTERS,* April 1993.
"A Silver Opportunity." *The Science Teacher,* April 1996, 23-25.
"Recycling Bottleneck." *Time,* September 14, 1992, 52-54.
"Two Strings of Pearls." *The Physics Teacher,* January 1991, 30.

# TEACHING SCHEDULE

| | TIME BLOCK 1 | TIME BLOCK 2 | TIME BLOCK 3 | TIME BLOCK 4 | TIME BLOCK 5 |
|---|---|---|---|---|---|
| Class Work | Read Intro +<br>A.1-A.3<br>Discuss Intro. to Petroleum Unit<br>YD A.1 | Discuss A.2<br>YD A.3<br>Prelab B.1 | LA B.1<br>Discuss A.3 | Discuss B.2<br>Prelab LA B.3<br>Getting Ready for LA B.3 | LA B.3<br>CQ A.6 |
| Homework | | SQ Part A<br>Getting Ready B.1 | Read B.2 | | CQ B.3<br>Read B.4-B.5 |
| | **TIME BLOCK 6** | **TIME BLOCK 7** | **TIME BLOCK 8** | **TIME BLOCK 9** | **TIME BLOCK 10** |
| Class Work | YD B.4<br>Discuss B.5<br>YT B.5 | LA B.7 | LA B.8 | Review Parts A& B | Quiz Parts A&B<br>YTs B.6 (2)<br>Discuss C.1<br>Plan YD C.1 |
| Homework | Read B.6 | YT B.7<br>Prelab LA B.8 | YT B.8 | SQ Part B<br>Read C.1 | Read C.2<br>YD C.1 interview |
| | **TIME BLOCK 11** | **TIME BLOCK 12** | **TIME BLOCK 13** | **TIME BLOCK 14** | **TIME BLOCK 15** |
| Class Work | YD C.1<br>Summary<br>YT C.2 (1st) | Discuss C.3<br>YT C.3 (1st)<br>Prelab LA C.5<br>Getting Ready C.5 | LA C.5 | Discuss C.6<br>YT C.6 | YT C.7 |
| Homework | YT C.2 (2nd)<br>Read C.3-C.5 | YT C.3 (2nd) | Read C.6 | Read C.7 | SQ Part C<br>Read D.1-D.3 |
| | **TIME BLOCK 16** | **TIME BLOCK 17** | **TIME BLOCK 18** | **TIME BLOCK 19** | **TIME BLOCK 20** |
| Class Work | Review Part C<br>Quiz Part C<br>Discuss D.1<br>LA D.2 | LA D.2 cont. | Discuss D.3-D.5<br>Prelab D.6 | LA D.6 | SQ Part D |
| Homework | Read D.3-D.5 | | | | |
| | **TIME BLOCK 21** | **TIME BLOCK 22** | **TIME BLOCK 23** | **TIME BLOCK 24** | **TIME BLOCK 25** |
| Class Work | Review Part D<br>Quiz Part D<br>Discuss E.1-E.2<br>YT E.1 (1st) | PIAT F.1-F.2<br>SQ Part E | PIAT F.1-F.2 | Review Petroleum unit | Exam Petroleum unit |
| Homework | YT E.1 (2nd) | | | | |

**LA**=Laboratory Activity; **CQ**=ChemQuandary; **YT**= Your Turn; **YD**=You Decide; **PIAT**=Putting It All Together; **SQ**=Summary Questions; One **Time Block**=Approximately 50 minutes

# SUPPLY LIST

## Expendable Items and Unusual Materials

| Quantity | Unit | Item | Sections Used |
|---|---|---|---|
| 450 | mL | 10W-40 oil | B.3 |
| 180 | mL | 2-propanol | B.1 |
| 50 | g | asphalt pieces | B.3 |
| 12 | g | boiling chips | B.1, D.6 |
| 12 | | candles | C.5 |
| 12 | mL | concentrated sulfuric acid | D.6 |
| 450 | mL | distilled water | B.3 |
| 12 | 12-oz | empty soft drink cans | C.5 |
| 12 | | hotplates | B.1 |
| 5 | lb | ice | C.5 |
| 12 | | index cards | C.5 |
| | | iodine crystals | B.1 |
| 450 | mL | kerosene | B.3 |
| | | matches | C.5 |
| 60 | 7-mm | metal beads | B.3 |
| 24 | mL | methanol | D.6 |
| 450 | mL | mineral oil | B.3 |
| 12 | sets | space-filling models | B.7, B.8, D.6 |
| 420 | mL | water | B.1 |
| 12 | | wooden splints | B.1 |

The Petroleum unit develops the concepts of resources and conservation, which started in the Water unit and continue through the Resources unit. Students are introduced to the properties and chemistry of petroleum and the central concepts of organic chemistry: naming, covalent bonding, hydrocarbon isomerism, and synthesis.

Our society is highly dependent on petroleum as an energy source, but petroleum is also vitally important as the source of key petrochemicals used as building blocks. The major theme of this unit is separating and altering the compounds found in petroleum to produce a wide variety of consumer products.

Although students are exposed daily to many organic-based materials ranging from sandwich bags to cassette tapes to clothing, most chemistry courses postpone any treatment of organic chemistry until the end of the course—if it is covered at all. In *ChemCom,* students gain an early view of organic chemistry through its introduction in this unit, and through its extension to important biomolecules in the section "Understanding Foods" in the Food unit.

Knowledge of organic chemistry can help citizens, voters, and lawmakers become wise managers of this important resource. Students are challenged to consider the choices involved in using petroleum for building or burning.

# LEARNING GOALS

Upon completion of this unit the student will be able to

1. Compare the benefits with the burdens of using petroleum for building and burning. [A.2]
2. Identify regions of high petroleum use and regions of petroleum reserves, and discuss the economic and political implications of petroleum supply and demand. [A.3]
3. Describe the chemical makeup of petroleum and how it differs from other natural resources. [B.1]
4. Explain the relationship between density and viscosity and the number of carbon atoms in a hydrocarbon molecule. [B.2, B.3]
5. Describe the process of fractional distillation; list the five major fractions of petroleum distillation and typical products manufactured from each fraction. [B.2, B.4]
6. Name the first ten alkanes and draw structural and electron-dot formulas for each. [B.5, B.7]
7. Describe the processes involved in covalent bonding. [B.6]
8. State and explain the effect of carbon chain length and sidechains on hydrocarbon boiling points. [B.7, B.8]
9. Define the term *isomer,* and draw structural formulas for the isomers of a given hydrocarbon. [B.8]
10. Trace the history of energy sources and consumption patterns in the United States, and account for major changes. [C.2]
11. Explain endothermic and exothermic reactions in terms of bond breaking and forming, and give examples of each type of reaction. [C.3]
12. Identify energy conversions in an automobile, and calculate the savings resulting from increased automobile efficiency. [C.3]
13. Define heat of combustion and specific heat, and calculate energies of various combustion reactions. [C.4–C.6]
14. Write balanced equations, incorporating energy terms, for the combustion of hydrocarbon fuels. [C.6]
15. Define the term *octane number*, state its relationship to grades of gasoline, and identify two ways of increasing octane number. [C.7]
16. Compare saturated with unsaturated hydrocarbons in terms of formulas, structures, and physical and chemical properties. [D.2, D.3]
17. Identify the functional groups for common alcohols, carboxylic acids, and esters. [D.4–D.6]
18. Describe polymerization; give one example each of addition and condensation polymerization reactions. [D.5]
19. Describe major sources of energy for the United States today and alternative sources of fuels for the future. [E.1, E.2]

# A. PETROLEUM IN OUR LIVES

Part A highlights our reliance on a wide variety of petroleum-based products and on the use of petroleum as a major source of energy. The worldwide interdependence of nations that produce and consume this nonrenewable resource is considered.

## A.1 YOU DECIDE: IT'S A PETROLEUM WORLD *(page 160)*

To emphasize petroleum as a chemical feedstock for "building," this activity illustrates a variety of products made from petroleum that students may take for granted. Figure 4, page 160, is a photograph containing several petroleum-based products. After students examine it, ask them to list items that are made from petroleum. When they have completed their lists, show the students the photograph in the Black Line Masters in the back of the Guide from which all petroleum-based products have been removed. They can decide which petroleum-based items they would be least willing and most willing to give up if a severe petroleum shortage occurred.

This opening activity should sensitize students to the fact that they use many petroleum-based products when they participate in enjoyable activities. Make clear how issues of supply and abundance of this important nonrenewable resource affect their lives and require them to think about priorities.

## A.2 PETROLEUM IN OUR FUTURE *(pages 161–162)*

This unit presents charts and graphs to describe the production and use of petroleum and discusses the effects of these on pricing. Future production and use are predicted using the graphs and charts.

## CHEMISTRY AT WORK/Oil's Well That Ends Well: The Work of a Petroleum Geologist

The petroleum geologist described here must have knowledge of a variety of scientific fields.

## A.3 YOU DECIDE: WHO HAS THE OIL? *(pages 164–165)*

The goal of this student activity is to compare world petroleum consumption and supplies in different geographic regions by examining graphical presentations of data concerning petroleum reserves, populations, and consumption of petroleum. Once again, students see that unequal distribution of a natural resource (and varying demand) leads nations to import-export relationships. Make sure students note the definition of known reserves and understand how to read the various figures.

### Answers to Questions

1. Middle East
2. Middle East (57% world's oil vs. 3% world's population)
3. Asia and Pacific (5% world's oil vs. 57% world's population )
4. North America (32% vs. 12%), Asia and Pacific (17% vs. 5%), Eastern Europe (18% vs. 9%), Western Europe (21% vs. 4%), Central and South America (6% vs. 5%)
5. Africa (3% vs. 8%), Middle East (3% vs. 57%)
6. Africa, Middle East
7. Same regions as in Answer 4
8. Middle East and North America, Middle East and Western Europe, Middle East and Eastern Europe, Middle East and Asia
9. Differences in political and economic systems
10. North America (32%), Western Europe (21%)

11. North America and Western Europe are highly dependent upon petroleum as fuel for transportation (have more private cars than other nations) and as raw materials for industry.

12.a. (1) North America has a larger population than the Middle East.

(2) The Middle East produces much more petroleum than it consumes, and it has a much smaller population than North America.

(3) North America consumes much more petroleum than it produces.

b. Strained political relationships between countries that import and export oil will likely result in the disruption of the flow of petroleum.

## PART A: SUMMARY QUESTIONS *(page 165)*

1. a. 18 million barrels $\times$ 0.13 = 2.34 (or 2.3) million barrels used to build (The text does not belabor precision in calculations. To simplify this and other calculations, encourage your students to have the same number of digits in their answer as in the data. This is a rough approximation of significant figures.)

   b. 18 million barrels $\times$ 0.87 = 15.66 (or 16) million barrels burned

   c. Many answers are possible, including drugs, films, clothing, soaps, detergents, vinyl records, synthetic-fiber rugs.

   d. Answers depend on items listed in Answer 1c.

2. a. Attempts to make up the shortfall in supply would increase expenditures and consumer costs to support more exploration for oil reserves and for higher-cost alternatives such as oil shale and tar sands.

   b. If the cost of oil increased 10% and the supply decreased, the short- and long-term costs to consumers would be greater than the initial 10% increase in cost.

3. a. A very good trade relationship is needed between North America and parts of the Middle East. The Middle Eastern countries, particularly OPEC countries, have in the past been free to raise their prices, and oil-importing countries have been forced to pay the increases. If it becomes too expensive to import oil from the Middle East, North America may turn to more locally available resources or search for petroleum substitutes. The United States now imports increasing amounts of petroleum from Mexico, South America, and Canada.

   b. As other nations increase their use of Middle East oil, the North America–Middle East trade relationship will depend on price increases, as mentioned in Answer 3a. Developing technology in other world regions might lead new nations to become petroleum exporters, providing another source of petroleum for North America, as well as increasing trade relations with those countries.

4. $10{,}000 \text{ mi} \times \frac{1 \text{ gal}}{27 \text{ mi}} \times \frac{42 \text{ gal}}{21 \text{ gal}} \times \frac{1 \text{ barrel}}{42 \text{ gal}} = 18 \text{ barrels}$

# B. PETROLEUM: WHAT IS IT? WHAT DO WE DO WITH IT?

Part B examines the differences between petroleum and metallic ores and describes the major components of petroleum, their separation, and their uses. Students explore not only the properties of petroleum fractions, but also the relationship between structure, bonding, and hydrocarbon boiling points through a series of discovery-oriented activities. The rationale for bonding is introduced, and students use models to explore the structures of common alkanes, their naming, and the existence of structural isomers.

## B.1 LABORATORY ACTIVITY: SEPARATION BY DISTILLATION

*(pages 166–169)*

### Pre-Lab Discussion

Indicate that this activity models the separation technique used in petroleum refineries: distillation. Remind students that petroleum is a mixture of many related substances. This distillation activity simulates the refining methods used with crude oil.

Point out that the unknown liquid solution that students will receive is a mixture of two of the four substances as shown in Table 1 (even though all students will have only a mixture of water and 2-propanol). Student observations will permit the identification of the two components in the mixture.

### Time

50 minutes

### Materials (for 24 students working in pairs)

24 100-mL beakers
12 100-mL graduated cylinders
12 125-mL Florence flasks
12 funnels
12 thermometers (-10 °C to 110 °C)
12 hot plates (do not substitute a burner)
24 ring stands
12 clamps
12 rubber stoppers with 2 holes to fit the tops of the Florence flasks
12 condensers
12 corks
12 wooden splints
600 mL of distillation solution prepared by mixing 420 mL water and 180 mL 2-propanol (iso-propanol)
boiling chips
glass tube to connect the flask to the condenser (<90° bend)
rubber tubing
iodine crystals

### Safety

Students should wear safety goggles and aprons throughout the laboratory activity. The unknown liquid is flammable. No flames should be allowed in the laboratory. **Point out the hazards of heating unknown liquids that might be flammable. Careful observation is necessary during the entire heating process. Never heat the flask to dryness.** Remind students that iodine must be treated with special care. Skin contact with iodine crystals can cause lesions, and iodine gas is intensely irritating to the skin. Students should not be permitted to use thermometers containing mercury.

## Advance Preparation

Construct and display a correctly assembled distillation setup to enhance the pre-lab discussion (see Figure 8 in the student text). Demonstrate how to insert the thermometer and glass tubes into the stopper. You may wish to assemble the thermometer/stopper/tube combinations to reduce accidents.

## Lab Tips

- As usual, protective eyewear is required.
- Table 1 lists four possible candidates for students to consider as components of the original mixture. **Even though other options are listed, do not use any distillation solution other than the 2-propanol–water system specified in the materials list. This system poses the least hazard and gives good results.** However, do not reveal the composition of the mixture to the students. Allow them to use their observations to identify the mixture's components.
- Check all connections in each distillation apparatus before students begin work. Check to see that the tubing from the condenser is in the sink so that a flood will not result. Remind students to keep the water pressure low to prevent the rubber tubing from separating from the condenser.

## Expected Results

The time necessary will vary, but the boiling liquid should reach two temperature plateaus. The first, the boiling point of the 2-propanol, will be about 82 °C; the second will be about 100 °C, the boiling point of water. At higher altitudes the boiling points will be lower. The temperature of a pure substance does not change during the boiling process. However, 2-propanol and water form constant boiling mixtures. This fact along with errors in reading the thermometers will cause the plateaus to vary. Iodine is more soluble in the first distillate.

## Answers to Questions

1. Graph of time/temperature data with a plateau at about 82 °C and another at about 100 °C.
2. Temperatures should correspond to the plateaus, and should correspond closely to the temperatures at which the first drops of distillate were collected.
3. The first distillate is 2-propanol, and the second is water.
4. Iodine is more soluble in 2-propanol.
5. It could be boiled again. If two plateaus occur, it is not a pure substance.

## Post-Lab Discussion

Have students compare their conclusions regarding the components of the mixture. The discussion can focus on the contrasting physical and chemical properties of water and 2-propanol. Discuss which property makes it possible to separate the mixture by distillation.

Compare the results of the laboratory activity with fractional distillation of petroleum. In both, a mixture is separated by difference in boiling point. The laboratory activity involved only two components, but crude oil contains hundreds of different compounds, many with similar boiling points. Fractional distillation of crude oil generally results in mixtures.

## B.2 PETROLEUM REFINING *(pages 169–170)*

The text and diagrams clarify several aspects of fractional distillation, which separates the crude oil into its components. The tall tower allows lower-boiling hydrocarbons to remain gases until they reach the top; higher-boiling substances, which condense more easily and require more energy to vaporize, stay near the bottom. The difference in temperature throughout the column promotes the evaporation, condensation, and subsequent separation of the various fractions.

## B.3 LABORATORY ACTIVITY: VISCOSITY AND DENSITY

*(pages 170–173)*

In this activity, students observe selected physical properties of a number of petroleum fractions and determine the relative density and viscosity of several petroleum components. Students find the relationship between the viscosity of a sample and the average number of carbon atoms in its molecules.

### Time

50 minutes

### Materials (for 24 students working in pairs)

60 36-mL culture tubes with caps or test tubes with corks
60 7-mm metal beads
12 25-mL graduated cylinders
450 mL mineral oil (light)
450 mL household lubricating oil (light oil)
450 mL kerosene
450 mL motor oil, 10W-40
150 g paraffin wax (bars)
50 g asphalt (pieces) from road surface
450 mL distilled water
balances
several digital timers, stopwatches, or watches that count seconds

### Safety

Students should wear safety goggles and aprons throughout the laboratory activity.

### Advance Preparation

Determine the average mass of an empty culture tube plus cap, and the average mass of a metal bead. Measure the total volume of water that is held by one culture tube. The volume occupied by the bead is not significant. Write on the chalkboard the values for mass of vial plus cap, mass of bead, and volume of liquid in vial for students to use to complete the density portion of this activity.

Label a set of five culture tubes for each team: mineral oil, kerosene, motor oil, household lubricating oil, and water. Add a 7-mm metal bead to each of 60 culture tubes. Fill each tube with the appropriate liquid, and cap tightly. Once prepared, the tubes can be stored for use in subsequent years.

### Lab Tips

- Metal beads can be purchased from suppliers of earth science materials.
- Asphalt can be found at the edge of a newly paved road, at a refinery, or from a roofing contractor. Student samples of solids should weigh about 10–20 g each.

### Pre-Lab Discussion

Briefly review the logic of the liquid density determination outlined on page 172. Also review the water-displacement method for determining the volume of irregular solids (such as asphalt and paraffin) as part of this discussion.

Help students understand viscosity by pouring common materials such as vinegar and honey and having students compare the viscosities.

Remind students to leave the sample tubes sealed at all times.

## Expected Results

| Sample | Physical state at room temperature | Density, g/cm³ | Average time of fall, s | Relative viscosity |
|---|---|---|---|---|
| Mineral oil | liquid | 0.90 | 1.91 | 4 |
| Asphalt | solid | 1.60 | | |
| Kerosene | liquid | 0.74 | 0.35 | 1 |
| Paraffin wax | solid | 0.70 | | |
| Motor oil | liquid | 0.88 | 2.76 | 5 |
| Household oil | liquid | 0.91 | 0.72 | 3 |
| Water | liquid | 1.01 | 0.47 | 2 |

*Note: The motor oil used for these data was 10W–40, and the household oil was 3-in-1 oil.*

### Answers to Questions

1. Liquid petroleum products are less dense than water and will float on its surface. Oil floating on water can catch fire and burn.
2. As the number of carbon atoms per hydrocarbon molecule increases, the viscosity increases.
3. a. The petroleum fractions are dissolved together, and solutions do not settle.
   b. If separation could be performed on the basis of viscosity at room temperature, the energy needed to operate a fractional distillation column could be saved.

### Extensions

To demonstrate viscosity, a eudiometer tube may be used, which allows a longer drop time and increased accuracy. Viscosity can also be compared at various temperatures.

Determine the viscosities of motor oils of different ASTM ratings: 10W, 30W, etc.

## CHEMQUANDARY *Gasoline and Geography* *(page 173)*

Because the ease of evaporation of a liquid (its volatility) depends on its temperature, the properties of gasoline must be carefully controlled when it is used in regions with extreme temperatures. To ensure adequate vaporization and ignition of the fuel, gasoline used at high average temperatures must contain a larger fraction of high molecular weight molecules than gasoline used when the average temperature is low.

## B.4 YOU DECIDE: CRUDE OIL TO PRODUCTS *(pages 173–175)*

Students are introduced to the variety of products obtained from petroleum fractions. The objective is to classify various branches shown in Figure 10 (page 174) as petroleum for "burning" and/or feedstock for "building." This activity establishes important connections between the refined fractions from the original crude oil mixture and the petroleum-based products we use in daily life.

### Answers to Questions

1. and 2. Answers depend on which fraction is used. Petrochemicals are usually "building" uses of petroleum. Fertilizers and ammonia are used in agriculture to "build" food. Residues produce materials used for building (paving, shingles, etc.). Lubricating oils are neither burned nor used for building, and in many cases they can be recycled. Almost every other source falls into the "burning for energy" category.

3. (This sample answer is based on the choice of intermediate distillates; other answers are possible for other branches.) If intermediate distillates were eliminated, homes could not be heated with oil furnaces; there would be no diesel fuel for operating heavy construction and farming equipment; there would be a decrease in the feedstock available for producing petrochemicals such as alcohols, medicines, plastics, and dyes.

4. Students might choose fuels for heating homes, petrochemicals used in the production of medicines, or fertilizers to grow food, among others. Probe for the reasons supporting specific student choices.
5. Some students might choose the building of new substances as more important, but they will probably still realize that an economy such as ours cannot maintain itself without fuels for transportation. (Other answers are possible, based on the students' own priorities.)

## B.5 A LOOK AT PETROLEUM MOLECULES *(page 175)*

Distillation separates petroleum into fractions because of differences in the boiling points of the substances from which it is composed. This leads to an examination of the boiling points of various hydrocarbons, and the intermolecular forces responsible for the differences in boiling points.

## YOUR TURN *Hydrocarbon* Boiling Points *(pages 175–176)*

Students have previously observed differences in physical properties of various hydrocarbon molecules in the laboratory. Chemists determine physical and chemical properties of substances and organize the data in ways that lead to the identification of useful patterns. These patterns often stimulate searches for explanations of such regularities.

In order to encourage students to recognize patterns, one chemistry text suggests changing the class seating pattern regularly and asking students to identify the system for assigning seats: alphabetical, by age, sex, height, etc. This preliminary activity would provide an easy introduction to this *Your Turn.*

Curiously, some students prefer to organize chemical information alphabetically. By having them do a few problems with tables arranged according to alphabetical or numerical rank, you can convince them that organizing according to boiling points provides a simpler method for classifying the properties of these compounds. Eventually, the naming system and its relationship to number of carbons and boiling points will reinforce this concept.

### Answers to Questions

1. a. The hydrocarbons in Table 1 are alphabetized by name.
   b. Such an alphabetical arrangement might be useful for locating a specific hydrocarbon in a handbook or index.
2. a. The boiling-point data would be better organized if the hydrocarbons were listed from high to low (or low to high) boiling-point values.
   b.

| Hydrocarbon | Boiling point (°C) |
|---|---|
| Methane | -161.7 |
| Ethane | -88.6 |
| Propane | -42.1 |
| Butane | -0.5 |
| Pentane | 36.1 |
| Hexane | 68.7 |
| Heptane | 98.4 |
| Octane | 125.7 |
| Nonane | 150.8 |
| Decane | 174.0 |

3. a. Methane, ethane, propane, butane. You may want to emphasize the point that substances that are gases at room temperature have already boiled at a lower temperature. Students often associate boiling with a high temperature. This mistaken notion can be countered by the demonstration idea found in the Enrichment Materials section.
   b. Pentane.

4. Decane has a higher boiling point than butane, indicating that the intermolecular forces between decane molecules are greater.
5. The greater the intermolecular attraction, the greater the viscosity and the higher the freezing point. Pentane =1, octane = 2, decane = 3.

## B.6 CHEMICAL BONDING *(pages 177–179)*

The bonding responsible for petroleum molecules is explained. Ionic bonding, involving the gain and loss of electrons, is contrasted with covalent bonding, which is introduced by examining hydrogen molecules and methane. Electron-dot and structural formulas enable students to draw pictures of the compounds and lead them to the model building in the next activity.

## B.7 LABORATORY ACTIVITY: MODELING ALKANES *(pages 179–182)*

In this activity, students build three-dimensional models of methane, ethane, and propane. They derive the general formula for alkanes, practice writing electron-dot representations, and begin to view the two-dimensional structural formulas of molecules from a more realistic three-dimensional perspective. Although only a few elements are involved in organic compounds, a huge number of compounds are possible because a carbon atom can bond to other carbon atoms to form long chains and rings. Constructing models and diagrams enables students to see clearly the differences in the three formulas.

### Materials (for 24 students working in pairs)

12 sets of space-filling or ball-and-stick models

### Pre-Lab Discussion

One reason organic chemistry is so extensive, in terms of the number of known compounds, is the unique bonding capability of carbon. Carbon forms four covalent bonds and is the only element capable of such repeated bonding to itself. Remind students that carbon needs to add four electrons, so it always forms four bonds.

### Time

50 minutes

### Answers to Questions

1. No question
2. a. 6 hydrogens

   b. 8 hydrogens

   c. 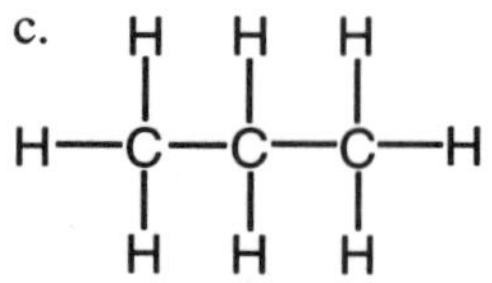

   propane

3. a. 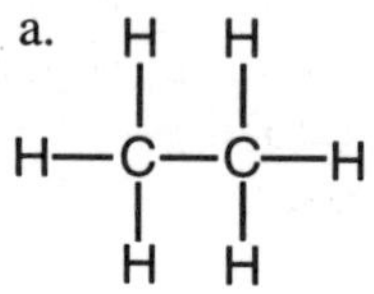

   ethane

   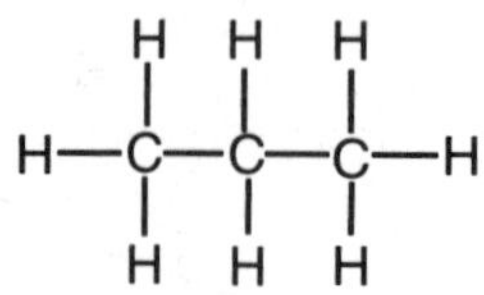

   propane

   b. $C_3H_8$
4. $CH_4$, $C_2H_6$, $C_3H_8$, $C_4H_{10}$, $C_5H_{12}$, $C_6H_{14}$, $C_7H_{16}$, $C_8H_{18}$, $C_9H_{20}$, $C_{10}H_{22}$

5.

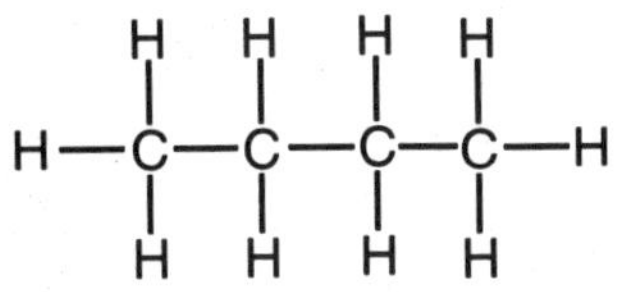

butane

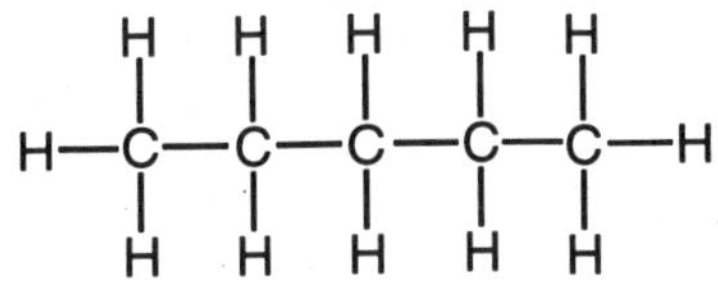

pentane

6. a. (1) Heptane (2) Nonane
   b. (1) $C_7H_{16}$ (2) $C_9H_{20}$
7. a. $C_{25}H_{52}$
   b. The short version takes much less time and space; the long version would not fit on this page!
8. a. Ethane, $C_2H_6$: 2(12) + 6(1) = 30 g/mol
   b. Butane, $C_4H_{10}$: 4(12) + 10(1) = 58 g/mol

## Post-Lab Discussion

The enormous number and variety of hydrocarbons demand an orderly system of nomenclature, which is introduced here. Students should be familiar with the Greek prefixes used in polygons in geometry (penta-, hexa-, etc.). Using the prefixes enables students to learn formulas for $C_5$–$C_{10}$ compounds easily. Repeated practice, with as many practical information tidbits as possible (cows "burp" methane, propane is in house trailer fuel, butane in lighters, and so on) should help students be able to name hydrocarbons if given the formula, or write a formula from a name.

Point out the general formula for alkanes and the alternate ways of writing structural formulas shown in Table 3, page 181. If students prefer the lines of structural formulas, do not force them to move to condensed formulas. Eventually, the ease of writing this style of formula will encourage them to change their practice.

## YOUR TURN *Alkane* Boiling Points: Trends *(page 182)*

Students are challenged to explain the observed boiling-point trend in the alkane series and predict boiling points of larger hydrocarbons. Discuss the chemical explanation for differences in boiling points, which is based on the notion of intermolecular forces. Longer (and less spherical) molecules have more surface area and points of contact for intermolecular attractions to occur.

## Answers to Questions

1. Approximately 30 °C
2. a. $C_{11}H_{24}$: 204 °C b. (actual: 196 °C)
   $C_{12}H_{26}$: 234 °C (actual: 216 °C)
   $C_{13}H_{28}$: 264°C (actual: 235 °C)
3. The greater the number of carbons, the stronger the attractive forces.

## B.8 LABORATORY ACTIVITY: ALKANES REVISITED *(pages 182–184)*

Students build models of $C_4H_{10}$ and $C_6H_{14}$ and find that in each case the atoms can be arranged in different ways (two and five, respectively), which introduces the notion of isomers. Lead students to deduce that isomers should exhibit different physical properties because they are different geometric arrangements of atoms, that is, different structures.

## Time

50 minutes

## Materials (for 24 students working in pairs)

12 space-filling or ball-and-stick model kits

## Lab Tips

Because building all five isomers of hexane would require more atoms than found in a typical kit, students will need to join into larger groups. The typical kit contains 12–14 carbon atoms, and so if each team (with one kit) joins with two others (for a total of three kits/six students), each team can be asked to build as many unique isomers as they can in a given time period.

## Answers to Questions

1. Two isomers of butane are possible.
2. a and b.

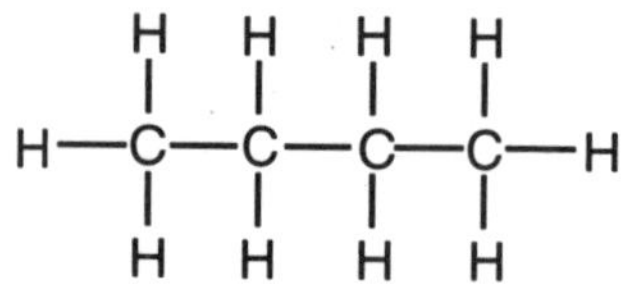

butane

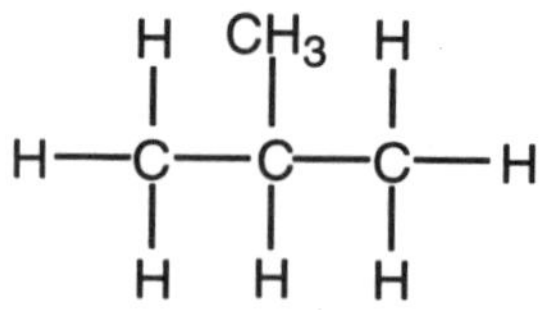

2-methylpropane

3. No question
4.

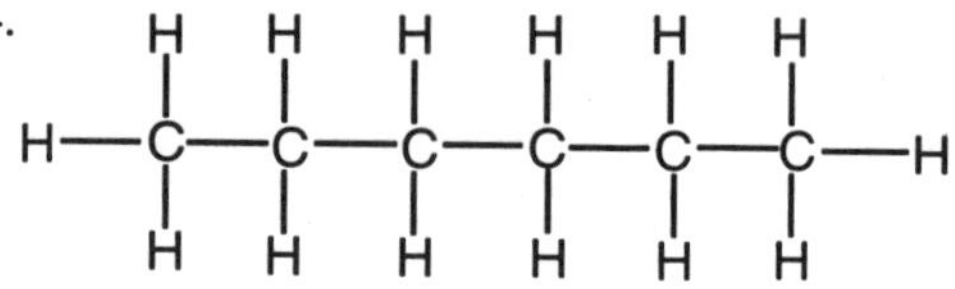

hexane

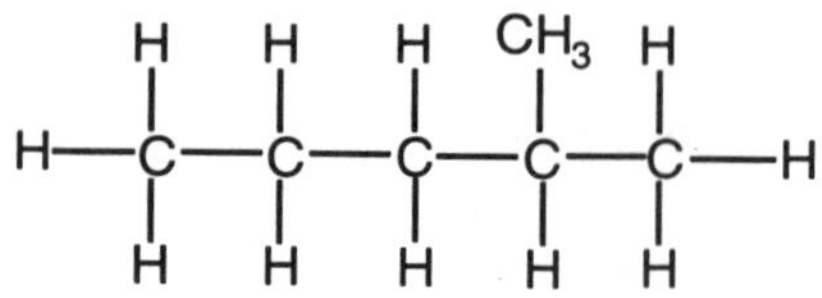

2-methylpentane

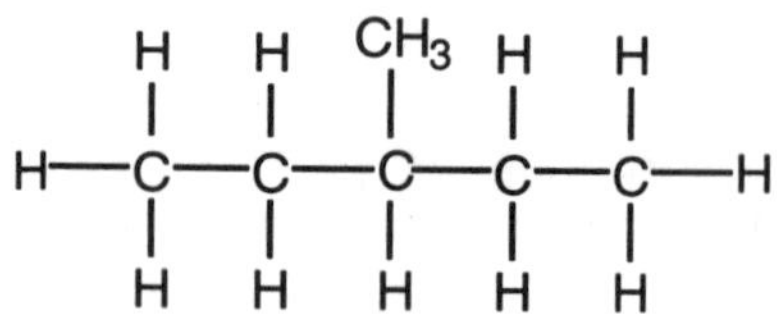

3-methylpentane

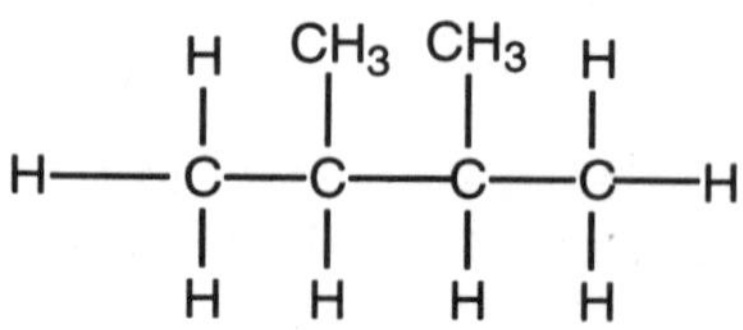

2,3-dimethylbutane

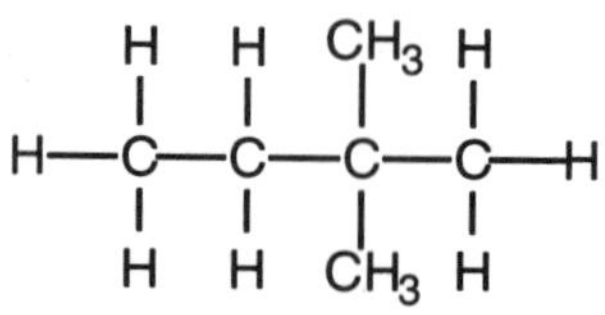

2,2-dimethylbutane

5. a. No answer required.
   b. 5 isomers of hexane

## YOUR TURN *Alkane Boiling Points*: Isomers *(page 184)*

The purpose of this activity is to apply and test the idea that increased branching causes hydrocarbon boiling points to decrease. The complications introduced by isomers of hydrocarbons are minimized here, as no nomenclature is required of the students. Instead, students see how isomer structure influences boiling points. Intermolecular attractions decrease as the molecular structure approaches a spherical shape (increased branching), minimizing surface area contact between molecules. Knowledge of this is an important introduction that pertains to altering fuels and octane numbers, covered in Part C.

## Answers to Questions

1. The boiling points decrease with increasing number of carbon sidechains within an isomer series.
2. a. 3-Methylhexane, boiling point (bp) = 92.0 °C
   b. 2,2-Dimethylpentane, bp = 79.2 °C
   c. Heptane, bp = 98.4 °C
3. a. One would expect its boiling point to be between 99.2 °C and 117.7 °C because it is intermediate in the number of sidechains (or branching). Its actual boiling point is 112 °C, lower than octane and 2-methylheptane, and higher than 2,2,4-trimethylpentane.
   b. One would expect it to have a higher boiling point than all the $C_5H_{12}$ isomers because of its larger size. (Some students may think that because it is branched, it might have a lower boiling point than unbranched pentane. Carefully compare the two models and point out that the $C_8H_{18}$ branched-chain isomer does have the greater surface. In fact, its boiling point is 112 °C, compared with 36.1 °C for pentane.)

## PART B: SUMMARY QUESTIONS *(page 185)*

1. a. Petroleum is similar to metallic resources in that it is nonrenewable and world demand is increasing.
   b. Petroleum is different in that it cannot generally be recycled (although used motor oil can be). Petroleum is a liquid and provides a much greater variety of substances than does any particular metallic resource because it is a complex mixture of many compounds that can easily be chemically altered.
2. a. 0.702 g/mL
   b. 17.6 g
3. a. Building and burning
   b. Building—petrochemical products (ethene or ethylene for polyethylene, alcohol solvents, drugs, etc.); Burning—fuels (methane/ethane in natural gas used to heat homes; octane and other liquid hydrocarbons in gasoline).
4. $C_{25}H_{52}$
5. a. The number of carbons (which is related to total molar mass) and the degree of branching within a hydrocarbon molecule help determine the relative boiling points.
   b. Boiling points increase as the number of carbons per molecule increase; there is more surface area for intermolecular contact. Boiling points decrease with increasing branching in the molecules because of less surface area for intermolecular contact.
6. a. Isomers are molecules having the same molecular formula but different structural arrangements (and thus different properties).
   b. Any structures of the nine possible $C_7H_{16}$ isomers will do. (Some students may mistakenly draw two versions of the same isomer and count it twice. To convince them that the two represent the same structure, have them build both chains, without hydrogen, with the model kit and compare them.)
7. Ask students to draw the isomers and count the bonds and hydrogens. Probably, in one case the student was not counting the C–C bonds.

In response to a common student question, the number of possible isomers increases rapidly as one moves to higher alkanes. There are 18 isomers for $C_8H_{18}$; the number rises to 1,858 for $C_{14}H_{30}$. How about $C_{20}H_{42}$? The claim is 366,319 isomer possibilities!

## EXTENDING YOUR KNOWLEDGE *(page 186)*

- Students investigating this problem will learn that the actual hydrocarbon blend used in gasoline varies according to seasonal temperatures and climate. In cold conditions, more high-volatility hydrocarbon molecules are used in gasoline to promote proper fuel vaporization in the engine. Under warmer conditions, the gasoline blend contains more low-volatility hydrocarbon fractions to avoid premature or too rapid vaporization.

A related effect is experienced with diesel fuels; summer-grade diesel fuel actually thickens to a gel if the temperature drops below about –15 °C (not unusually cold for northern climates). "Winterized" diesel fuel is required to prevent automobiles from stalling under these conditions. This fuel contains a higher fraction of low-viscosity, more volatile, low molar mass hydrocarbons.

- Butane, normally a gas at room temperature, is a liquid when held under pressure in a closed volume. This increased pressure raises butane's boiling point. When the valve is opened to the lower pressure of the atmosphere, the boiling point of butane becomes lower than room temperature, and the liquid vaporizes. When the pressure is increased, the boiling point is elevated.
- Salt water promotes corrosion. Thus, although salt will lower the freezing point of radiator fluid, it will contribute to the cooling system's early demise. Ethylene glycol, an organic molecular substance, is much less reactive with the radiator components.

Knowing the differences among properties of petroleum hydrocarbons enables chemists to isolate useful fractions. With a better understanding of petroleum chemistry, students are ready to tackle a burning issue—petroleum as a source of energy (Part C).

# C. PETROLEUM AS AN ENERGY SOURCE

The focus in this part of the unit is on the burning of petroleum products. Students learn some history of how petroleum has affected our lives. Historical perspective is provided when students interview senior citizens about energy and resource use. A *Your Turn* shifts the focus to how U.S. energy sources have changed over the years.

Energy released from burning petroleum is described in terms of bond breaking and bond making. The energy released in hydrocarbon burning is measured by students in a simple calorimetry laboratory activity. The unit concludes by examining how chemists and chemical engineers crack and isomerize molecules to convert petroleum fractions in lesser demand into fractions more in demand.

## C.1 YOU DECIDE: THE GOOD OLD DAYS? *(pages 188–189)*

Interviewing senior citizens about energy and use of resources provides historical insights that are fascinating and revealing to young people. The groundwork for involving families in *ChemCom* was already established in the water use diary in the Water unit. The simplest proposal is for students to contact older family members. Many families, however, do not have grandparents nearby, so encourage students to interview an older person in the neighborhood. If your community has a retirement home, it is often possible to bring senior citizens to the school for the class to interview. You may wish to enlist the interest/help of a social studies teacher or run the study as a joint effort. Be sure that students record the specific time period(s) being discussed by the person being interviewed. Some students may wish to videotape an interview and present it to the class.

The class should decide as a group on 10–12 questions that will help clarify the connection between lifestyles and energy use. The class should then be divided into several teams. Students first practice their interview plan by questioning other team members. Their answers are recorded to form a basis for comparing lifestyles today with those of earlier times.

Ask the students to devise some convenient way for the total class interview data to be summarized and displayed. A bulletin board, poster, or even a length of brown wrapping paper taped to the wall can serve this purpose.

Conduct a class discussion focusing on the three questions found on page 189. Be prepared to address other concerns or findings students have identified in this information-gathering activity.

Point out that to make the best use of petroleum today, it is prudent to examine the energy sources that we have used over the past century. This leads to a closer look at past, present, and future energy sources.

### Answers to Questions

1.

| Energy Uses | Early 1900s | Today |
|---|---|---|
| a. Home heating | Wood, coal | Gas, electricity |
| b. Lighting | Oil lamps, gas | Electricity |
| c. Public transportation | Trains, street cars | Buses, subways, planes |
| d. Private transportation | Horse, buggy | Cars |
| e. Cooking | Wood, coal stoves | Gas, electric, and microwave ovens |
| f. Food packaging | Brown paper | Plastic |
| g. Laundry | Washboards, clotheslines | Electric washers and dryers |
| h. Clothing | Cotton, wool | Variety of synthetic and natural fabrics |

2. Positive features: slower-paced lifestyle, possibly less pollution, etc. Negative features: higher death rate, inferior medical treatment, smaller variety of consumer items and foods, etc.
3. If alternative energy sources (including conservation) are not adequately developed in time, we could potentially arrive at a situation far worse than the "good old days." This topic will be discussed further in section F.

## C.2 ENERGY: PAST, PRESENT, AND FUTURE *(pages 189–192)*

This section identifies the sun as the primary source of energy and speculates how the discovery of fire probably influenced the development of civilization. This narrative provides an introduction to the *Your Turn* activity, which draws attention to a graph representing types of energy use for the past 140 years.

### YOUR TURN *Fuel Sources over the Years* *(page 190)*

In this activity, based on the graph in Figure 13 (page 190), students trace the growth of energy use and associate it with economic factors and the availability of particular types of energy sources. Note that basic graph-reading skills are practiced and enhanced in this activity.

#### Answers to Questions

1. a. Up to about 1880, wood supplied more than 50% of our energy needs.
   b. Trains were the chief form of long-distance transportation during this period. Horses were used for short distances.
2. a. The decline in local supplies of wood and development of the internal combustion engine (a coal-burning device) contributed to the demise of wood as a primary energy source.
   b. Coal was the next important energy source.
3. Oil burned in lamps was the most important use of petroleum prior to 1910.
4. a. Oil became an important energy source in the 1920s.
   b. Petroleum use increased as the number of automobiles increased, spurred by the Ford assembly line with its mass production approach.
5. a. The most recent energy source to enter the picture is nuclear energy.
   b. To generate electrical power

### YOUR TURN *Energy Supplied by Various Fuels* *(pages 191–192)*

This graph illustrates how energy used in the United States has increased and how the sources have changed since 1850.

#### Answers to Questions

1. a. Our energy use overall has accelerated since 1850.
   b. The increase was due to rising population, increased mobility, and greater consumption of processed or manufactured products per person.
2. a. 22%
   b. 14%
3. a. Using the graph, find the consumption of energy from natural gas in 1970 ($67 \times 10^{18}$ J $- 45 \times 10^{18}$ J) and the energy consumption in 1990 ($74 \times 10^{18} - 57 \times 10^{18}$ J). The difference is $5 \times 10^{18}$ J.
   Divide the answer by $22 \times 10^{18}$ J, the energy consumption in 1970.
   b. Decreased
   c. The natural gas consumption is less. The oil is used for transportation, which has increased in the 20 years from 1970 to 1990.

## C.3 ENERGY AND FOSSIL FUELS *(pages 192–194)*

This section introduces the chemistry of energy storage and release in petroleum molecules. Chemical reactions are discussed in terms of bond breaking and bond making. Emphasize that the information about reactions occurring in stages is speculative but helps us in thinking about the energy of reactions. Both reactions illustrate that in an exothermic reaction the energy necessary to break the bonds is greater than that used to make bonds. Both the combustion of methane and the formation of water are exothermic overall. If a reaction is exothermic, then the reverse reaction is endothermic. Thus the decomposition of water is endothermic. The two *Your Turn* activities focus on conversions of energy types, and savings that result from more fuel-efficient cars.

## Demonstration Idea

The traditional electrolysis of water, producing hydrogen and oxygen, demonstrates that the decomposition of water is endothermic. Collecting a small quantity of hydrogen and oxygen in a test tube and igniting with a splint demonstrates that the combination of hydrogen and oxygen is an exothermic reaction. The combustion of natural gas (mainly methane) in a Bunsen burner also demonstrates an exothermic process.

## YOUR TURN *Energy Conversion* *(page 194)*

The excess thermal energy of exothermic reactions can be harnessed in various ways to do useful work. (Students are invited to examine the efficiency of such energy conversions in the upcoming *Your Turn*.)

Emphasize that the automobile is a system consisting of a series of interconnected and coordinated energy conversion devices, typically, the following:

- The defroster converts electrical energy into thermal energy.
- The generator converts mechanical energy into electrical energy.
- The engine converts chemical (potential) energy into thermal energy into mechanical energy.
- The gasoline in the tank represents chemical (or potential) energy.

## YOUR TURN *Energy Conversion Efficiency* *(pages 196–197)*

In Figure 18, students consider energy losses associated with various energy conversions in the automobile. Representative calculations illustrate the monetary and resource losses resulting from inefficient energy conversion.

## Anwers to Questions

1. a. $200\,\frac{\text{mi}}{\text{wk}} \times 52\,\frac{\text{wk}}{\text{yr}} = 10{,}400\,\frac{\text{mi}}{\text{yr}}$

   b. $10{,}400\,\frac{\text{mi}}{\text{yr}} \times \frac{1\text{ gal}}{27.0\text{ mi}} = 385\,\frac{\text{gal}}{\text{yr}}$

2. $385\,\frac{\text{gal}}{\text{yr}} \times \frac{\$1.35}{\text{gal}} = \frac{\$520}{\text{yr}}$

3. a. $385\,\frac{\text{gal}}{\text{yr}} \times 0.75 = 289\,\frac{\text{gal}}{\text{yr}}$

   b. $\frac{\$520}{\text{yr}} \times 0.75 = \frac{\$390}{\text{yr}}$

4. a. $200\,\frac{\text{mi}}{\text{wk}} \times \frac{1\text{ gal}}{40.0\text{ mi}} \times \frac{52\text{ wk}}{\text{yr}} = 260\,\frac{\text{gal}}{\text{yr}}$

   b. 385 gal – 260 gal = 125 gal

   c. $\text{gal} \times \frac{\$1.35}{\text{gal}} = \$169$

5. a. $\frac{10{,}400\text{ mi}}{\text{yr}} \times \frac{1\text{ gal}}{50\text{ mi}} = \frac{208\text{ gal}}{\text{yr}}$;

$$\frac{385\text{ gal}}{\text{yr}} - \frac{208\text{ gal}}{\text{yr}} = \frac{177\text{ gal}}{\text{yr}}\text{ saved}$$

b. $$\frac{177\text{ gal}}{\text{yr}} \times \frac{\$1.35}{\text{gal}} = \frac{\$239}{\text{yr}}$$

## C.4 THE CHEMISTRY OF BURNING *(page 197)*

The chemical reaction with the addition of the energy term is introduced to describe the process of combustion. To make decisions about the wise use of fuels, one needs to know how to measure the amount of energy released by different fuels. Laboratory Activity C.5 illustrates a method of measuring and calculating the heat of a combustion reaction.

## C.5 LABORATORY ACTIVITY: COMBUSTION *(pages 198–201)*

Students will use a simple calorimeter to trap the energy from a burning candle and calculate the heat of combustion in kJ/g and kJ/mol.

### Time

50 minutes

### Materials (for 24 students working in pairs)

12 12-oz, empty soft drink cans (with pull tabs)
12 ring stands
12 ring clamps
12 thermometers
12 stirring rods
12 index cards
balances
12 paraffin ($C_{25}H_{52}$) candles
5 lb ice
matches

### Safety

Students should wear safety goggles and aprons throughout the laboratory activity. Remind students that they should stir the water with a stirring rod, not a thermometer, to avoid breakage.

Caution students not to inhale carbon soot. If carbon soot gets on skin (it may be an irritant), wash off immediately.

### Advance Preparation

Ask students to bring empty aluminum soft drink cans from home a day or two before the activity is scheduled. Flip up tabs on cans so that glass stirring rods can pass through (see Figure 19, page 199.) This minimizes heat loss during heating, but makes stirring and adding the water a little more difficult.

Each class will need chilled water. If the tap water is not cool enough, add ice until the temperature is approximately 10° C cooler than room temperature. Remove any unmelted ice from the water.

Provide the students with a place on the chalkboard, poster board, or overhead transparency to record their value for the heat of combustion of candle wax in kJ/g and kJ/mol.

### Pre-Lab Discussion

Introduce students to the energy unit, the joule (J). To raise the temperature of 1 g of water 1 °C, 4.2 J are required. To provide additional perspective, a 100-watt light bulb requires 6,000 J (6 kJ) of energy to operate for one minute.

A properly assembled apparatus (similar to Figure 19) on the front bench makes a good Pre-Lab discussion prop. It will also be useful in the Post-Lab discussion.

## Lab Tips

Most student-quality thermometers can be read only to 0.5 °C.

## Answers to Questions

1. For perspective, the actual heat of combustion for butane (about 49.3 kJ/g) is higher than that for paraffin (approximately 42.0 kJ/g). This use of a relatively crude calorimeter introduces many errors into the determination. Do not be surprised if the results vary greatly from team to team. The values determined in the laboratory activity will most likely be lower than the actual values.
2. The actual molar heat of combustion is greater for paraffin (approximately 14,800 kJ/mol) than for butane (2,859 kJ/mol). Student results should reflect a similar ordering of these two values, although their results will be lower than these accepted values because of heat losses.
3. Because each paraffin "molecule" on average contains 24 carbon–carbon bonds, while a butane molecule contains three carbon–carbon bonds, it might be expected that more energy would result from burning one mole of paraffin than from burning one mole of butane. Each additional -$CH_2$ unit produces one molecule of $CO_2$ and one of $H_2O$ and additional heat.
4. On a gram-for-gram basis, butane is a slightly better fuel. However, mole-for-mole (or molecule-for-molecule), paraffin is a much better fuel.
5. a. This is not a good assumption. Some thermal energy released during hydrocarbon burning will be lost to the surroundings directly from the flame, through the walls of the aluminum can, and from the water surface. This energy will not be accounted for in the calculations—thus the experimental heat of combustion will be lower than the true value.
   b. Other experimental errors include inaccuracy in weighing, the burning wick's small contribution of energy to the total heat of combustion (its heat of combustion is probably not equal to that of paraffin), and incomplete combustion of paraffin. These factors will tend to lower the total quantity of heat measured.

## Post-Lab Discussion

The class discussion can focus on the posted student data. Some of the student results may differ substantially from those of classmates. This provides an excellent opportunity to suggest checking calculations carefully and looking for sources of experimental error. Observation of the setup (Figure 19) is helpful. It is particularly useful to look for heat loss that is not measured, such as that absorbed by the can and that given off to the surrounding air.

This activity can and should be related to the more theoretical discussion of energy storage in chemical compounds and bond breaking and bond making in section C.3, Energy and Fossil Fuels.

If alcohol burners are available, this activity might be extended to determine the heat of combustion of a liquid fuel such as hexane, methanol, or ethanol. Straight-chain hydrocarbons produce a lot of soot when an alcohol burner is used. This would reinforce the comparison between liquid and solid fuels explored in the questions.

## C.6 USING HEATS OF COMBUSTION *(pages 201–202)*

Heats of combustion are further described, and students are given an opportunity to review balancing equations.

## YOUR TURN *Heats of Combustion* *(pages 202–203)*

Students establish quantitative and qualitative relationships between the heat of combustion and the first eight members of the alkane series. Point out that because the major use of petroleum is for burning as a fuel, the heat produced per gram is a primary factor in choosing appropriate types of hydrocarbons for burning.

## Answers to Questions

1. Answer is worked in student text as an example.
2. a. $CH_4(g) + 2\ O_2(g) \rightarrow CO_2(g) + 2\ H_2O(g) + 890\ kJ$

   b. $2\ C_4H_{10}(g) + 13\ O_2(g) \rightarrow 8\ CO_2(g) + 10\ H_2O(g) + 5{,}718\ kJ$
3. a. The molar heats of combustion increase as the number of carbons increases (this is a linear relationship), while the heat of combustion per gram of hydrocarbon decreases slightly.

   b. The molar heat of combustion increases about 660 kJ per carbon atom. Decane's molar heat of combustion should be 5,450 kJ/mol + 2(660 kJ/mol) = 6,770 kJ/mol. The heat of combustion per gram should be slightly less than that of octane. The exact number cannot be determined owing to the nonlinear relationship.

   c. The heat of combustion in kJ/mol was easier to predict owing to the linear relationship.
4. a. $2\ \text{mol}\ C_8H_{18} \times \frac{5{,}450}{\text{mol}} = 10{,}900\ \text{kJ}$

   b. $\frac{2{,}660\ \text{g}\ C_8H_{18}}{\text{gal}} \times \frac{47.8\ \text{kJ}}{\text{g}\ C_8H_{18}} = \frac{127{,}000\ \text{kJ}}{\text{gal}}$

   c. $20\ \text{gal} \times \frac{2{,}660\ \text{g}}{\text{gal}} \times \frac{47.8\ \text{kJ}}{\text{g}} \times 0.16 = 410{,}000\ \text{kJ}$
5. a. $C(s) + O_2(g) \rightarrow CO_2(g) + 394\ kJ$

   b. Octane (47.8 kJ/g) releases more energy per gram than carbon:

   $$\frac{394\ \text{kJ}}{\text{mol}} \times \frac{1\ \text{mol C}}{12.01\ \text{g C}} = \frac{32.8\ \text{kJ}}{\text{g C}}$$

   c. Carbon could be used in combustion reactions that do not require the vapor phase, such as removing oxygen from metals in ore refining.

   d. Coal would not work in the automobile engine, which requires the fuel to be in the vapor phase for rapid combustion.

## C.7 ALTERING FUELS *(pages 204–206)*

The process of cracking or breaking large hydrocarbons into smaller molecules is presented. Fuel fractions can also be isomerized to produce branched-chained hydrocarbons that burn more satisfactorily in car engines. This section documents ways in which chemists have been able to engineer hydrocarbon molecules to fulfill particular needs. The older method of adding tetraethyl lead to boost octane ratings is contrasted with adding oxygen-containing fuels such as methanol and methyl-*t*-butyl ether (MTBE).

On gasoline pumps there is sometimes a statement that the octane ratings were obtained by the (R+M)/2 method. This means that the octane rating is an average of two different test values: (1) the research octane number (RON) for a test engine run under conditions simulating city driving; and (2) the motor octane number (MON) for the engine run under highway conditions. If the RON is 93 and the MON is 91, then the octane number is 92.

In addition to being an octane enhancer, MTBE is also involved in attempts to reduce carbon monoxide in urban atmospheres. The Clean Air Act of 1992 requires that all oxygenated gasolines sold in at least four winter months in the 41 most populated metropolitan areas (New York, Los Angeles, Washington, D.C., etc.) must contain 2.7% oxygen by weight. Putting MTBE into gasoline brings the oxygen percentage of the mixture up to the required weight percent. Because the MTBE-enhanced fuel is already partially oxygenated, it burns cleaner, producing carbon dioxide directly rather than carbon monoxide.

Methyl-*t*-butyl ether is one of a class of compounds represented by the general formula, $R\text{-}O\text{-}R^1$, where R and $R^1$ are hydrocarbon chains.

## YOUR TURN A *Burning* Issue *(page 206)*

### Answers to Questions

1. Methanol and ethanol can be considered partially combusted molecules because oxygen is already included in their structures. Therefore, less energy should be produced when they are burned.
2. The gasohol mixture should decrease miles per gallon because there is less energy per gallon of gasoline to propel the car.

## PART C: SUMMARY QUESTIONS *(pages 207–208)*

1. a. Development of the internal combustion engine (automobiles and airplanes)
   b. Increasing industrial technology (refining, processing of crude oil, and manufacturing)
   c. Increasing population and demand for electricity (heating and cooling)
2. Wood is a solid renewable resource, not easily transported (a solid). Petroleum is a nonrenewable resource and easily transported (a liquid or gas can be piped). It burns more cleanly and efficiently than wood.
3. The energy stored in petroleum hydrocarbons was originally captured via photosynthesis in plants. Animals consume the plants. Both the plants and the animals die, eventually forming layers of the underground petroleum deposits we tap today.
4. chemical (gasoline) → thermal (engine) → mechanical (piston) → electrical (generator) → chemical (battery) → electrical (sunroof motor) → mechanical (sunroof retracting)
5. Energy efficiency represents the quantity of available energy that is converted into useful work by a machine. Two ways to increase energy efficiency are to decrease the number of conversions between the original fuel source and the end use, and to increase the efficiency of each step in the conversion scheme chosen.
6. a. $132{,}000 \text{ kJ} \times 0.75 = 99{,}000 \text{ kJ}$ wasted
   b. This energy is lost
      (1) as heat through the exhaust pipe and radiator,
      (2) as friction in moving parts,
      (3) in compressing air in the cylinders, and
      (4) in running accessories (radio, wipers, heater, lights, etc.).
7. $C_3H_8(g) + 5\ O_2(g) \rightarrow 3\ CO_2(g) + 4\ H_2O(g) + 2{,}200 \text{ kJ}$
8. a. The molar heat of combustion of methane (890 kJ/mol) is higher than that of water gas (525 kJ/mol).
   b. $10 \text{ mol water gas} \times \frac{525 \text{ kJ}}{\text{mol}} = 5{,}250 \text{ kJ}$
9. a. $\frac{2{,}512 \text{ kJ}}{2 \text{ mol } C_2H_2} = \frac{1{,}256}{\text{mol}}$
   b. $\frac{1{,}256 \text{ kJ}}{\text{mol}} \times 12 \text{ mol} = 15{,}000 \text{ kJ}$
10. Heat of combustion (kJ/g), molar heat of combustion (kJ/mol), boiling point, density, as well as others
11. a. Octane rating is a scale for measuring the antiknock characteristics of gasoline. Heptane is rated as 0 (poor antiknock behavior), and isooctane (2,2,4-trimethylpentane) is rated as 100 (good antiknock behavior).
    b. The gasoline has the same antiknocking behavior as a mixture of 93% isooctane and 7% heptane.
    c. The octane rating of gasoline can be improved by adding tetraethyl lead or by increasing the percent of branched-chain alkanes (which burn more smoothly) or aromatic hydrocarbons via isomerization processes.

## EXTENDING YOUR KNOWLEDGE *(page 208)*

Aluminum and magnesium are poor substitutes for petroleum. They are both more expensive per gram than petroleum, requiring large quantities of energy in their reduction from ores. Automobile engines are simply not designed for the use of such solid fuels.

# D. USEFUL MATERIALS FROM PETROLEUM

Part D returns to the "building" theme for petroleum. It describes alkenes, including *cis-trans* isomerism, and introduces the concept of functional groups and aromatic compounds. This leads to considering how polymers such as plastics are produced. Students are also given an opportunity to be synthetic chemists, observing the change in properties when an alcohol and an acid produce an ester.

## D.1 BEYOND ALKANES *(pages 209–210)*

This section introduces alkenes, hydrocarbons that contain at least one double bond. It is important for students to see the changes in nomenclature and structure that go with unsaturation. The model-building activity introduces *cis-trans* isomerism. Students are also introduced to the possibility of including other elements in carbon compounds, which is demonstrated in the next model-building activity.

## D.2 LABORATORY ACTIVITY: THE BUILDERS *(pages 210–212)*

Students work with models to explore the bonding and characteristics of alkenes. The purpose of the lab is to help students recognize structural details of common builder molecules. Model systems must be able to represent double bonds—if they cannot the old gumdrop/toothpick standby is preferred. As the students construct alkene molecules, call attention to the planar relationship of the groups on the double bond. Students are led to construct cyclic compounds and compounds containing elements in addition to carbon and hydrogen. Do not dwell excessively on nomenclature during this activity.

### Time

50 minutes

### Materials (for 24 students working in pairs)

12 sets of space-filling or ball-and-stick models

### Part 1. Alkenes

Electron-dot structures, structural formulas, condensed formulas, and systematic nomenclature are used to describe molecules with multiple bonds.

### Answers to Questions

1. See molecular structures in text.
2. $C_nH_{2n}$
3. Students should note that alkene molecules resist rotation about the double bond.
4. a. Four arrangements of $C_4H_8$ are possible, as shown by the structural formulas on page 59, as well as cyclobutane.

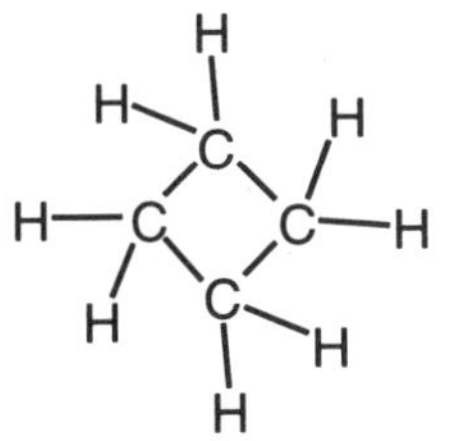

cyclobutane

Point out that cyclobutane is an isomer, but it does not have a double bond in the molecule.

b. Yes. Students should recognize the butene isomer; ethylpropene is less easily recognized.

5. Each pair alone represents the same substance (not isomers). Some students may correctly point out, however, that the substance represented by the first pair (1-butene) is an isomer of the substance represented by the second pair (2-methylpropene).
6. Only one isomer is possible for propene.
7. Isomers; same formula ($C_5H_{10}$) but different structures.
8. a. H–C≡C–H (ethyne or acetylene)
   b. $CH_3$–C≡C–$CH_3$ (2-butyne)

## Part 2. Another Kind of Isomerism: Which Side Are You On?

By this time, students are familiar with structural isomers and can apply this idea to *cis-trans* isomerism. They will recognize that the rigidity of the double bond prevents rotation and the two isomers are not superimposable. This concept is important in the discussion of fats in the Food unit.

### Answers to Questions

1. a. Yes
   b.

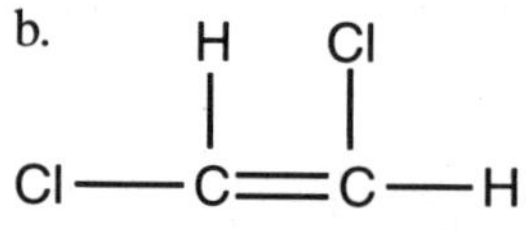


*trans*-1,2-dichloroethene

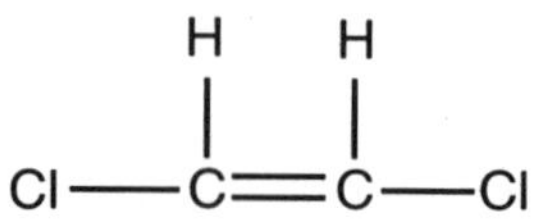


*cis*-1,2-dichloroethene

2. No, all possible ways of drawing the formulas are superimposable.

## Part 3. Compounds of Carbon, Hydrogen, and Singly Bonded Oxygen

Students should find that the two isomers of $C_2H_6O$ are easy to make and identify. They will recognize that the molecule with the -OH group will be more soluble in $H_2O$. If necessary, remind them that they have already tested this compound, ethanol, in the Solvents Laboratory Activity in the Water unit. The higher boiling point (80 °C) of ethanol compared to that of dimethyl ether (-24 °C) is explained by hydrogen bonding.

### Answers to Questions

1. a. Two distinct structural isomers of $C_2H_6O$, are possible:
   b. $CH_3$–$CH_2$–O–H and $CH_3$–O–$CH_3$
   Ethanol — Dimethyl ether
   c. Yes, they have the same formula, but the atoms are arranged in a different order.
2. a. Ethanol is more soluble in water than is dimethyl ether.
   b. Ethanol should have the higher boiling point.

## D.3 MORE BUILDER MOLECULES *(pages 212–213)*

Cycloalkanes, which students may have asked about when making isomers in Part B, and aromatic compounds are presented. The special nature and reactivity of the benzene ring structure are described. Point out that even though cycloalkanes are in a ring, like aromatics, cycloalkanes have more in common (relative unreactivity) with the linear alkanes. You may wish to allow some time for students to build models.

## D.4 BUILDER MOLECULES CONTAINING OXYGEN *(page 214)*

The functional groups responsible for alcohols, acids, and esters are diagrammed. It might be helpful for students to see that both alcohols and acids have the –OH functional group, but in esters the hydrogen is replaced by another carbon group. Point out that the specific functional group(s) found in a given organic compound determine many of its physical (solubility, boiling point, etc.) and chemical properties. (Remind students that with the exception of combustion, unsubstituted alkanes are fairly unreactive.) Molecules that contain the same functional group(s) share similar family properties.

If you have samples of some of these compounds, a safe way to test odors is to add a few drops of the liquid to a piece of cotton and then place the cotton in a stoppered test tube. The tubes can be circulated around the room with little risk of spilling or creating odors.

## D.5 CREATING NEW OPTIONS: PETROCHEMICALS *(pages 214–220)*

The extensive variety of petrochemicals presented should give students some sense of the importance of the chemical industry. Alkanes are relatively unreactive, but the unsaturated bonds in alkenes are quite reactive, and therefore alkenes are used as starting materials for making new molecules. Two reactions of ethene provide examples. Both alkenes and aromatics (another important class of builder molecules) occur naturally in petroleum but can also be obtained via cracking and reforming reactions. Once obtained, builder molecules can often be polymerized into macromolecules via addition or condensation reactions. Some of the products and their corresponding properties are discussed. Students need not memorize names, formulas, and reactions, but they do need to understand that chemists can create many diverse substances from petrochemicals.

## CHEMISTRY AT WORK/Manufacturing Safely and Efficiently: It's No Accident *(page 219)*

A regulatory affairs chemist makes sure that her company is in compliance with all laws and regulations. This segment again emphasizes the relationship between properties of materials and their uses.

## D.6 LABORATORY ACTIVITY: PETROCHEMICALS *(pages 220–221)*

The synthesis of methyl salicylate illustrates the observable changes produced when chemical reactions occur. The familiar odor of the product of this reaction makes this activity a pleasing experience.

### Time

30 minutes

### Materials (for 24 students working in pairs)

12 test tubes (15 × 125 mm)
12 10-mL graduated cylinders
12 150-mL beakers
12 test tube holders
12 test tube brushes
12 Bunsen burners, rings, ring stands, and wire screens
12 dropping bottles for dispensing of organic reactants
12 mL concentrated sulfuric acid (18 M $H_2SO_4$)
1 reagent bottle for dispensing 18 M $H_2SO_4$
24 mL methanol
6 g salicylic acid
boiling chips

### Safety

Students should wear safety goggles and aprons throughout the laboratory activity. Demonstrate the appropriate way to smell chemicals. **Remind students to exercise extreme caution when using 18M sulfuric acid. It is caustic and corrosive.**

### Pre-Lab Discussion

Discuss the general condensation reaction for forming esters, and once again point out the role of chemists as molecular architects in designing molecules. The flavor and perfume industries are heavily dependent on petrochemically synthesized esters, which are either unavailable in nature or available only by expensive extraction from dilute sources. Many low molecular weight esters have pleasant fruity odors/tastes and are often responsible for the natural odors/flavors of ripe fruit, flowers, and other plant substances. The continued search for new odors/flavors and for a theory capable of correlating structure with odor/flavor is a potential area for future chemical careers. See Post-Lab Discussion as well.

## Lab Tips

- The students will probably recognize the odor from its use in flavoring candy, perfumes, etc., but caution them not to ingest any of the product.
- Remind students to avoid the direct inhalation of any fumes—they should be gently fanned toward the nose.
- The boiling chip in the test tube minimizes erratic boiling, which can cause liquid to erupt out of the test tube. The hot water bath needs only to be above the boiling point of methanol (56 °C) and does not need to boil vigorously.
- Many other combinations can be used if you have the available chemicals. Smaller quantities may also be used. Sample procedure: combine the specified drops of acid and alcohol in a test tube. Add two drops of concentrated sulfuric acid. Then heat 5–10 minutes in the hot water bath. Cool and note the odor. ( See *Source Book*, Volume 3, *Organic Chemistry*, pp. 5–10, ChemSource, Inc., 1994.)

| Acid | Alcohol | Odor |
|---|---|---|
| 20 drops acetic acid | 28 drops *n*-butyl alcohol | pear |
| 10 drops acetic acid | 20 drops octyl alcohol | orange |
| 20 drops acetic acid | 30 drops pentanol | banana |
| 20 drops propanoic acid | 15 drops ethanol | rum |
| 0.4 g benzoic acid | 20 drops ethanol | "fruity" |

## Answers to Questions

1. a. Methanol: $CH_3OH$. Salicylic acid (2-hydroxybenzoic acid): $C_7H_6O_3$
   b. $CH_3OH + C_7H_6O_3 \rightarrow C_8H_8O_3 + H_2O$
2. Amyl acetate: $CH_3COOC_5H_{11}$ (pentyl ethanoate)
3. a. Alcohol (1-butanol)
   b. Ester (methylpropanoate)
   c. Acid (3-methylpentanoic acid)
   d. Acid (ethanoic acid)

## Post-Lab Discussion

The Post-Lab Discussion can continue the discussion of the reaction of the functional groups in the organic acids and alcohols that produced esters. Emphasize that a seemingly small difference in molecular structure can make a very noticeable difference in properties. Have students note the contrasting odors of reactants and products. To reinforce this point, you may want to synthesize several other esters (see suggestion in Lab Tips). Be sure to pass samples around the classroom and write out the specific reaction(s):

alcohol + organic acid + acid catalyst → ester + water

Be sure to distinguish organic acids (RCOOH) from common inorganic (or mineral) acids such as HCl, $H_2SO_4$, and $HNO_3$. Organic acids are typically weak acids and were originally obtained by isolation from plant and animal ("organic") sources. For example, formic acid (HCOOH/methanoic acid) was first obtained from the destructive distillation of ants, and acetic acid ($CH_3COOH$/ethanoic acid) from the aerobic fermentation of a mixture of cider and honey.

Students have probably smelled butyric acid ($CH_3CH_2CH_2COOH$/butanoic acid) in rancid butter or body odor. Today commercially used organic acids are typically synthesized from petroleum or coal tar, as are the alcohols with which they are reacted.

## PART D: SUMMARY QUESTIONS *(page 222)*

1. a. Ethene, $C_2H_4$. Common name ethylene. This colorless, odorless gas is used as an agent to speed ripening of fruits and in the synthesis of polyethylene, polystyrene, and other plastics. (Other answers are possible.)
   b. Benzene, $C_6H_6$, a clear, colorless, highly flammable liquid used as a solvent and in the synthesis of a wide variety of products including medicines, dyes, and plastics (Other answers are possible.)
2. $C_6H_{14}$, $C_3H_8 + C_3H_6$ or $C_4H_8$ and $C_2H_6$ (several possibilities)
3. a., b., c.
   (1) Ethanol, $CH_3CH_2OH$
   (2) Acetic acid,

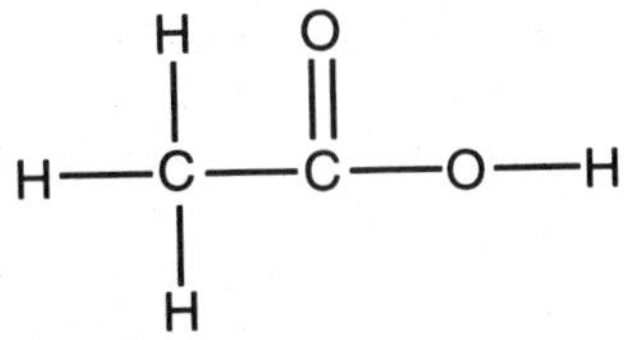

   (3) Methyl formate,

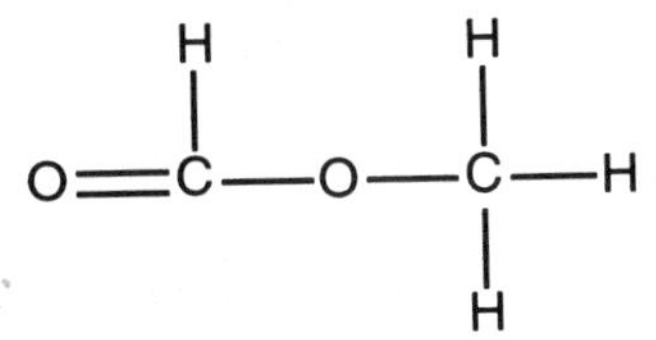

4. Benzene is more stable and less reactive than cyclic hydrocarbons.
5. (1) Carbon can bond with itself or other atoms to form single, double, or triple bonds; and (2) can form straight-chain or cyclic compounds and a variety of isomers.
6. Polymerization is the chemical process of joining together many small molecules to make a very large molecule. Examples include ethene (or ethylene), which joins to make polyethylene.
7. a. $Cl_2$
   b. $H_2O$ (in presence of an acid catalyst)
   c. NaOH

# E. ALTERNATIVES TO PETROLEUM

Part E briefly surveys various petroleum substitutes presently under consideration for use in place of the dwindling supply of nonrenewable fossil fuels. The promise of chemical technology is that economically and environmentally suitable alternatives to petroleum for both building and burning will be developed. Conservation not only provides scientists and engineers with time to develop such alternatives, but also is a wise course of action environmentally. Insights gained here lead to the culminating activity in Part F—the examination of what life would be like in a world without petroleum.

## E.1 ALTERNATIVE ENERGY SOURCES *(pages 223–227)*

After calculations concerning the present usage and future availability of petroleum and other fossil fuels, the text presents several suggestions to improve our energy future: offshore oil drilling, the extraction of petroleum from oil shale, coal liquefaction, and petroleum plantations. Biomass conversion, various forms of solar energy, and nuclear fission and fusion are options for replacing petroleum's role as an energy source at fixed site locations. Several topics raised here, such as nuclear energy and the greenhouse effect, will be covered in subsequent units.

### YOUR TURN *Energy Dependency* *(pages 223–225)*

This activity reestablishes the need to search for petroleum substitutes as fuel for our transportation needs. Students are asked to interpret the pie chart in Figure 23 (page 224), as well as data contained in Tables 6 (page 224) and 7 (page 225). It is not important for students to remember the actual numbers from the problems, but they should understand the logic of such calculations (the idea of using data on known reserves and extrapolating to predict the present use). Remind students that these are estimates and do not include unexplored resources such as the Arctic Natural Wildlife Reserve.

#### Answers to Questions

1. Petroleum
2. 88.3% (petroleum, natural gas, and coal)
3. 4.0% (hydro/solar/wind/geothermal, and wood)
4. a. 100 quads/year × 4 years × 0.402 = 161 quads used
   b. 1,000 quads −161 quads = 839 quads remaining
5. Assuming that the percent use by fuel type in Figure 23 remains constant:
   a. 400 quads × 0.248 = 99.2 quads used
      1,700 quads − 99 quads = 1,601 quads remaining
   b. 400 quads × 0.238 = 95 quads used
      40,300 quads − 95 quads = 40,205 quads remaining
6. 100% − 63% = 37%
7. Coal will more readily substitute for petroleum at stationary sites. Transportation vehicles are engineered to use liquid fuels.

### YOUR TURN *Prized Petroleum* *(page 227)*

Students are asked to critique a comment made by Daniel Yergin about petroleum.

#### Answers to Questions

1. It probably means "driving force" or "vital."
2. Fueled machinery to produce goods, provided starting materials for a huge variety of products, permitted people to be mobile. (A variety of answers are possible.)
3. The Gulf War
4. The United States is the largest consumer of petroleum. To keep its economy rolling, it needs to import petroleum.

5. a. If it were not for petroleum, transportation would be very different, and civilization as we know it would not exist.
   b. Most of the trappings of civilization—plastics, fabrics, dyes—have their beginnings in petroleum.
6. The role of petroleum might be less important if we learn to substitute other fuels or building block materials.

## E.2 BUILDER MOLECULE SOURCES *(pages 227–229)*

Even though some petroleum currently used as an energy source can be diverted to producing petrochemicals, the world's supply of petroleum will eventually be consumed. Industrialized nations are turning to coal first as a source for needed raw materials. Biomass conversion is second. Finally, because all of the mentioned alternatives for building and burning uses of petroleum have disadvantages, it is wise to consider conservation goals for extending present supplies through efficiency measures, recycling/reuse, and gradual changes in consumer behavior. This factual information could be enhanced by asking students to make a chart listing each alternative source, its advantages, drawbacks, and renewability.

## PART E: SUMMARY QUESTIONS *(page 229)*

1. a. Petroleum, a nonrenewable resource, supplies nearly half our energy needs. Our high rate of consumption, compared with estimated future supplies, poses a serious threat to our way of life. Transportation accounts for more than half of our use of petroleum for burning; industry uses 10% of the total oil per year for building products that support our lifestyles.
   b. As petroleum supplies dwindle, we are faced with the choice of to build or to burn—or with making decisions regarding the relative amounts of building and burning to be supported.
2. a. Research and development regarding renewable sources of energy give some hope that new technologies will be in place to serve as substitutes for dwindling supplies of petroleum and other nonrenewable resources. This policy will help to avert drastic and sudden changes in lifestyles.
   b. It is least likely that hydroelectric, solar, or wind power will be used for future transportation purposes, if by "transportation" we mean what the term signifies today (emphasis on privately owned motor vehicles).
   c. The community may have large towers for wind turbines, large fields of solar cell arrays, and/or individual residences with solar collectors and wind-powered generators.
3. Students may choose any one of the answers below. It is necessary to support the choice with reasons.
   a. Coal is in plentiful supply, but environmental problems are associated with its widespread use. The cost of converting coal to liquid fuel is currently greater than the cost of producing the same quantity of fuel from petroleum.
   b. Petroleum can be extracted from oil shale and tar sand, the crude oil condensed from kerogen vapor. These shales contain a large supply of pre-oil. However, the technology is not adequately developed, access to federally owned land containing the shale has not yet been secured, and conversion processes require large amounts of water—already scarce in the western states, where the oil shale and tar sands are found.
   c. Petroleum can be extracted from some plants, implying the development of "petroleum plantations," yet larger-scale technology supporting this idea has not yet been demonstrated. Also, there would be the ethical question of whether it is justifiable to devote large tracts of land to "growing petroleum" when much of the world's population is starving.

4. a. The burning of petroleum is most likely to be curtailed before building. The petrochemicals needed to produce detergents, drugs, plastics, and fibers will likely be given higher priority than certain uses of petroleum as fuel.
   b. (1) Burning uses much more petroleum. It would be less difficult to reduce the large number.
   (2) At least some of the building petroleum can be recycled, where the burning of petroleum results in a loss that cannot be recycled.
5. Desirable compounds would be obtained quickly, with low expense in terms of energy to process the compounds, and with low environmental impact—no air or water pollution. The compounds would also not require the use of resources used to meet other societal needs, such as food.
6. Support: Oil has made modern civilization possible, both in terms of transportation and consumer products.
   Refute: Many other terms could (and have been) used, such as information age, computer age, etc., each with good supporting arguments.

7. a. $150 \times 10^9 \text{ barrels} \times \dfrac{1 \text{ day}}{8 \times 10^6 \text{ barrels}} \times \dfrac{1 \text{ year}}{365 \text{ days}} = 51.4 \text{ years}$

   b. $249 \times 10^6 \text{ people} \times \dfrac{24 \text{ barrels}}{\text{person/year}} = 6.0 \times 10^9 \text{ barrels/year}$

   $\text{oil shale} = \dfrac{150 \times 10^9}{6.0 \times 10^9} = 25 \text{ years}$

   c. The numbers used in the calculations are estimates.

# F. PUTTING IT ALL TOGETHER: LIFE WITHOUT GASOLINE

## F.1 CONFRONTING PROSPECTS *(pages 230–231)*

A 1977 article by Isaac Asimov gets students to think about a life without gasoline as well as the difficulty in making predictions and extrapolating from data. Before this exercise, students probably knew Asimov primarily as a writer of science fiction.

### Answers to Questions

1. a. Gasoline prices have climbed. He might have based this on rising prices at the time and extrapolated—much as we have extrapolated from graphs.
   b. Cars have not disappeared and we still have fuel to operate furnaces in houses. Perhaps Asimov intended to exaggerate to shock readers into action. Perhaps some conservation measures have decreased the use of oil (not so likely, though). Maybe these scenarios will still occur, but the timetable is stretched out.
2. Student stories. Provide some mechanism for students to share their stories if they wish. You might borrow a technique from English teachers and help the students publish their stories in a magazine. In evaluating the stories, look not only at the literary merit, but also at whether the science is plausible and reflects what the students now should know about fuels.

## F.2 LOOKING BACK AND LOOKING AHEAD *(page 231)*

This brief transitional section to the next unit suggests an unexpected connection between petroleum use by society and food use by individuals—in both cases we confront building/burning issues. Using a resource for energy and for building is an elegant lead-in to the Food unit, which will develop this insight in detail.

# ENRICHMENT MATERIALS

***Teacher Background***

Some Important Commercial Polymers and Their Uses

| Polymer | Repeating unit(s) in molecule | Monomer(s) | Uses |
|---|---|---|---|
| Polyethylene | $-CH_2-CH_2-$ | $CH_2=CH_2$ | Films for toys and housewares, plastic bottles, pipe fittings |
| Polyvinyl chloride | $-CH_2-CH-Cl$ | $CH_2=CH-Cl$ | Pipe fittings, film and sheeting, wire insulation, auto parts, flooring, adhesives, coatings |
| Polystyrene | $-CH_2-CH-C_6H_6$ | $CH_2=CH-C_6H_6$ | Packaging and containers, toys, disposable food containers and utensils, brush handles |
| Polyester | $-C-O_2-CH_2-$ | $C-CO_2H$ and $CH_2OH$ | Textile fibers for clothing, sheets, curtains, towels, carpets |
| Polypropylene | $-CH_2-CH-CH_3$ | $CH_2=CH-CH_3$ | Fibers for carpeting, auto parts, rope nets |
| Polyacrylonitrile | $-CH_2-CH-CN$ | $CH_2=CH-CN$ | Acrylic fiber, carpets, sweaters, single- and double-knit clothing |
| Polyamide | $-C(CH_2)4-C-$ $ONH(CH_2)6-NH-$ | $H_2N(CH_2)_6NH_2$ and $HO_2C(CH_2)_4CO_2H$ | Textile fibers, bristles, carpets, upholstery fabrics, yarn, tire cord |

Source: "Plastics: Utilizing the Properties of String-like Molecules," by JCE staff, *Journal of Chemical Education*, (56), 42, 1979.

## PART E

**Synthetic fuels:** To obtain any fuel in a useful form requires a certain amount of energy for building the necessary equipment and for mining, pumping, and processing the fuel. Given present technology and oil prices, all the synthetic fuels have lower net energy values than the 50:1 ratio for domestic oil.

**Coal-derived fuels:** Ordinary coal is composed primarily of carbon, with only about one hydrogen atom for every 16 carbons. Because liquid and gaseous hydrocarbon fuels contain a much higher percent hydrogen (methane, 4H/1C; octane, 2.25H/1C; heavy fuel oil, 1H/6C), temperature and pressure are used to break down the chemical structure of coal in order to increase the hydrogen/carbon ratio.

In World War II, when Germany was unable to import oil, coal-derived gasoline was used to fuel planes and tanks. More recently, South Africa's Sasol I plant has been producing oil from coal since 1960. The net energy value for coal-derived fuels is about 17:1.

**Shale oil:** Of the 1.8 trillion barrels of shale oil in Colorado, Wyoming, and Utah, about 600 billion are considered recoverable. Unfortunately, there are a number of problems with its extraction: (1) Much energy is needed to extract kerogen, the oil-like component. Temperatures of approximately 500 °C result in a net energy value of only about 6.5:1. (2) One to four barrels of water is needed to produce one barrel of oil. (3)

Much waste is generated (1.5 tons of shale generated per barrel of oil produced). Thus, at the present time, shale oil is expensive from both economic and environmental perspectives.

**Alcohols:** As partially oxidized hydrocarbons, alcohols have a lower energy content than an equal volume of gasoline. The chief advantages of using alcohols are that they are obtained from renewable resources and can be mixed with regular gasoline up to concentrations of 10–15% without the need for engine redesign. Methanol is made from wood and, given methanol's high affinity for water, it is commonly used as a fuel-drying additive. Ethanol is made from grain (wheat or corn) and is less toxic and corrosive than methanol.

## CHEMQUANDARIES

### Section B.3

Ask students under what conditions the water-displacement method will not work without modification. Consider solids such as sugar and wood. (If the solid dissolves in or reacts with water, it is necessary to use another liquid, such as oil or mercury. If the solid is less dense than water, it is necessary to use another liquid of lower density than water, or sink it by force with a solid of known volume. If an object is too big to fit into a graduated cylinder, an overflow can may be used. (You may wish to demonstrate use of the overflow can.)

### Section B.5

1. Why would butane be a poor substitute for propane in the tanks of portable stoves used by campers?
   **Answer:** Butane has a higher boiling point ( 0.5 °C vs. –42.1 °C) than propane. At winter temperatures (and increased pressure), butane might liquefy. Since the stove works on the basis of pressurized gas, this would be a poor fuel choice. Better substitutes might include either ethane or methane because their boiling points are lower than propane.

2. If one looks at a see-through butane lighter, one sees that at least some of the butane is in the liquid state. Explain how this is possible if the lighter is at room temperature. (See also the second Extending Your Knowledge item, page 186.)
   **Answer:** The boiling point of a liquid depends on pressure. Apparently, the butane is under pressure higher than 1 atm (where it boils at -0.5 °C), and its boiling point at this pressure is greater than room temperature. Therefore, some of the butane is a liquid.

### Section B.7

You are given three bottles. One contains a viscous liquid, the other a waxlike solid, and the third a liquid so volatile that it begins to evaporate rapidly when the stopper is removed from the bottle. You are told that each bottle contains one of the following compounds: $CH_4$, $C_2H_6$, $C_4H_{10}$, $C_7H_{16}$, $C_{14}H_{30}$, or $C_{22}H_{46}$. Identify the three substances.

**Answer:** Viscous liquid, $C_{14}H_{30}$; waxlike, $C_{22}H_{46}$; gas under pressure, $C_4H_{10}$.

### Section C.9

Imagine that you are planning a backpacking trip and wish to take along one of the first eight alkanes as a hydrocarbon fuel.

a. Which form of the data (kJ/mol or kJ/g) would be more useful in helping you pick the best fuel? Explain.
b. Ignoring the mass of the fuel container, which fuel would be most appropriate for your trip? Why?
c. In reality, what other factors would you need to consider if you selected one of the gaseous fuels?

**Answers:**

a. Energy per mass (kJ/g) is the most important consideration because you have to carry it.

b. Methane is the best choice because it releases the most energy per gram of substance.

c. Gaseous fuels such as methane must be kept in pressurized containers to be carried in practical quantities. Safety is an important factor to consider.

## Part E

Grain alcohol (ethanol) has been proposed as a substitute for imported oil. A typical American car traveling 10,000 mi/yr at 15 mi/gal would require an amount of grain-produced alcohol equal to 14,700 lb of grain. For comparison, a subsistence diet for a typical adult male is the equivalent of about 400 lb of grain per year. Assuming that it takes approximately 1 acre of land to produce 1,890 lb of grain:

a. How many acres of land must be set aside to grow the grain needed for one car?

b. How many people could be fed on the yearly diet of an alcohol-driven car?

**Answers:**

a. $14{,}700 \text{ lb} \times \frac{1 \text{ acre}}{1{,}890 \text{ lb}} = 7.78 \text{ acres}$

b. $14{,}700 \text{ lb} \times \frac{1 \text{ person}}{400 \text{ lb}} = 37 \text{ persons}$

# DEMONSTRATION IDEAS

## Section B.5

Dry Ice (bp –78 °C) can be used to liquefy propane. A convenient setup is to attach a piece of copper tubing, with one end coiled or looped, to a piece of rubber tubing attached to a propane tank. If the coil is placed in a container of Dry Ice/acetone, with the other end outside, liquid hydrocarbon can be collected in a beaker and observed to boil as it warms to room temperature. If liquid nitrogen (bp -196 °C) is used, methane or ethane from the laboratory gas jet can be liquefied.

## Section C.7

1. Consider having students demonstrate the meaning of the term *cracking* with a large molecule created with molecular model kits.
2. A catalyst is a substance that can be used to alter (typically accelerate) the rate of a chemical reaction (by lowering the activation energy) without becoming consumed in the process. The following demonstration illustrates using a catalyst in the cracking process. This demonstration should be conducted beneath a fume hood.

Melt about 1 $cm^3$ of petroleum jelly in a test tube and add a small amount of vermiculite for support. When the mixture is semisolid, clamp the test tube in a horizontal position and place several pieces of broken porcelain, which act as a catalyst, in the test tube and fit it with a delivery tube. Place the end of the delivery tube in another test tube immersed in a beaker of cold water. Heat the porcelain chips strongly until red hot. Then gently heat the petroleum jelly. As the gas passes over the hot catalyst, it is cracked. The smaller fragments are collected as a liquid in the vertical test tube. Test the liquid for flammability, and point out that the liquid phase indicates that the molecules are smaller. The attractive forces are weaker between the smaller molecules. Reference: "Burning and Bonding," *Prime Science*, Kendall/Hunt, Dubuque, Iowa, 1997.

## Section D.5

1. Students are always fascinated by the breakdown of polystyrene in acetone. Fill a transparent container, such as a large beaker or graduated cylinder, one-third full of acetone and begin adding Styrofoam® pellets. You will certainly have no trouble

getting volunteers to help. The vast quantity that can be broken down with bubbling occurs because the pellets are air-filled foam.

2. Synthesis of nylon: *Nylon* is a generic term used to represent synthetic polyamides. Nylons made from diamines and dibasic acids or diacid chlorides are designated by two numbers, the first representing the number of carbon atoms in the diamine, and the second the number of carbon atoms in the dibasic acid. The history of the development of nylon and its use in World War II make it a particularly interesting library research topic.

• ***Caution:*** The diamine, dichloride, and sodium hydroxide are all irritating to the skin, eyes, and respiratory system; using a hood is advised. It is also advisable to wear plastic or rubber gloves. The demonstration is quite safe if properly performed, but do not directly handle the product until it has been washed with ethanol or water.

• The reaction does not depend on the exact stoichiometry of the reactants, but the concentrations should be close to those indicated for optimal results.

• The aqueous diamine solution may be stored, but the acid chloride solution is somewhat unstable because of hydrolysis (only make up as much as you need on a given day). The pure acid chloride should be stored in a brown bottle and out of strong light under a hood.

• **Solution 1:** Gradually heat a reagent bottle of 1,6-diaminohexane (also called hexamethylenediamine, melting point (mp) = 42 °C) in a warm water bath until the solid begins to melt. Decant 2.9 g directly into a 100-mL beaker and dilute to 50 mL with 0.5 M NaOH solution (20 g NaOH per 1 L).

• **Solution 2:** Prepare a 0.25 M diacid chloride solution by diluting 2.3 g of adipoyl dichloride to 50 mL with hexane (density of hexane is 0.66 g/cm$^3$).

• Using equal volumes of the two solutions (as little as 5 mL of each can be effective), slowly pour Solution 2 down the side of the beaker, on top of Solution 1 (Solution 2 is less dense than Solution 1), so as to minimize agitation between the immiscible solutions.

• Use forceps or a bent paperclip to hook the polymer film that forms at the interface of the two solutions, and slowly pull it from the center of the beaker. Continue pulling out the nylon "thread" until the solutions are exhausted. A thread 10–15 feet long is possible. If it breaks, you may be able to hook a new thread. Longer threads are possible if you avoid contact with the sides of the beaker.

• Any excess solution remaining should be placed in a solvent waste container or stirred, and the resulting solid washed and disposed of in a solid waste container.

## Reaction

diamine + diacid chloride → polyamide + 2 HCl

$$H_2N(CH_2)NH_2 + Cl{-}\overset{\overset{O}{\|}}{C}{-}(CH_2)_8{-}\overset{\overset{O}{\|}}{C}{-}Cl \xrightarrow{NaOH} {-}[\overset{\overset{H}{|}}{N}{-}(CH_2)_6{-}\overset{\overset{H}{|}}{N}{-}\overset{\overset{O}{\|}}{C}{-}(CH_2)_8{-}\overset{\overset{O}{\|}}{C}]{-} + 2\,HCl$$

# SUPPLEMENTAL READINGS


"Special Report on Energy." *National Geographic,* February 1981. [Whole issue.]

"Methane." *CHEM MATTERS,* February 1983.

"Hot & Cold Packs." *CHEM MATTERS,* February 1987.

"Instant Hot Pack." *CHEM MATTERS,* February 1987.

"Liquid Crystals." *CHEM MATTERS,* December 1983.

"Across Australia by Sun Power." *National Geographic,* November 1983, 600-607. [Solar powered car.]

"Race Car Rally." *Science and Children,* February 1994, 26-29.

"Polymers." *CHEM MATTERS,* April 1986. [Several articles: "Polymers" (4-6), "Tyvek" (8-10), "Polysaccharides" (12-14), and "Silly Putty" (15-17).]

*Polymer Chemistry,* National Science Teachers Association, 1986. [Book available from the American Chemical Society, Education Division. Developed by teachers in cooperation with Arco, DuPont, and Shell, the first half of the book provides background information on natural polymers; chapters on polymer structure, synthesis, economic value, and a minitext for students. The second half of the book is demonstrations, lab exercises, games, and other activities. The accompanying disk is a student tutorial for use on the Apple II computer.]

"The Interrupted Party." *CHEM MATTERS,* October 1984.

"Oil: Production, Consumption, and Reserves: The Role of Natural Resources in the Gulf War." *The Science Teacher,* April 1991, 12-18.

"The Blackened Bucket." *CHEM MATTERS,* December 1984.

"Polymer Samples for College Classrooms." *Journal of Chemical Education,* February 1984, 161-163. [Presents a lists of companies, contact persons, and the type and form of polymer that can be provided gratis.]

"The Missing Warning." *CHEM MATTERS,* October 1985.

"Flash Point!" *CHEM MATTERS,* December 1986.

"Natural Dyes." *CHEM MATTERS,* December 1986.

*Playing with Energy.* National Science Teachers Association, 1981, 106 pp. [Games and simulations for grades 9-12 selected from the Project for Energy-Enriched Curriculum learning packets.]

"Dissolving Plastic." *CHEM MATTERS,* October 1987.

"Exploring the Solubility of Polyvinyl Alcohol (PVA) Film." *CHEM MATTERS,* October 1987, 15.

"Non-Safety Glass." *CHEM MATTERS,* October 1987.

"Dissolving Plastic Contest." *CHEM MATTERS,* April 1988.

"The Exploding Tire." *CHEM MATTERS,* April 1988.

"What's Best for Your Car." *Consumer Report,* February 1987, 88-94. [Discusses and compares the viscosity, additives, etc., found in major brands of motor oil.]

"The Smell of Danger." *CHEM MATTERS,* October 1988.

"Burning: Is Torching the Most Benign Way to Clear Oil Spilled at Sea?" *Science News,* October 2, 1993, 220-223.

"Killing for Oil." *CHEM MATTERS,* October 1988.

"Fireside Dreams." *CHEM MATTERS,* December 1988.

"Geochemistry in the Search for Oil." *Chemical & Engineering News,* February 10, 1986, 28-34. [Special report reprint: fairly technical, more for teachers than students.]

"PET Recycling." *CHEM MATTERS,* October 1994.

"Bubblegum." *CHEM MATTERS,* October 1994.

"Alkenes and Their Derivatives: The Alchemists' Dream Come True." *Journal of Chemical Education,* August 1989, 670-672.

"The Fruits of Ethylene." *CHEM MATTERS,* April 1989.

"Fifty Years of Nylon Stockings." *CHEM MATTERS,* February 1990.

"Nylon." *CHEM MATTERS,* December 1990.

"Stick to It." *CHEM MATTERS,* December 1993.

"Diesel Under Pressure." *CHEM MATTERS,* February 1991.

"Rudolph Diesel's Engine." *CHEM MATTERS,* December 1990.

"Firing Up Fuel Cells." *Science News,* November 13, 1993, 314-316.

*Energy Facts,* 1984 (First Edition). U.S. Government Printing Office, 1985, 56 pages. (S/N 061-003-00453-9). [Published by the Energy Information Administration, provides a quick reference to a broad range of domestic and international energy data over the time period 1973-1984 through the use of graphs and charts. Also lists other sources of information.]
"Nonleaded Gasoline: Its Impact on the Chemical Industry." *Journal of Chemical Education,* September 1987, 773-776.
"Alcohol in Your Tank." *CHEM MATTERS,* December 1988.
"Brazil's Program To Use Ethanol as Transportation Fuel Loses Steam." *Chemical & Engineering News,* October 18, 1993, 13-15.
"Ice That Burns." *CHEM MATTERS,* October 1995.
"Fuel for Thought: Examining Alternatives." *ChemEcology,* July/August 1993, 2-5.
"Health Studies Indicate MTBE Is Safe Gasoline Additive." *Chemical & Engineering News,* September 20, 1993, 9-12.
"Electric Cars." *CHEM MATTERS,* April 1993.
"Solar Chemistry." *CHEM MATTERS,* February 1991.
"Salt-Gradient Solar Ponds." *CHEM MATTERS,* December 1989.

# TEACHING SCHEDULE

| | TIME BLOCK 1 | TIME BLOCK 2 | TIME BLOCK 3 | TIME BLOCK 4 | TIME BLOCK 5 |
|---|---|---|---|---|---|
| Class Work | Read Introduction<br><br>Discuss Intro. to Foods | Discuss A.1-A.3<br>YD A.1<br>Prelab B.1<br><br>Getting Ready B.1 | LA B.1 | YT B.2<br>YD B.3 | YT B.4 (2) |
| Homework | Read A.1-A.3 | SQ Part A<br>YD A.3 | Read B.2 | Read B.4 | Read B.5 |

| | TIME BLOCK 6 | TIME BLOCK 7 | TIME BLOCK 8 | TIME BLOCK 9 | TIME BLOCK 10 |
|---|---|---|---|---|---|
| Class Work | Review Parts A&B<br>SQ Part B | Quiz Parts A&B<br>Discuss C.1-C.2<br><br>YT C.2 (1st) | Discuss C.3<br>YT C.3 (1st) | Review Part C | Quiz Part C<br>Discuss D.1<br><br>Prelab LA D.2 |
| Homework | Read C.1-C.2 | YTs C.2 (2)<br>Read C.3 | YT C.3 (2nd) | SQ Part C<br>Read D.1 | YT D.1<br>Getting Ready D.2 |

| | TIME BLOCK 11 | TIME BLOCK 12 | TIME BLOCK 13 | TIME BLOCK 14 | TIME BLOCK 15 |
|---|---|---|---|---|---|
| Class Work | LA D.2 | Discuss C.3<br>YT D.3<br><br>Prelab D.4 | LA D.4 | Discuss D.5<br>Prelab D.6<br>LA D.6 | CQ D.6<br>YD D.7 finish |
| Homework | Read D.3 | | Read D.5 | YD D.7 collect labels | |

| | TIME BLOCK 16 | TIME BLOCK 17 | TIME BLOCK 18 | TIME BLOCK 19 | TIME BLOCK 20 |
|---|---|---|---|---|---|
| Class Work | Review Part D | Quiz Part D<br>YT E.1 | YD E.2 | Review Food unit | Exam Food unit |
| Homework | SQ Part D<br>Read E.1-E.3 | | | | |

**LA**=Laboratory Activity; **CQ**=ChemQuandary; **YT**= Your Turn; **YD**=You Decide; **PIAT**=Putting It All Together; **SQ**=Summary Questions; One **Time Block**=Approximately 50 minutes

# SUPPLY LIST

## Expendable Items and Unusual Materials

| Quantity | Unit | Item | Sections Used |
|---|---|---|---|
| 120 | mL | ammonia, household or 3M | D.6 |
| | | assorted beverages | D.2 |
| 1 | bottle | blue food coloring | D.6 |
| 12 | 6-inch-long | chromatography paper, 1 inch wide | D.6 |
| 12 | size 20 | corks | B.1 |
| 30 | pieces | filter paper, Whatman No. 1 | D.4 |
| 1 | bottle | green food coloring | D.6 |
| 33 | mL | hydrochloric acid, conc | D.4 |
| 5 | lbs | ice | B.1 |
| 12 | | index cards | B.1 |
| 1 | g | iron(III) chloride hexahydrate | D.4 |
| 1 | tbs | laundry starch, liquid | D.2 |
| 1 | pkg | M & M candies | D.6 |
| | | matches | B.1 |
| 12 | | paper clips | B.1 |
| 50 | | peanuts | B.1 |
| 0.053 | g | potassium iodate | D.2 |
| 1 | g | potassium iodide | D.2 |
| 1.9 | g | potassium thiocyanate | D.4 |
| 1 | bottle | red food coloring | D.6 |
| several | 25-g | samples of food | D.4 |
| 1 | pkg | Skittles candies | D.6 |
| 12 | oz | Soft drink cans, empty | B.1 |
| 50 | mL | sulfuric acid, 0.50 M | D.6 |
| 6 | mL | sulfuric acid, 3 M | D.2 |
| 36 | | toothpicks | D.6 |
| 1 | 100 mg | vitamin C tablet | D.2 |
| 1 | bottle | yellow food coloring | D.6 |

In approaching the study of food, our most important renewable resource, we examine food for energy and food for structure—an organization that parallels the Petroleum unit's theme of burning and building. In using this unit, students recognize that all food is made of chemicals. They examine chemical structures and behavior of carbohydrates and fats (as energy sources) and proteins (as builder molecules). Vitamins and minerals are included to round out the nutrition story, and students are introduced to the pervasive role of additives.

Underlying the information concerning food chemistry is the need to become informed decision makers about food choices and diet. In the culminating activity, students choose between two frozen dinners, based on their knowledge of Calories and the roles of the nutrients described on the labels.

# LEARNING GOALS

Upon completion of this unit, the student will be able to:

1. Compare the uses of food in terms of "building" and "burning." [Introduction, B.1]
2. Distinguish malnutrition from undernutrition, and identify parts of the world where these problems are most acute. [Introduction, A.3]
3. Define calorie and joule, and calculate energy changes from calorimetry data. [B.1]
4. Correlate weight gain or loss with caloric intake and human activity. [B.2]
5. Compare and contrast mono-, di-, and polysaccharides in terms of structural formulas and properties. [B.3]
6. Identify key functional groups in carbohydrates and fats, and write an equation for the formation of a typical fat. [B.3, B.4]
7. Distinguish between saturated and unsaturated fats, relating the consumption of each to health. [B.4]
8. Define and illustrate the concept of a limiting reactant, using biochemical examples and calculations. [C.2]
9. Describe how functional groups in amino acids interact in protein formation. [C.3]
10. Describe five functions of proteins in the body. [C.3]
11. Discuss the concepts of essential amino acids, complete proteins, and complementary proteins with respect to a balanced diet. [C.3]
12. Separate and measure proteins and carbohydrates in nonfat milk, and calculate a sample's caloric value. [C.4]
13. Distinguish water-soluble from fat-soluble vitamins (with specific examples of each), and discuss the implications of these differences in terms of dietary needs. [D.1]
14. Analyze the vitamin C content of foods by performing titrations. [D.2]
15. Identify minerals used in the body, and distinguish between macrominerals and trace minerals. [D.3]
16. Determine the iron content of foods by colorimetry. [D.4]
17. Discuss the relative risks and benefits of various food additives in terms of their purposes, and provide specific examples. [D.5]
18. Discuss the roles of the Food and Drug Administration and federal regulations in ensuring food safety. [D.5]
19. Compare and contrast menus from several cultures in terms of calories and nutritional balance, and analyze the nutritional quality of food recorded in a personal food diary. [E.2]

# A. GETTING STARTED

The necessity of food for life is discussed, along with its dual roles of burning and building. These topics lead to a comparison of our diets with those of other cultures. Different cultures have different ways, developed mostly by trial and error, of meeting the body's needs for "burning" and "building."

Most areas of the United States provide opportunities to sample a wide range of ethnic foods. You may wish to have students contribute information about foods with which they are familiar and compare the different ways of meeting the same nutritional needs.

In the United States, food labels help us to make choices. This unit begins by having students look at the information about nutrition that is readily available to everyone who takes the time to read food labels. After the food pyramid is introduced, students are asked to keep a food diary and analyze it by referring to the food pyramid. The difference between being malnourished and undernourished is explained.

Food labels are used as a focus for discussion about food and nutrition. Table 1 provides nutritional information about sources of calories from nutrients found in one food.

## A.1 YOU DECIDE: LABELS AND DAILY VALUE *(page 236)*

This section asks students to examine the label reproduced in Table 1.

### Answers to Questions

1. a. Sugar and starch
   b. Calcium and iron
2. No, the percent protein is not given.
3. (Answers will vary.)
4. An unsaturated fat must have at least one double bond in the molecule.
5. Something with a lot of carbohydrates, sugar, and fat; it might be snacks such as cookies or rolls.

## A.2 THE FOOD PYRAMID *(pages 236–237)*

The food pyramid is introduced as a guide for planning food choices that will ensure a healthful diet. The current recommendations for a healthful diet, approved in 1991, are based on a pyramid relationship among the various foods we should eat. The food pyramid incorporates the traditional basic four food groups of the 1958 pie chart, but with much different emphases. In particular, the pyramid calls for eating habits that increase the proportions of bread, cereal, grains, and pasta (at the base of the pyramid) and decrease fats, oils, and sweets (near or at the top).

## A.3 YOU DECIDE: KEEPING A FOOD DIARY *(pages 237–238)*

Once again, *ChemCom* makes chemistry relevant to students' lives—this time, asking them to look at their eating habits. Young people spend a lot of time thinking about what they eat and how this affects how they look. Eating disorders related to concern about appearance are common among both boys and girls.

This activity highlights two dimensions of nutrition: too little energy in the diet (undernutrition) and insufficient quantities of certain molecular building blocks (malnutrition). Many teenagers consume large quantities of food but insufficient quantities of nutrients. Others, especially those concerned about weight, may actually consume too few calories for good health.

This activity reveals information about personal nutrition that will probably surprise some students. Assign different days to different students. Be sure to include weekend days. To encourage precise reporting about quantities, you may want to bring a typical meal (or typical food samples) into the classroom to help students practice the estimates they will need to make in collecting their data. A chart is a useful way of recording data. The chart can be filled out as homework, but be sure to allow class time for comparison and group analysis of the charts. Collect the food diaries to use in B.3 *You Decide*.

### Answers to Questions

1. (No answer)
2. (Individual answers)
3. (Individual answers)
4. (Individual answers)
5. The food pyramid is intended to be followed over a longer period, not just for a single day.

## PART A: SUMMARY QUESTIONS *(page 238)*

1. Carbohydrates, fats, proteins, vitamins, minerals.
   a. Carbohydrates
   b. Proteins
2. Fats, oils, sweets
3. a. Undernourishment is associated with an insufficient number of calories for burning, and implies malnourishment as well. Malnourishment means inadequate supplies of needed nutrients.
   b. Yes, one can be malnourished even if sufficient numbers of calories are ingested.
   c. No, being undernourished means insufficient calories, which means that the body will not have sufficient materials for building either.
4. a. Eating a variety of foods increases the likelihood of an adequate balance of all needed nutrients.
   b. Calorie-counting alone is inadequate, since the body needs food for building as well as burning.

# B. FOOD AS ENERGY

Part B highlights the energy aspect of foods by introducing students to food calories through a laboratory activity similar to the one used in the Petroleum unit. The energy units of the dietitian (Calorie) and the chemist (calorie) are defined and related quantitatively. Students get considerable practice in doing Calorie calculations and finding the percent of calories from fat in a food.

Students examine the basic structures and properties of fats and carbohydrates, including the difference between saturated and unsaturated fats and the concept of *cis-trans* isomers. A typical student fast-food meal is analyzed according to the guidelines in this section.

## B.1 LABORATORY ACTIVITY: FOOD ENERGY IN A PEANUT

*(pages 239–242)*

Students will use the same calorimeter that was used in the Petroleum unit, C.5 Laboratory Activity: Combustion.

### Time

50 minutes

### Materials (for a class of 24 working in pairs)

12 12-oz empty soft drink cans (with pull tabs)
12 ring stands
24 ring clamps
12 thermometers
12 stirring rods
12 index cards
balances
peanuts
5 lb ice
matches
12 paper clips
12 corks, size 20

### Safety

Students should wear safety goggles and aprons throughout the laboratory activity. Students should use non-mercury thermometers. To avoid breakage, remind students to handle the thermometer carefully, especially when stirring the water. Caution students to handle the can carefully to avoid cuts from the metal.

### Advance Preparation

If you have saved the calorimeters from the Combustion activity in the Petroleum unit, the preparation time for this activity is minimal. You will need to provide the class with chilled water. If the tap water is too warm, use ice to cool a large quantity about 10 °C below room temperature. Be sure that no ice is present in the can.

### Pre-Lab Discussion

Students should review the Combustion activity in the Petroleum unit. It is particularly helpful to review the physical setup and the sources of experimental error.

### Lab Tips

"Dry" peanuts do not work well. Sometimes it is difficult to light the peanut. Longer fireplace matches or even wooden splints work better than book matches. If a certain peanut does not light after several tries, it will save time to discard that peanut and try a different one. Remember that students will have to find the mass of the new one before beginning.

## Sample Data and Calculations

| | |
|---|---|
| Mass of peanut burned | 0.52 g |
| Mass of water in can | 97.90 g |
| Initial temp. of water | 2.0 °C |
| Final temp. of water | 25.0 °C |

$$\Delta H = 97.90 \text{ g} \times (25.0 - 2.0)\ °\text{C} \times \frac{1 \text{ cal}}{\text{g} \times °\text{C}} = 2{,}300 \text{ cal}$$

$$\frac{2{,}300 \text{ cal}}{0.52 \text{ g}} \times \frac{1 \text{ Cal}}{1{,}000 \text{ cal}} = \frac{4.4 \text{ Cal}}{\text{g}}$$

The label on the peanut jar is used for calculating the Cal/serving and the percent difference.

## Calculations

1. $250 \text{ mL} \times \frac{1.0 \text{ g}}{\text{mL}} = 250 \text{ g}$

   $250 \text{ g} \times \frac{1 \text{ kg}}{1{,}000 \text{ g}} = 0.250 \text{ kg}$

2. $78.4\ °\text{C} - 22.2\ °\text{C} = 56.2\ °\text{C}$

3. $250 \text{ g} \times 56.2\ °\text{C} \times \frac{1 \text{ cal}}{\text{g}\ °\text{C}} = 14{,}050 \text{ cal}$

4. a. $14{,}050 \text{ cal} \times \frac{1 \text{ cal}}{1{,}000 \text{ cal}} = 14 \text{ Cal}$

   b. 14 Cal

5. (Answers will vary.)

6. $\frac{230 \text{ Cal}}{38 \text{ g}} = \frac{6.0 \text{ Cal}}{\text{g}}$

7. (Answers will vary.)

## Post-Lab Discussion

The calculations may give students difficulty. You may wish to walk them through the calculations described in the student text. If necessary, repeat with another set of data before asking students to use their own numbers. It is best to approach the calculations stepwise as described in the student text. Highlight the fact that one Calorie (Cal), the food unit, is different from the chemist's calorie (with a lower-case "c"). The food Calorie is the same as a kilocalorie (kcal), 1,000 calories. After all calculations are complete, have the students put the results of the calculation of cal/g of peanut on the board for comparison. Additional practice problems are found in the Enrichment Materials section at the end of this unit.

No more than two significant figures should be reported in the results. Student results from combustion activities carried out like this one often lack precision.

After comparison of the results, ask students how this activity is like the one in the Petroleum unit and how it is different.

## B.2 FOOD FOR THOUGHT—AND FOR ENERGY *(pages 242–243)*

The source of the energy stored in food by photosynthesis is sunlight. In the *Your Turn*, students have an opportunity to practice calculations using calories. A marginal note reminds them that, while BTUs are used in the United States and Great Britain, most of the world uses joules as a unit of energy.

## YOUR TURN *Calorimetry* *(page 243)*

This activity reinforces how the energy content of foods can be determined from calorimetry data. The calculations continue to move students away from formula usage and toward quantitative reasoning. Beginning with the definition of a calorie in terms of mass of water and temperature change, help students grasp the algorithm of multiplication by asking several questions, such as: How many calories to raise 1 g of water 5 °C? 1 g of water 10 °C? 2 g of water 5 °C? 5 g of water 10 °C? (Twice as much water or change in temperature, twice as much energy.) An additional ChemQuandary is found in the Enrichment Materials section at the end of this unit.

### Answers to Questions

1. $\frac{1 \text{ cal}}{\text{g } °\text{C}} \times 160 \text{ g} \times (60 - 22)\ °\text{C} = 6{,}080 \text{ cal}$

   $6{,}080 \text{ cal} \times \frac{1 \text{ Cal}}{1{,}000 \text{ cal}} = 6.08 \text{ Cal } (6.08 \text{ kcal})$

2. a. $\frac{1 \text{ cal}}{\text{g } °\text{C}} \times 6 \text{ glasses} \times \frac{250 \text{ g}}{\text{glass}} \times (37 - 0)\ °\text{C} = 55{,}500 \text{ cal}$

   b. $55{,}500 \text{ cal} \times \frac{1 \text{ Cal}}{1{,}000 \text{ cal}} = 55.5 \text{ Cal}$

   $240 \text{ Cal} \times \frac{1{,}000 \text{ cal}}{1 \text{ Cal}} = 240{,}000 \text{ cal}$

   c. $240{,}000 \text{ cal} \times \frac{16 \text{ glasses}}{55{,}500 \text{ cal}} = 25.9 \text{ glasses}$

   d. No, this is not an efficient way to diet. One would need to consume very large quantities of water, although health food advocates suggest this technique to assist in a diet program. It is considered important for general health to drink eight glasses of water daily.

## B.3 YOU DECIDE: ENERGY IN—ENERGY OUT *(pages 244–246)*

Students keep diaries of personal activity for a 24-hour period and calculate the total amount of energy expended. This activity emphasizes links among the caloric values of foods, human activity, and weight gain or weight loss. Since all that we do uses energy, every minute counts. Encourage students to consider broad categories, such as 5 or 6 hours sitting at school, 1/2 hour (or more) driving, 15 minutes of standing doing dishes or housework, and so on. The total time should equal 24 hours, which will require fairly accurate accounting of activities.

The calculations of energy use are similar to the Water Use Diary: basically students multiply the number of hours by the energy use per hour, given in Table 2. You may have some students who are more active in sports such as wrestling or basketball. It is safe to project a higher caloric output, 1,100 to 1,200 kcal/h, for those activities.

Students compare the total energy consumed in food with the total energy expended during physical activity. One of three conditions results: maintenance of body weight, weight gain, or weight loss.

If you give the assignment as homework, the diary probably will be more accurate and the answers to the questions can be completed in less than one class period.

### Answers to Questions

1. a., b. (Individual student answers)

   c. (Individual answers; generally *ChemCom* students, except for those involved in sports, have been below average in energy use.)

2. a. Calories = 250 + 125 = 375 Cal

   (i) $375 \text{ Cal} \times \frac{1 \text{ h}}{375 \text{ Cal}} = 1.0 \text{ h}$

(ii) $\frac{3.5 \text{ mi}}{300 \text{ Cal}} \times 375 \text{ Cal} = 4.38 \text{ mi}$

(iii) $375 \text{ Cal} \times \frac{1 \text{ h}}{520 \text{ Cal}} = 0.72 \text{ h (crawl)}$

b. $375 \text{ Cal} \times \frac{1 \text{ lb}}{4{,}000 \text{ Cal}} = 0.094 \text{ lb}$

c. $\frac{3 \text{ sundaes}}{\text{wk}} \times 16 \text{ wk} \times \frac{0.094 \text{ lb}}{\text{sundae}} = 4.5 \text{ lb}$

d. (Students will express their own opinions here.)

3. a., b. One could, for example, lower the intake of some other high-energy foods in their regular diet to compensate for the energy in the ice cream. Some students may even suggest maladaptive methods such as skipping meals, drinking lots of water, or regurgitating the food.

## B.4 FATS: STORED ENERGY WITH A BAD NAME *(pages 246–250)*

This section begins by distinguishing between the definition of fat in common language and the definition of fat as used by scientists. It continues by defining saturated fats, polyunsaturated fats, and triglycerides. A *Your Turn* gives students an opportunity to do some calculations from the label on a package of butter. Finally *cis* and *trans* fatty acids and questions about nutritional safety are discussed.

This section is more interesting to students if it is closely tied to healthful diets and prevention of diseases related to high consumption of fats.

### YOUR TURN *Calories From Fat* *(pages 249–250)*

The calculations done by students in this *Your Turn* point out that nearly all the Calories in both butter and margarine are Calories from fat.

### Answers to Questions

1. $\frac{7.2 \text{ g}}{10.9 \text{ g}} \times 100\% = 66\%$

2. a. There must be some mono-unsaturated fats in butter
   b. $10.9 \text{ g} - 0.4 \text{ g} - 7.2 \text{ g} = 3.3 \text{ g}$

3. $\frac{10.9 \text{ g}}{14.0 \text{ g}} \times 100\% = 78\%$

4. a. $10.9 \text{ g} \times \frac{9 \text{ Cal}}{\text{g}} = 98.1 \text{ Cal}$
   b. $\frac{98.1 \text{ Cal}}{100 \text{ Cal}} \times 100\% = 98.1\%$

5. $10 \text{ g} \times \frac{9 \text{ Cal}}{\text{g}} = 90 \text{ Cal}$
   $\frac{90 \text{ Cal}}{90 \text{ Cal}} \times 100 = 100\%$

### YOUR TURN *Fats in the Diet* *(page 251)*

In this *Your Turn* the fat content of alternative and regular doubleburgers is compared.

### Answers to Questions

1. Doubleburger: $\frac{26 \text{ g}}{215 \text{ g}} \times 100\% = 12\%$

   Alternative: $\frac{10 \text{ g}}{206 \text{ g}} \times 100\% = 4.9\%$

Yes. The alternative burger has less fat.

2. Doubleburger: $26\text{ g} \times \dfrac{9\text{ Cal}}{\text{g}} = 234\text{ Cal}; \dfrac{234\text{ Cal}}{500\text{ Cal}} \times 100 = 46.8\%$

Alternative: $10\text{ g} \times \dfrac{9\text{ Cal}}{\text{g}} = 90\text{ Cal}; \dfrac{90\text{ Cal}}{320\text{ Cal}} \times 100\% = 28.1\%$

a. The alternative burger meets the guideline.

b. Doubleburger: $16\text{ g} \times \dfrac{9\text{ Cal}}{\text{g}} = 144\text{ Cal}$; this is greater than 10%.

Alternative: $5\text{ g} \times \dfrac{9\text{ Cal}}{\text{g}} = 45\text{ Cal}$; this is greater than 10%.

3. The doubleburger. It has 10 g unsaturated fat, compared with 5 g for the alternative burger.

## B.5 CARBOHYDRATES: ANOTHER WAY TO COMBINE C, H, AND O *(pages 251–253)*

Carbohydrate structure brings students back to photosynthesis as the energy-storing process in foods. Different types of saccharides, including polysaccharides, are presented, along with some nutritional advice.

Sugars in the chain form, such as glucose, link to make the ring form, which is more stable. This difference in bonding between starch and cellulose (Figure 9, page 253) is small but critical. In digestible starch, both rings are attached to the connecting oxygen (this is similar to an ether functional group) so that the oxygen is below both rings (this is called the alpha linkage). In cellulose, the oxygen is above one ring and below the other (the beta linkage). Human enzymes can break down the more exposed bonding in the starch molecule, but only certain animals have enzymes capable of digesting cellulose.

Additional teacher background is found in the Enrichment Materials section at the end of this unit.

## PART B: SUMMARY QUESTIONS *(pages 254–255)*

1. Three -OH groups, two C=O groups, and one C=C double bond.
2. (No answer)
3. (No answer)
4. $\dfrac{1\text{ cal}}{\text{g °C}} \times 1{,}000\text{ g} \times (62 - 22)\text{ °C} \times \dfrac{1\text{ Cal}}{1{,}000\text{ cal}} = 40\text{ Cal}$
5. Typical example for a 150-lb person:

6:00 PM–11:00 PM—light housework

$$\frac{180\text{ Cal}}{\text{h}} \times 5\text{ h} = 900\text{ Cal}$$

11:00 PM–7:00 AM—sleeping

$$\frac{80\text{ Cal}}{\text{h}} \times 8\text{ h} = 640\text{ Cal}$$

7:00 AM–8:00 AM—showering, etc.

$$\frac{140\text{ Cal}}{\text{h}} \times 1\text{ h} = 140\text{ Cal}$$

Total = 1,680 Cal that need to be replaced

6. a. $2.5 \times 10^9\text{ lb} \times \dfrac{4{,}000\text{ Cal}}{\text{lb}} = 1.0 \times 10^{13}\text{ Cal}$

b. $\dfrac{2{,}500\text{ Cal}}{\text{day}} \times \dfrac{365\text{ days}}{\text{y}} = \dfrac{912{,}500\text{ Cal}}{\text{y}} = \dfrac{9.13 \times 10^5\text{ Cal}}{\text{y}}$ for each person

$$1.0 \times 10^{13}\ \text{Cal} \times \frac{1\ \text{person}}{9.13 \times 10^5\ \text{Cal}} = 1.1 \times 10^7$$ persons, or slightly more than 10 million

7. a. $\frac{2{,}200\ \text{Cal}}{\text{day}} \times 0.30 \times \frac{1\ \text{g}}{9\ \text{Cal}} = 73\ \text{g}$

   b. $\frac{3{,}000\ \text{Cal}}{\text{day}} \times 0.30 \times \frac{1\ \text{g}}{9\ \text{Cal}} = 100\ \text{g}$

8. a. (i) $34\ \text{g} \times \frac{4\ \text{Cal}}{\text{g}} = 136\ \text{Cal};\ \frac{136\ \text{Cal}}{410\ \text{Cal}} \times 100\% = 33\%$

   (ii) $20\ \text{g} \times \frac{9\ \text{Cal}}{\text{g}} = 180\ \text{Cal};\ \frac{180\ \text{Cal}}{410\ \text{Cal}} \times 100\% = 44\%$

   b. (i) $46\ \text{g} \times \frac{4\ \text{Cal}}{\text{g}} = 184\ \text{Cal};\ \frac{184\ \text{Cal}}{410\ \text{Cal}} \times 100\% = 46\%$

   (ii) $22\ \text{g} \times \frac{9\ \text{Cal}}{\text{g}} = 198\ \text{Cal};\ \frac{198\ \text{Cal}}{400\ \text{Cal}} \times 100\% = 50\%$

   c. $80\ \text{g} \times \frac{4\ \text{Cal}}{\text{g}} = 320\ \text{Cal}$

   d. More than 30% of the calories come from fat, but less than 60% come from carbohydrates.

   e. Probably most of them

## EXTENDING YOUR KNOWLEDGE *(page 255)*

a. $\frac{30\ \text{lb}}{60\ \text{days}} \times \frac{4{,}000\ \text{Cal}}{\text{lb}} = \frac{2{,}000\ \text{Cal}}{\text{days}}$

b. 3,000 Cal – 2,000 Cal = 1,000 Cal

c. This is not a sensible way to lose weight. A large reduction in Calories should not be undertaken without doctor's orders.

# C. FOODS: THE BUILDER MOLECULES

Part C focuses on food nutrients, primarily proteins, as builder molecules. The concept of a limiting reactant and how it applies to reactions in living systems is introduced. Amino acids, enzymes, and their relationship to proteins are discussed. The laboratory activity testing the nutritional value of milk is found in the Enrichment Materials section at the end of this unit.

## C.1 FOODS AS CHEMICAL REACTANTS *(page 256)*

This section introduces biochemical reactions with an examination of the breakdown of glucose in the body. It also explains the function of enzymes, which will be studied more thoroughly in the Personal Chemistry and Choices unit.

## CHEMISTRY AT WORK/Food for Thought

It may be somewhat surprising for urban students to discover that the use of high technology—computers and even satellites—is part of modern agriculture. The couple described use these and many other tools to reduce costs and increase production.

### Answers to Questions.

1. Malthus was not able to predict the introduction of fertilizers and other agricultural methods for increasing food supplies. (The increase in productivity probably does not follow his arithmetical formula.)
2. Even with increased production, the population is in danger of outgrowing the food supply.
3. Students will suggest a variety of techniques ranging from planting crops closer together to utilize more high rise housing to using lawn space to grow vegetables.

## C.2 LIMITING REACTANTS *(page 258)*

The concept of limiting reactants is introduced through something familiar to many students—a recipe. The concept is then applied to balanced chemical equations and, finally, to nutrients and living systems. Demonstration and homework suggestions are found in the Enrichment Materials section at the end of this unit.

## YOUR TURN *Limiting Reactants* *(pages 258–259)*

This activity gives students practice in predicting the amounts of products obtained when one reactant is limited. Although none of the systems is "chemical", the calculations correspond closely to those required in chemical systems. This early practice is deliberately limited to concrete, macroscopic systems, so that the central idea will be easier to understand. If students have difficulty with the concept, suggest that they find the possible products for each reactant and compare the results.

### Answers to Questions

1. a. Seven cakes
   b. Flour is the limiting reactant.
2. a. Napkins are used up first.
   b. Napkins
   c. 19 lunch boxes are possible.
3. a. 20 booklets
   b. The pages are the limiting reactant.
   c. 60 staples and 20 covers are left over.

## YOUR TURN *Limiting Reactants: Chemical Reactions* *(page 260)*

This extension of the previous activity applies the concept of a limiting reactant to moles and molecules. The importance of limiting reactants in the fertilizer industry provides a real-world example.

## Answers to Questions

1. a. 1 mol nitrogen, 3 mol hydrogen, 2 mol ammonia
   b. 28 g nitrogen, 6 g hydrogen, 34 g ammonia
2. a. Nitrogen is the limiting reactant.
   b. Hydrogen is found in excess.
   c. 3 kg
   d. 34 kg of ammonia

## YOUR TURN *Limiting Reactants in Plants and Humans* *(pages 261–262)*

The concept of limiting reactants is applied to the nutrients needed to grow common agricultural crops.

## Answers to Questions

1. Boron is not an essential nutrient for humans.
2. Sodium, chromium, cobalt, fluorine, iodine, nickel, and selenium are not essential for plants.
3. a., b. (Answers will vary, depending on the product examined.)
4. a. Nitrogen, phosphorus, potassium
   b. It would be difficult to grow adequate quantities of food.

## C.3 PROTEINS *(pages 263–264)*

This section introduces proteins and their building blocks, amino acids, as well as the condensation reaction that produces the peptide bond. The structures of amino acids and their conversion to peptide units may need some clarification. Formation of the peptide bond may be illustrated with a ruler having a red tail (for carboxyl group) and a blue tail (for amino group). Point out that amino acids always combine so that red (carboxyl) reacts with blue (amino); the product still has an amino group and a carboxyl group free to react. Also point out that there are two ways to unite the amino acids, producing two different peptides. Additional teacher background and a ChemQuandary are found in the Enrichment Materials section at the end of this unit. In this edition, the Laboratory Activity: Milk Analysis is also found in the Enrichment Materials section.

## YOUR TURN *Molecular Structure of Proteins* *(pages 264–265)*

Students examine and practice writing structural formulas for amino acids and proteins. Consideration of how many formulas can be written with only a few amino acids helps them realize how a staggering number of different protein structures can be based on a limited number of amino acids.

## Answers to Questions

1. a.

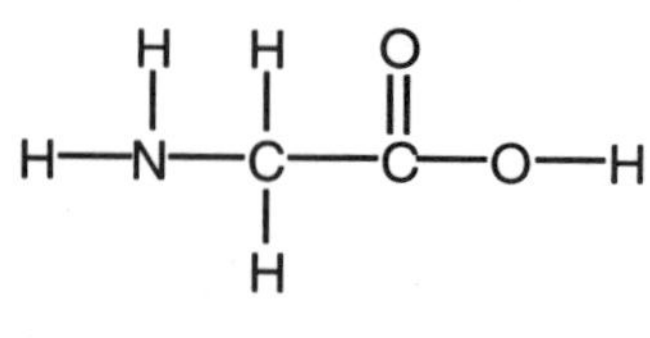

glycine

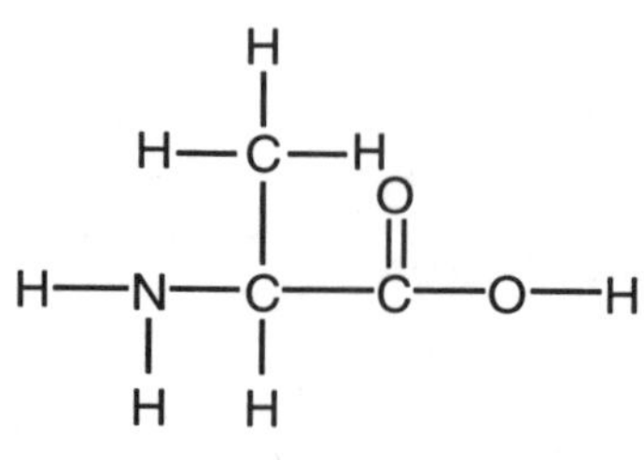

alanine

   b. Functional groups: both molecules contain the amino group ($NH_2$) and the carboxyl or carboxylic acid group (COOH).
   c. Alanine has a methyl ($CH_3$) group in place of the hydrogen atom on glycine.

2. a.

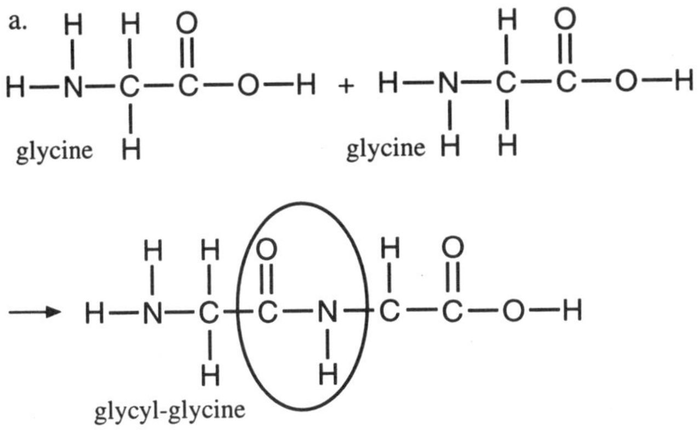


b.

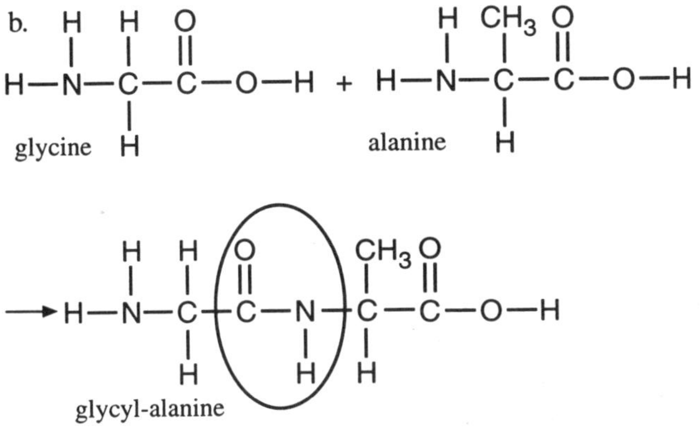


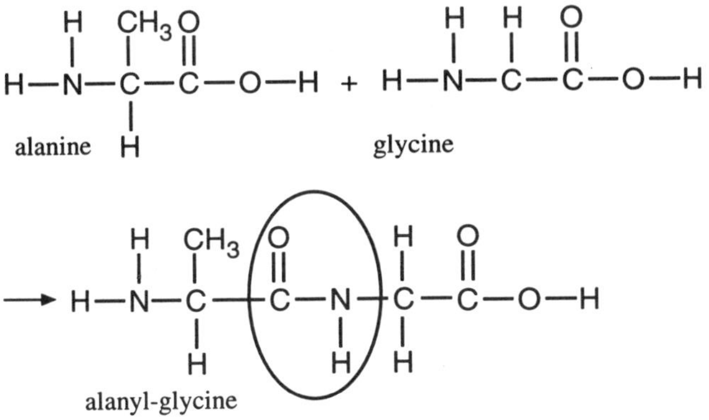


3. (No answer required)

4. a. abc, bac, cab, acb, bca, cba-6 (3 × 2 × 1)
   b. 27 (3 × 3 × 3)
   c. 24 (4 × 3 × 2 × 1)
   d. Millions-plus
   e. With such a vast number of arrangements, unique "genetic codings" would be possible at the molecular level.

## YOUR TURN *Protein in the Diet* *(pages 267–268)*

Since the body cannot store excess protein, it is important that we take in a continual supply of these biomolecules for body-maintenance "building" and "rebuilding" purposes. Here students examine and apply the Recommended Daily Dietary Allowances (RDAs) for protein, as summarized in Table 8 on page 267.

### Answers to Questions

1. 45 g of protein. Student answers will depend on their age and gender: 44 g for the 15- to 18-year-old female students; 59 g for the 15- to 18-year-old males.

2. Infants are more rapidly building new tissues.
3. a. Probably lower than 50 g
   b. With lower body weight, protein needs are less.
4. a. Mother's milk
   b. Nursing (lactating) women require higher supplies of protein to produce breast milk.

## PART C: SUMMARY QUESTIONS *(page 269)*

1. a. The proper amounts and kinds of foods are important both for "building" and for "burning." Certain nutrients are needed for the efficient use of food for energy. Any nutrient lacking or in short supply can become a limiting reactant.
   b. The same is true for plant nutrients.
2. a. Essential amino acids are needed by the body, but cannot be synthesized.
   b. Deficiencies in these amino acids can adversely affect body chemistry by inhibiting protein synthesis and the rebuilding of body structures.
3. a. Carbon dioxide
   b. Ammonia
   c. $36 \text{ g } CO_2 \times \frac{34 \text{ g } NH_3}{44 \text{ g } CO_2} = 27.8 \text{ g } NH_3$ required

   $34\text{g } NH_3 - 28 \text{ g } NH_3 = 6 \text{ g } NH_3$ left over

   d. $36 \text{ g } CO_2 \times \frac{60 \text{ g } CON_2H_4}{44 \text{ g } CO_2} = 49 \text{ g } CON_2H_4$
4. The body requires other nutrients in addition to the amino acids found in proteins. Fats, carbohydrates, minerals, and vitamins all have essential roles to play. In fact, excessive protein intake can cause health problems.

5. 

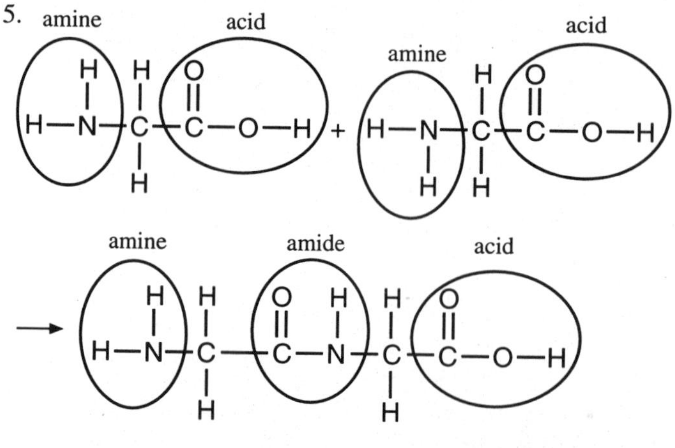


6. a. 44 g (Students' answers may be slightly higher.)
   b. $44 \text{ g} \times \frac{16 \text{ oz}}{1 \text{ lb}} \times \frac{1 \text{ lb}}{23.5 \text{ g}} = 30 \text{ oz}$ (white meat)
   c. $44 \text{ g} \times \frac{1 \text{ cup}}{8.03 \text{ g}} = 5.5$ cups (whole milk)

7. a. A complete protein contains all of the essential amino acids; an incomplete one does not.
   b. One needs all the essential amino acids to continue protein synthesis in the body. Complementary proteins, such as nuts and grains, should be included in the diet.
8. a. Burning 1.0 mol of methane produces more thermal energy than burning 1.0 mol CO and 1.0 mol $H_2$.
   b. (2) is an amino acid.

# D. OTHER SUBSTANCES IN FOODS

In Part D we discuss the roles of vitamins, minerals, and other food additives. Students learn why vitamins and minerals are essential and they learn the importance of the RDAs. In the laboratory they determine the quantity of vitamin C by titration and iron by colorimetry, and they analyze food colors by paper chromatography. Finally, students briefly examine the benefits and risks of some common food additives.

## D.1 VITAMINS *(pages 270–271)*

RDAs for selected vitamins are examined in this section, as well as the health effects of consuming too much or too little of particular vitamins. Additional teacher background is found in the Enrichment Materials section at the end of this unit.

## YOUR TURN *Vitamins in the Diet* *(page 272)*

This activity provides students with practice in interpreting data from tables and applying it to their own situations.

### Answers to Questions

1. a. $B_{12}$ (cobalamin) deficiency results in pernicious anemia and exhaustion; D (calciferol) deficiency results in rickets.
   b. By taking vitamin supplements
2.

| Vegetable | Your RDA | | No. Servings to Supply Your RDA | |
|---|---|---|---|---|
| *(1-Cup Serving)* | $B_1$ | C | $B_1$ | C |
| Green peas | | | | |
| Males | 1.5 | 60 | 3.88 | 1.03 |
| Females | 1.1 | 60 | 2.84 | 1.03 |
| Broccoli | | | | |
| Males | 1.5 | 60 | 25.9 | 0.73 |
| Females | 1.1 | 60 | 19.0 | 0.73 |

   a. Yes, vitamin $B_1$
   b. Since there is no perfect food, variety in the diet assures that the proper kinds of vitamins are consumed. Green peas are a better source of $B_1$ compared to broccoli, but are not as good a source of vitamin C.
   c. Even if the caloric intake is adequate, vitamin deficiencies from the lack of variety in foods will limit body building and burning processes.
3. Fat-soluble vitamins: they can be stored and thus accumulate in the body, while excess water-soluble vitamins are more likely to be excreted.
4. a. Water-soluble vitamins
   b. Long cooking can dissolve and often destroy these vitamins from vegetables. Raw or lightly steamed vegetables retain maximum concentrations of vitamins.

## D.2 LABORATORY ACTIVITY: VITAMIN C *(pages 273–275)*

Students use microtitration techniques and calculations to determine the amounts of vitamin C in several beverages. The use of Beral pipets and wellplates enables teams of students to analyze several different samples in 50 minutes. The macroscale version of this laboratory activity is found in the Enrichment Materials section at the end of this unit.

### Time

50 minutes

## Materials (for a class of 24 working in pairs)

12 10-mL graduated cylinders
12 Beral pipets
12 Beral pipets containing iodine solution
12 Beral pipets containing 1% starch solution
vitamin C solution, 0.100 g/100 mL water
Beral pipets containing beverage samples
toothpicks for stirring

## Safety

Students should wear safety goggles and aprons throughout the laboratory activity.

## Advance Preparation

**Iodine solution**: Dissolve 1.0 g of potassium iodide and 0.053 g potassium iodate in 25 mL water. Add 6 mL of 3 M sulfuric acid. Add water to make total volume 100 mL. Fill 12 pipets shortly before class time and provide a capped bottle for refills. The iodine pipets cannot be cleaned, so use the same pipets for every class. The reaction in the iodine solution: $IO_3^- + 5\,I^- + 6\,H^+ \rightarrow 3\,I_2 + 3\,H_2O$

**Starch solution:** Dissolve 1 teaspoon of liquid laundry starch in 100 mL hot (not boiling) water. Fill 12 pipets and provide a capped bottle for refills.

**Vitamin C solution:** Crush a 100-mg vitamin C tablet and dissolve in water. Dilute to 1000 mL. (This is much more than needed, but it is difficult to work with smaller quantities of the tablet.) The solution is somewhat unstable and should be used within a week of preparation.

**Beverage samples:** Choose five or more beverages to test. The color of dark juices and sodas obscures the endpoint of the reaction, so they are best avoided. If colored juices are used, suggest that all five samples titrated be of the same color—Tang, Orange Juice, Hi-C, and so on. Remove any pulp from juices by filtering through cheesecloth before placing in pipets.

Beverages vary considerably in vitamin C content. Some are listed below.

| Vitamin C content | Beverage |
|---|---|
| Very high | Pineapple or orange juice |
| High | White grape juice, grapefruit juice, or Hi-C |
| Medium | V-8 juice |
| Low | Apple juice, milk, or Tang |
| Very low or none | Gatorade, Sprite |

## Pre-Lab Discussion

Students have probably heard of vitamin C both as a component of various juices and soft drinks and as a cold preventive. Discuss the types of foods which are high in vitamin C. Students will probably be surprised to find that significant amounts of vitamin C are found in broccoli and green and red peppers. (See *Chem Matters*, April 1995.)

Demonstrate the use of the pipet and wellplate in titration. Emphasize the importance of holding the pipet at the same angle each time it is used.

## Answers to Questions

1. a., b. (Answers will vary. Some may be surprised that Gatorade does not contain vitamin C.)
2. Potatoes, broccoli, red and green peppers

## Post-Lab Discussion

Students may have some difficulty with calculations. Depending on their mathematical ability, they may need help calculating the conversion factor. Working through a sample calculation in step 2.5 may be necessary.

Though student data may not exhibit great precision, it should allow them to rank-order the beverages quite reliably in terms of vitamin C levels.

Relate the laboratory activity to the text materials by discussing how much of each drink would be necessary to acquire the recommended RDA for vitamin C.

### Extension

If time permits, some teams can investigate the amount of vitamin C in other foods. Use a blender to obtain the juice. Another interesting investigation is to see what heating or microwaving does to the vitamin C content.

## D.3 MINERALS: AN IMPORTANT PART OF OUR DIET *(pages 275–277)*

This brief section emphasizes that minerals are also essential to health, as cofactors in biochemical catalysis and as component parts of the body's structure. Minerals are classified as either macrominerals or trace minerals. Although only small quantities of trace minerals are needed, they can act as limiting reagents. Health problems occur with either too much or too little of any needed mineral.

## YOUR TURN *Minerals in the Diet* *(pages 277–278)*

Students use the RDAs to examine the variety of ways to obtain the required minerals.

### Answers to Questions

1. $12 \text{ mg Fe} \times \frac{1 \text{ slice}}{0.8 \text{ mg Fe}} = 15 \text{ slices (male)}; 15 \text{ mg Fe} \times \frac{1 \text{ slice}}{0.8 \text{ mg Fe}} = 19 \text{ slices (female)}$

2. $1{,}200 \text{ mg Ca} \times \frac{1 \text{ cup}}{288 \text{ mg Ca}} = 4.2 \text{ cups}$

3 a. Iron

b. $\frac{27 \text{ mg Ca}}{1{,}200 \text{ mg Ca}} \times 100\% = 2.25\%; \frac{0.4 \text{ mg Fe}}{12 \text{ mg Fe}} \times 100\% = 3.33\% \text{ (male)}$

$\frac{0.4 \text{ mg Fe}}{15 \text{ mg Fe}} \times 100\% = 2.67\% \text{ (female)}$

4. a. $\frac{1{,}200 \text{ mg}}{\text{day}} \times \frac{365 \text{ days}}{\text{y}} \times \frac{1 \text{ g}}{1{,}000 \text{ mg}} = 438 \text{ g/year}$

b. Bones and teeth contain large amounts.
c. Ca: milk, dairy products; P: animal protein
d. Ca: rickets, osteomalacia, and osteoporosis; P: no deficiency
e. Yes; infants and women

5. a. Prevention of goiter
b. Seafoods

## D.4 LABORATORY ACTIVITY: IRON IN FOODS *(pages 278–279)*

This activity demonstrates one way to determine the mineral content of foods. Students use colorimetry to determine the relative levels of iron in several common foods.

### Time

50 minutes

### Materials (for a class of 24 working in pairs)

24 crucibles, porcelain
24 clay triangles

12 ring stands
12 ring clamps
12 Bunsen burners
12 tongs
12 test tubes
12 funnels
12 50-mL or larger beakers
12 rubber stoppers
2 burets for dispensing HCl and KSCN
30 filter papers (Whatman no.1, 11 cm diameter)
200 mL 2 M hydrochloric acid (HCl) (33 mL conc. HCl/200 mL solution)
200 mL 0.10 M potassium thiocyanate (KSCN) (1.9 g/200 mL)
10 mL 1.0% $Fe^{3+}$ ($FeCl_3$) stock solution for preparing $Fe^{3+}$ color standards
several 2.5-mg samples of foods (see Lab Tips)

## Safety

Students should wear safety goggles and aprons throughout the laboratory activity. If possible, use a hood for ashing. Caution students to handle hot apparatus with care.

## Advance Preparation

The iron standards procedure will provide you with six sets of color standards for a range of $Fe^{3+}$ concentrations for each bench. Label six test tubes with the concentration labels in the shaded portion of the table below. Use test tubes that are the same size the students will use.

Dissolve a 0.50-g sample of iron(III) chloride hexahydrate in 9.5 mL of distilled water, producing a 1.0% $Fe^{3+}$($FeCl_3.6H_2O$) standard.

Use 250-mL beakers to make the dilutions according to the table below.

| To prepare concentration | Use | Add water |
|---|---|---|
| 0.1% | 10 mL 1% | 90 mL |
| 0.01% | 0 mL 0.1% | 90 mL |
| 0.005% | 50 mL 0.01% | 50 mL |
| 0.0025% | 50 mL 0.005%` | 50 mL |
| 0.00125% | 50 mL 0.0025% | 50 mL |
| 0.0010% | 10 mL 0.01% | 90 mL |
| 0.00050% | 50 mL 0.0010% | 50 mL |
| 0.00025% | 50 mL 0.00050% | 50 mL |

Pour 5 mL of the solutions described in the shaded portion of the table into separate appropriately labeled test tubes, and add 5 mL of 0.10 M KSCN. Invert each tube once to mix. You should have enough of each solution to make six standards. Place a set of standards (one of each concentration) in a rack, and place them where two pairs of students can use them.

## Pre-Lab Discussion

The discussion should include the use of color-intensity methods in quantitative analysis. Making and diluting Kool-Aid illustrates the use of color to estimate concentration.

Using a magnetic stirrer, you can extract iron from breakfast cereals such as Total. Fill a 1,000-mL beaker with dry cereal, add water and a magnetic stirring bar, and stir. After the mixture is suitably mixed, remove the magnet. Iron filings should be present on the stirring bar. You might wish to discuss with your students the ability of the body to absorb elemental iron, perhaps by adding the stirring bar to a beaker of dilute HCl. Stomach acid is approximately 0.10 M acid.

## Lab Tips

Some foods that work well in this activity are parsley, raisins, spinach, cauliflower, broccoli, canned kidney beans, dry breakfast cereals. Approximate iron contents can be found in the Appendix to the Student Text.

Large pieces of food should be crushed or ground, to prevent making a hard product that is difficult to dissolve.

If color standards need to be stored, they should be placed in a dark place overnight. The color standards cannot be kept and used from year to year.

Remind students to adjust the burner so that it will produce the hottest flame. Demonstrate the proper technique for handling crucibles, and caution students about their fragility. Some foods may produce an unacceptable amount of smoke. Crucible lids can help control this problem. After the food is charred, remove the lid until the food has completely turned to ash. (Dry breakfast cereals do not produce much smoke.)

## Answers to Questions

1. No, the iron(III) concentration values found by color comparison in the solutions do not represent the percent iron in the original food samples—they just represent the percent iron in the water solution. More calculations would be necessary to find this value.
2. See student values from Post-Lab summary. For a comparison with accepted values, see the Appendix.
3. Macrominerals such as calcium and sodium, and trace minerals such as chromium, zinc, and fluorine

## Post-Lab Discussion

Have students put their results on the chalkboard or poster paper and discuss them with the class. Discuss any discrepancies in the order.

## Extension

If a spectrophotometer is available, you can demonstrate how it uses color to determine concentration more precisely.

# D.5 FOOD ADDITIVES *(pages 279–281)*

This section briefly surveys the benefits (and some associated risks) involved in modern use of food additives.

# D.6 LABORATORY ACTIVITY: FOOD COLORING ANALYSIS *(pages 281–283)*

Students use paper chromatography to separate dyes in candies and compare them to dyes in food coloring. They use the data on the chromatography to calculate Rf values.

## Time

50 minutes

## Materials (for a class of 24 working in pairs)

12 8-oz gas-collecting bottles
12 rubber stoppers to fit gas bottles
12 24-well wellplates
36 micropipets
chromatography paper, 1-inch-wide strips
toothpicks
1 package M&M's candies
1 package Skittles candies
red food coloring
blue food coloring
yellow food coloring
green food coloring

sulfuric acid ($H_2SO_4$), 0.05 M (2.8 mL conc. $H_2SO_4$/1.0 L solution)
ammonia ($NH_3$), 3 M or household ammonia

## Safety

Students should wear safety goggles and aprons throughout the laboratory activity. If possible, handle the ammonia in a hood.

## Advance Preparation

Cut 12 strips of chromatography paper, each 6 inches long.

## Pre-Lab Discussion

Solely for the sake of appearance, a food product may contain a number of varied artificial dyes. Paper chromatography is a way to separate these dyes. The various dyes in a mixture respond differently to the paper and the solvent. Some are held more firmly to the paper than others. It is difficult to compare results unless all conditions are carefully controlled. Therefore, a standard is generally run with an unknown, and the Rf values are compared.

## Lab Tips

This activity is not difficult for students. The most likely error is allowing the spots to be submerged in the solvent. If this happens, the dye merely dissolves in the solvent, and no spots appear on the chromatography paper.

## Expected Results

The Rf values are given for red candy and food coloring. Do not be concerned if student values are slightly different. The spots will be in the same order, but the actual distances may differ. Many things affect the results: temperature of the room, small changes in concentration of the solvent, and even the paper used.

| Candy | Dye Color | Distance Dye Spot Moved | Distance Solvent Front Moved | Rf Value |
|---|---|---|---|---|
| M&M's | red | 30 mm | 68 mm | 0.44 |
| Skittles | red | 25 mm | 68 mm | 0.37 |
| | yellow | 37 mm | 68 mm | 0.54 |
| Food Coloring | pink | 22 mm | 68 mm | 0.332 |
| | red | 45 mm | 68 mm | 0.66 |

## Answers to Questions

1. The dyes in the ink will also be separated by paper chromatography.
2. (Answers will vary with student results.)
3. (Answers will vary with student results. The dye with the smallest Rf value.)
4. (Answers will vary with student results. The dye with the largest Rf value.)
5. a. (Answers will vary.)
   b. Any dye that shows only one spot may be made of a single dye. However, it could be that more than one dye has the same Rf value.
6. a., b., c. (Answers will vary with student results.)

## Post-Lab Discussion

The post-lab discussion can focus on two topics: the use of dyes as food additives and the technique of paper chromatography. The answers to the questions can be used as guidelines in discussing the technique of paper chromatography. Get student opinions on the appropriateness of using dyes to improve the appearance of foods. Looking at labels will reveal a somewhat surprising list of foods that contain dyes. The discussion leads naturally into the following sections on food legislation and the *ChemQuandary* on nitrite additives. An additional *ChemQuandary* and a Demonstration are found in the Enrichment Materials section at the end of this unit.

## Food Legislation *(pages 284–285)*

In this section we discuss the laws that enable monitoring food safety and purity in the United States. Discussion of the Delaney Clause, which requires prospective food additives to be tested on laboratory animals, leads to the *ChemQuandary* on weighing risks and benefits of using nitrites to preserve meats. In 1996 the part of the Delaney Clause relating to pesticide residues was repealed. It was replaced with a new standard of "a reasonable certainty of no harm." The artificial sweetener aspartame and its relationship to phenylketonuria are also discussed.

## CHEMQUANDARY *Nitrite Additives* *(page 285)*

This *ChemQuandary* probes the benefits and risks of nitrites as food additives. Here are some questions that students might ask. (Other questions, of course, are equally acceptable.)

### Some Possible Questions

1. How conclusive is the evidence that nitrosamines cause cancer in laboratory animals and in humans?
2. How deadly is the botulin toxin? Are there antidotes for it?
3. How often does botulism occur and how often does nitrite cause cancer?
4. What are some possible substitutes for sodium nitrite as a preservative?
5. Have substitutes been used? What are the risks involved with them?

## D.7 YOU DECIDE: FOOD ADDITIVE SURVEY *(page 287)*

The purpose of this activity is to make students more aware of food additives found in products that they consume everyday. Students survey five or six packaged foods and select six additives to describe and investigate.

This activity will take about 30 minutes to complete. (Answers will vary.) Students may collect the data from the labels as homework, but the activity should be discussed in class. Emphasize the risks and benefits of using food additives in the various products the students have selected.

## CHEMISTRY AT WORK/Keeping Food Supplies Safe for Consumers *(page 288)*

A research chemist manages a group who work to understand and prevent food toxins.

## PART D: SUMMARY QUESTIONS *(page 289)*

1. a. Water solubility: you need more water-soluble vitamins on a daily basis.
   b. Toxicity: too high a dose of fat-soluble vitamins can be toxic.
   c. Cooking precautions: overcooking can break down fat-soluble vitamins and leach out water-soluble vitamins.
2. Macrominerals are needed in large amounts, but microminerals are also important because they can act as limiting reactants. Some examples are that lack of calcium in the diet (from dairy products) causes bone problems and lack of iodine in the diet (from seafood) causes problems with the thyroid gland such as goiter.
3. a. Calcium: dairy and eggs
   b. Iron: meats
   c. Vitamin C: fruits and vegetables
   d. Vitamin A: fruits, vegetables, and dairy
4. a. $\frac{60\text{ mg}}{\text{day}} \times \frac{1\text{ cup orange juice}}{124\text{ mg vitamin C}} = 0.48\text{ cups orange juice}$

   b. $\frac{60\text{ mg}}{\text{day}} \times \frac{1\text{ cup}}{2.29\text{ mg}} = 26.2\text{ cups}$
5. Some food additives are natural substances, such as vitamin C and minerals needed by the body. All foods may be viewed as chemicals.
6. a. Least essential—coloring agents; have no impact on nutritive value

b. Most essential—preservatives that prevent bacterial growth; prevent the spread of disease and illness

7. If the benefit is deemed high, and the risk is judged to be low, and no better substitute is available, then the wisest course is to continue to use the additive.

8. a. Delaney Clause—"No additive shall be deemed safe if it is found to induce cancer when ingested by man or animal."

   b. The Delaney Clause prevents the use of additives that may cause cancer in humans.

   c. The Delaney Clause may prevent using additives that reduce other diseases, more deadly than cancer caused by the additive.

# E. PUTTING IT ALL TOGETHER: WHICH FOOD IS BETTER?

Students apply what they have learned in the unit to the very practical activity of using food labels to make informed decisions about nutrition. In the first activity, students calculate their personal daily requirements of nutrients; in the second, they use this information to make a choice between two possible frozen dinners.

## E.1 FOOD LABELS *(pages 290–291)*

The 1994 changes in food labels are highlighted. Students use Figure 18 and values from food labels to calculate personal values.

## YOUR TURN *Finding Your Daily Values* *(pages 291–292)*

The answers to the questions are given for a 165-lb male. Students will have different values, but the methods of calculation are the same.

Calculation of Factor: $\frac{3{,}375 \text{ Cal}}{2{,}000 \text{ Cal}} = 1.69$

### Answers to Questions

1. 300 g carbohydrates × 1.69 = 507 g carbohydrates
2. a. 65 g total fat × 1.69 = 110 g total fat

   b. 20 g saturated fat × 1.69 = 33.8 g saturated fat
3. $\frac{23 \text{ g}}{507 \text{ g}} \times 100 = 4.5\%$
4. $\frac{113 \text{ g dinner}}{10\%} \times 4.5\% = 50.8 \text{ g dinner}$
5.

| | Daily Calories Level |
|---|---|
| Total fat | About 110 g or less |
| Saturated fat | About 33.8 g or less |
| Total carbohydrate | About 507 g or less |
| Dietary fiber | About 42.3 g or less |
| Protein | About 60 g or less |

## E.2 YOU DECIDE: WHAT KIND OF FOOD AM I? *(pages 292–293)*

Students use the values calculated in question 5 of the previous *Your Turn* to answer some of the questions.

### Answers to Questions

1. a. 110 g − 8 g = 102 g

   b. 45 g + 8 g = 53 g; $\frac{53 \text{ g}}{110 \text{ g}} \times 100 = 48\%$
2. a. The Chicken Delight contains a smaller percent of the daily value for fat, saturated fat, cholesterol, and sodium. These are all nutrients you want to reduce in your diet. It also contains a higher percent of the daily value for vitamin A and iron.

   b. The Chicken Supreme has lower total Calories and higher percents of vitamin C and calcium.
3. 15 mg Fe × 1.69 = 25.3 mg Fe; 15 mg × 0.15 = 2.25 mg

   $\frac{2.25 \text{ mg}}{25.3 \text{ mg}} \times 100 = 8.89\%$

4. $\frac{23\text{g}}{60\text{ g}} \times 100 = 38\%$

5. The friend should eat only $\frac{8\text{ g}}{15\text{ g}} \times 100 = 53\%$.

6. Energy is obtained primarily from carbohydrates and fats. Vitamins are present in only small amounts.
7. Everything in the food is "chemical." When people say "no chemicals", they often mean no additives, but this is incorrect.

## E.3 LOOKING BACK AND LOOKING AHEAD *(page 293)*

As the text suggests, this marks the half-way point in the full *ChemCom* program. Four additional units complete the 1-year course. Use this as an opportunity to look back with students at the range of topics and issues already covered and to look forward to the remaining four units.

# ENRICHMENT MATERIALS

## TEACHER BACKGROUND

### Section B-5

The body is able to metabolize various classes of biomolecules to meet its energy needs:

1. **Carbohydrates** (4 Cal/g): The body's preferred fuel is ultimately converted into glucose, which circulates in the blood and provides energy to cells. Excess glucose is converted into glycogen, which is stored in the liver and muscle cells or, if these "banks" are full, as fat stored in fat cells.
2. **Fats** (9 Cal/g): Of the fats in food and the human body, 95% are triglycerides (the other 5% are phospholipids), which consist of a glycerol ($C_3H_8O_3$) backbone with three attached fatty acids. Under conditions of low to moderate activity, both components are burned along with glucose to fuel voluntary muscles. The red blood cells and nerve cells are unable to utilize fatty acids for fuel. Also, under conditions of strenuous activity, even the muscles shift to using only glucose (from stored glycogen), since oxygen cannot be supplied fast enough to support fat oxidation. By serving as a fuel under normal conditions, and storing glucose that is in excess of glycogen storage capacity, fats serve a vital energy role.
3. **Protein** (4 Cal/g): If the body is supplied with adequate calories from carbohydrates and fats, protein is not used as a fuel, as it has a more important "building" function. Under starvation conditions, the body will break down protein to fuel the brain, nerve cells, and red blood cells, and to maintain vital organ functions. Only when reserves of carbohydrates and readily available fat have been exhausted does cellular metabolism turn to burning proteins.
4. **Alcohol (ethanol)** (7 Cal/g): Alcohol provides empty calories and lacks nutrients. It can displace more nutrient-dense foods and contribute to malnutrition and obesity, as well as act as a central nervous system depressant.

### Section C.3

From a base of 26 letters, we can form thousands of unique words by varying the number, type, and sequence of letters. However, some combinations lack meaning and are gibberish (zxq or hjt). A similar argument can be made for proteins. (For instance, sickle cell anemia is caused by one mistake in a sequence of approximately 300 amino acid units.) Given that proteins (words) contain from 50 to 10,000 amino acid units (letters), the usual accuracy of protein synthesis is remarkable—relatively few "misspellings" occur.

Just as different countries have different languages, different species have their own characteristic protein languages. In fact, individual humans each have a unique set of more than 30,000 different proteins. (Of course, similarities in protein language among humans is greater than between individual humans and other mammals, which is greater than between mammals and nonmammals.) This biochemical uniqueness is most evident in the action of antibodies and the immune system.

### Complete and Complementary Proteins

Most animal sources of protein (such as meat, fish, poultry, milk, and eggs) are considered complete, in that they closely match the needs of the human body for all the essential amino acids. Efficient use of protein intake requires that the entire set of essential amino acids be eaten within about a 4-hour period. This can be accomplished without eating meat if one chooses nonmeat options that complement each other. In general, combinations of legumes, such as nitrogen-fixing beans and nuts (deficient in methionine), and grain (deficient in tryptophan and lysine) are effective. Examples include peanut butter and whole wheat bread, black beans or soybeans and rice, beans and corn, and so on. Vegetarian diets can be quite healthful for adults, especially if dairy products are included. In any case, given the overconsumption of protein in the typical American diet, the presence of fat in most meat, and the negative health impact of saturated fats and cholesterol, eating less meat and more vegetables and fruits is a good idea.

### Section D.1

For a substance to qualify as a vitamin, its absence must cause a specific deficiency disease that is cured when the substance is resupplied. There are 13 known vitamins required by humans; four are fat soluble (A, D, E, and K) and nine are water soluble (C and eight B-complex).

No reliable scientific evidence supports the notion that vitamin supplements have beneficial effects for the average person eating a balanced diet. Vitamin deficiency is rare unless a person's diet is extremely unbalanced. Yet, a $3 billion-a-year industry has as many as 40% of all adult Americans buying vitamin or mineral supplements. Many advertisements are designed to inspire insecurity and are often based on the deception that "if some is good for you, more is better."

Most nutritionists prefer the balanced nutrient intake provided by a varied diet to popping pills. If people choose to take supplements as an added "insurance" along with a balanced diet, nutritionists recommend products that contain no more than about 100% of the RDAs for a wide range of vitamins and minerals. Taking excessive amounts of one or more nutrients can adversely affect the body's absorption, utilization, and excretion of others. As an example, high zinc intakes seem to impair copper's function, and many Americans get too little copper anyway. High doses of fat-soluble vitamins and megadoses (on the order of $100 \times \text{RDA}$ ) of the water-soluble ones can have serious toxic effects.

Advertisements sometimes promote the idea that natural vitamins are better than synthetic ones. Actually, individual cells cannot tell the difference between the two. From a chemical point of view, vitamin C synthesized in the laboratory is identical to vitamin C from rose hips.

The actual vitamin content of a given food depends to a large extent on how the food was processed, stored, and made ready for eating. In general, the fresher and less processed the food, the higher its vitamin content. In particular, fresh fruits and vegetables are slightly better than frozen ones, which are significantly better than canned. In cooking, the less time and/or water used, the better. Steaming, pressure cooking, or microwaving are preferred over boiling in water, which can destroy some fat-soluble vitamins and leach out many of the water-soluble ones. If food is boiled, nutritionists recommend using the drained-off water in making soups and gravies.

## CHEMQUANDARIES

### Section B.2

- In general, why is a weight-loss plan based only on exercise (without controlling caloric intake) not likely to be adequate for anyone who is more than a little overweight?

**Answer:** Physical activity typically accounts for only about 30% of the body's energy output: the other 70% is to maintain basal metabolism. Even the most vigorous physical activity burns off only about 1,000 Cal/h, and 1 lb of body fat is equivalent to about 4000 Cal. Thus, if one wanted to lose 1-2 lb/wk (the maximum recommended without medical supervision) using exercise alone, one would need to set aside 4-8 h/wk for vigorous physical exercise. Such a plan would probably be unhealthful or even impossible for a very overweight person. A balanced combination of low-calorie diet and exercise would be more effective in both the short and long term.

- A hiker wishes to minimize the quantity of food that she needs to carry with her on a short trip in a cold climate. If she also wishes to maximize her caloric intake, would she be better advised to take along high-carbohydrate or high-fat foods?

**Answer:** Fats, since they deliver more Cal/g than carbohydrates.

## Section C.3

- Some athletes load up on protein for "extra energy" by eating steak before a competition. Is this a wise choice?

**Answer:** No. The body's preferred fuels in order of decreasing use are carbohydrates, fats, and proteins. The athletes would be wiser to consume pre-game meals high in carbohydrates.

- What are the logic and myth behind high protein diet formulations?

**Answer:** The logic behind trying to lose weight by eating only protein is that the body does not store excess protein (or even excess amino acids). However, after excreting the nitrogen, the body does convert excess protein by-products into fat. Another problem is that the metabolism of excess protein puts a strain on the liver and kidneys. For the most part, the $70-million-a-year protein supplement industry is supported by mistaken ideas about protein requirements and utilization. The average American consumes almost twice as much protein as needed.

- As most animal sources of protein are "complete", one could build a healthful diet by excluding other food types. True or false?

**Answer:** False. The word "complete" in this context refers only to needed amino acids, not to nonprotein nutrients. An adequate diet must also contain carbohydrates, fats, vitamins, and minerals.

## Section D.6

- Maraschino cherries are bleached with sulfur dioxide and then dyed red with an organic food coloring. What are the benefits and risks of such cherries relative to the "natural" product? What problems would a manufacturer face in switching to the natural?

**Answer:** The benefits are cosmetic and esthetic—people have come to expect such cherries to be bright red. The risks include the possibility of a reaction in people sensitive to sulfur dioxide and the possibility that, at some future date, the particular red dye used may come under suspicion as a health threat. A manufacturer would need to change people's image of what a good cherry is supposed to look like.

- Would you expect to find golden raisins in an organic food store? Why or why not?

**Answer:** Probably not, since they have been bleached with sulfur dioxide.

- Why does it make sense to ban additives that have been shown to be possible causes of cancer, if carcinogens also occur naturally in food?

**Answer:** Although risks from either may be quite small, the former is an avoidable risk, while the latter is not. In general, the greater the total exposure to carcinogens, the more likely one will develop cancer.

# DEMONSTRATIONS

## Section C.2

This demonstration provides a chemical example of the limiting reactant concept, involving easily observed chemical changes. Use a shield during this demonstration. Take care that no flames are used when the balloons are emptied.

## Materials

8 culture tubes
8 round balloons (uninflated diameter 4 inches)
50 mL 3 M hydrochloric acid
55 cm magnesium ribbon

## Demo Tips

Fill the tubes with the proper volumes of hydrochloric acid, clearly label them by number, and place them in test tube racks. Fold each piece of magnesium ribbon so it will fit inside an uninflated balloon.

**Part A** Constant amount of HCl, variable amount of Mg
Tube 1: 5 mL 3 M HCl + 2.5 cm Mg ribbon (3 M HCl: 12.5 mL conc. HCl/50 mL solution)
Tube 2: 5 mL 3 M HCl + 5.0 cm Mg ribbon
Tube 3: 5 mL 3 M HCl + 10.0 cm Mg ribbon
Tube 4: 5 mL 3 M HCl + 15.0 cm Mg ribbon

**Part B** Constant amount of Mg, variable amount of HCl
Tube 5: 2 mL 3 M HCl + 5.0 cm Mg ribbon
Tube 6: 4 mL 3 M HCl + 5.0 cm Mg ribbon
Tube 7: 6 mL 3 M HCl + 5.0 cm Mg ribbon
Tube 8: 10 mL 3 M HCl + 5.0 cm Mg ribbon

Work with one tube at a time. (The procedure that follows is one you should practice before going public with your students. The manipulations are not difficult—just a bit awkward.) Hold the tube at an angle to prevent the ribbon from prematurely falling into the acid. Carefully stretch the neck of the balloon containing the magnesium strip over the lip of the culture tube. Then stand the tube upright, so that the magnesium ribbon falls into the acid. Note the relative volume of hydrogen gas generated (as indicated by the size of the balloon) and how much of the original sample of magnesium ribbon is consumed. Repeat for each tube in the set.

Finally, display and compare all the tube-balloon systems. Invite students to interpret the results in terms of the limiting reactant concept. Ask students—after they have observed all tubes carefully—which reactant apparently limited the production of hydrogen in the Part A tubes (1-4). Then invite them to analyze the Part B tubes in a similar fashion. Also help students sense how this demonstration helps to clarify the limiting-nutrient ideas developed in the unit.

## Section D.6

Prepare 100 mL of approximately 0.2 M sodium sulfite (2.5 g $Na_2SO_3$/100 mL). Dip a fresh slice of apple into the sulfite bath for one minute. Remove it and place it on a towel near an untreated "control" apple slice. Continue with your lesson, and view the slices later in the class period.

The treated slice will continue to look fresh long after the control slice begins to darken. Ask students which apple slice they would be more likely to choose at a salad bar (assuming they were unaware of the sulfite treatment). Ask students if they think the use of this additive is justified.

# OTHER ACTIVITIES

## Section B.1 Problems

1. How much energy is released by a burning peanut if a 60-mL sample of water is heated from 15 °C to 38 °C?

   (1 cal/g °C) × (60 mL) × (1 g/mL) × (38 – 15) °C = 1,380 cal = 1.4 kcal = 1.4 Cal

2. How much energy is released by burning 0.2 g of candle wax if the temperature of a 80 g sample of water at 20 °C is heated to 45 °C?

   (1 cal/g °C) × (80 g) × (45 – 20) °C = 2,000 cal = 2.0 kcal = 2.0 Cal

3. How much water could be heated by burning 1 g of paraffin (heat of combustion is about 10,000 cal/g) from 0 °C to body temperature (37 °C)?

   (10,000 cal/g) × (1/37 °C) × ( 1 cal/g °C) = 270 g

4. What would be the final temperature of a 250-g sample of water, beginning at 20 °C, if 1.5 g of propane were burned to heat the water? Heat of combustion of propane is 12.0 kcal/g.

   Energy = 1.5 g × 12.0 kcal/g = 18 kcal = 18,000 cal

   18,000 cal = (250 g) × (temp. change) × (1 cal/g °C) temp.

   change = 72 °C

   72 °C = $T_f$ – 20 °C; $T_f$ = 92 °C

## Section C.2 Homework Extension

Challenge your students to try some at-home experiments on the concept of limiting reactant. Some options to explore include: (a) Vary the amount of baking powder used in making biscuits, and note differences in the characteristics of the final product, and (b) add different amounts of baking soda to a constant vinegar/soap solution and note the amounts of bubbles produced—or keep the amount of baking soda constant and vary the concentration of vinegar.

## Student Reports

1. In developing nations, vitamin A deficiency (xerophthalmia/night blindness) is second only to general malnutrition in its negative effects. Investigate the effects of the deficiency and efforts being made to combat it.
2. Examine the controversy surrounding health claims made about vitamin C and colds (or cancer), or vitamin E and breast lumps, or vitamin $B_6$ and premenstrual syndrome.
3. Research the benefits and risks associated with various sugar substitutes: cyclamates, saccharin, aspartame, and polyhydroxy alcohols (sorbitol and xylitol).
4. Research the controversy surrounding the use of antibiotics in animal feed and their subsequent occurrence in foods.
5. Investigate the rise in the use of refined sugar in the United States between about 1900 and today.

# MACROSCALE LABORATORY ACTIVITIES

Additional instructions for the teacher are found in the corresponding sections of this guide. For teachers preferring the macroscale activity, the student instructions are printed on separate pages, which are ready for duplication. Larger quantities of substances and solutions are often needed for the macroscale activity; in addition, some of the macroscale laboratory activities also require significantly more time than the small-scale versions. Questions for students to answer are found in the student text following the small-scale version of the activity.

## C.4 LABORATORY ACTIVITY: MILK ANALYSIS

### Teacher's Directions

In this activity, students conduct an analysis of the main components of skim milk. They determine the percents of protein, carbohydrate, and water in a sample and calculate its energy content.

### Time

90 minutes

### Materials (for a class of 24 working in pairs)

*Day 1*

12 50-mL beakers
12 150-mL beakers
12 25-mL graduated cylinders
12 short-stemmed funnels
12 ring clamps
12 10-cm dropping pipets (medicine droppers)
12 ring clamps

12 ring stands
12 spatulas
12 dropping bottles for dispensing acetic acid
12 wash bottles
paper towels
48 filter papers (Whatman no.1, 11-cm diameter)
180 mL nonfat milk
50 mL concentrated acetic acid (approximately 17.4 M)

***Day 2***

| Group 1—Protein | Group 2—Carbohydrate/water |
|---|---|
| 24 test tubes | 6 evaporating dishes |
| 6 test tube racks | 6 glass stirring rods |
| 6 100-mL graduated cylinders | 6 10-mL graduated cylinders |
| 6 250-mL beakers | 6 250-mL beakers |
| 6 ring clamps | 6 ring clamps |
| 6 ring stands | 6 ring stands |
| 6 wire gauze | 6 wire gauze |
| 6 Bunsen burners | 6 Bunsen burners |
| 6 dropping bottles | boiling chips |
| 25 mL Molisch reagent | |
| 25 mL concentrated sulfuric acid ($H_2SO_4$) | |
| 50 mL Biuret reagent | |

## Safety

Students should wear safety goggles, aprons, and gloves throughout the laboratory activity. Concentrated acetic acid should be used only with direct teacher supervision. Caution students to swirl carefully (Part 1, Step 4). Part 2, Step 6, using concentrated sulfuric acid, should be done as a teacher demonstration.

## Advance Preparation

Both the Molisch and Biuret reagents may be ordered as prepared solutions. The Biuret reagent may also be available from a neighboring biology teacher.

*Molisch reagent:* Dissolve 1.25 g 1-naphthol (alpha-naphthol) in 100 mL of 95% ethanol. Should be freshly prepared.

*Biuret reagent:* Dissolve 0.25 g copper(II) sulfate pentahydrate ($CuSO_4{\cdot}5H_2O$) in 100 mL solution, producing 0.01 M $CuSO_4$. Also prepare 250 mL of 10.0 M sodium hydroxide (100 g NaOH/250 mL). This solution gets very hot. It should be prepared in a pyrex container sitting on a cooling ice bath. (Store in a plastic bottle to avoid glass-etching.) Prepare the Biuret solution when needed by adding 6–7 mL of the copper sulfate solution (with stirring) to the 250-mL sodium hydroxide solution.

Do not substitute whole milk because separation of the fat from whole milk is time consuming. Sharing class data and finding a class average will simplify the comparisons with standard values for milk at the end of this activity.

## Pre-Lab Discussion

*Exercise extreme caution* when students use concentrated acetic acid and sulfuric acid. Place concentrated sulfuric acid in a suitably labeled dropping bottle or reagent bottle. Instruct students to keep the dropper in a vertical position, so the acid cannot run back into the rubber bulb, and to add the acid slowly to avoid splattering. The test tube should be held at a slight angle, not pointing at anyone. If your students are not especially mature, you may wish to add the sulfuric acid for them.

## Lab Tips

Remove the filter papers containing protein from the funnels after about two hours. Set them on a paper towel to dry overnight. Try not to leave the protein sample in one large clump (low surface area retards drying). Spread the protein thinly on the filter paper with a spatula, as you would butter bread. Leave the appropriate beaker and filtrate next to each filter paper. Although air-drying will not remove all the water from the protein, the mass of water remaining is not large enough to affect the student results significantly.

### Post-Lab Discussion

The post-lab discussion should reexamine the statement that milk is a nearly perfect food in light of student results in this activity. Sharing the class data and finding an average will improve the agreement with given average values.

### Answers to Questions

1. Students should find that their results compare favorably with the accepted values.
2. The milk sample contains other substances such as minerals and vitamins.
3. The calculations in this activity assume that the milk sample is composed of only three components. Thus, the sum should equal 100%.
4. Protein, formed in the same manner as the protein was separated from the carbohydrates in this activity
5. a. $8.0\text{ g} \times \frac{9\text{ Cal}}{\text{g}} = 72$ Cal from fat; $8.0\text{ g} \times \frac{4\text{ Cal}}{\text{g}} = 32$ Cal from protein

   $11.0\text{ g} \times \frac{4\text{ Cal}}{\text{g}} = 44$ Cal from carbohydrates

   Total Cal = 72 Cal + 32 Cal + 44 Cal = 148 Cal

   b. 80 Cal for nonfat milk, which is 68 Cal less than the whole milk sample analyzed. Note that this is almost entirely due to the fat content of whole milk.

## D.2 LABORATORY ACTIVITY: VITAMIN C Teacher's Directions

(Note: The teacher's directions for the Small-Scale Version on pages 141–142 and the Macroscale Version are the same except for Materials, Advance Preparation, and part of the Pre-Lab Discussion, which are given below.)

### Materials (for a class of 24 working in pairs)

12 150-mL beakers
12 25-mL graduated cylinders
12 125-mL Erlenmeyer flasks
12 ring stands
12 buret clamps
12 burets
12 dropping bottle for dispensing 1% starch solution
450 mg vitamin C (ascorbic acid)
200 mL 1% starch solution

### Advance Preparation

*Iodine solution:* Dissolve 10.0 g of potassium iodide and 0.53 g potassium iodate in 250 mL water. Add 60 mL of 3 M sulfuric acid. Add water to make a total volume of 1 L. The equation for the reaction is $IO_3^- + 5\ I^- + 6\ H^+ \rightarrow 3\ I_2 + 3\ H_2O$.

*Starch solution:* Dissolve 2 teaspoons of liquid laundry starch in 200 mL water. Fill 12 dropper bottles.

*Vitamin C solution:* Crush a 100-mg vitamin C tablet and dissolve in water. Dilute to 1,000 mL. The solution is somewhat unstable and should be used within a week of preparation.

*Beverage samples:* Choose five or more beverages to test. The color of dark juices and sodas obscures the endpoint of the reaction, so they are best avoided. If colored juices are used, suggest that all five samples titrated are the same color—Tang, Orange Juice, Hi-C, and so on. Remove any pulp from juices by filtering through cheesecloth.

### Pre-Lab Discussion

Demonstrate the technique of titration, including the reading of the buret.

## C.4 LABORATORY ACTIVITY: MILK ANALYSIS

### Getting Ready

In this laboratory activity, you will determine the percents of protein, carbohydrate, and water in nonfat milk. Once the composition of milk has been determined, you will calculate its food energy value. You will then compare your laboratory results with accepted values for nonfat milk.

The milk analysis will be carried out during two class sessions. You will first separate protein from nonfat milk by precipitating protein and filtering off the solid. Then the quantity of water in nonfat milk will be found by evaporating the water and weighing the remaining milk residue. You will calculate the percent carbohydrate by difference; that is, once you know the percents of protein and water, you will assume the rest is carbohydrate.

### Procedure, Day 1

**Part 1: Removal of Milk Protein**

1. Carefully weigh an empty 50-mL beaker. Record the mass.
2. Add 15 mL of nonfat milk to the beaker. Weigh the beaker and milk. Record the mass.
3. Using the results of Steps 1 and 2, calculate the mass of your milk sample.
4. Add 30 drops of concentrated acetic acid to the beaker containing the nonfat milk. (*Caution:* Keep your nose and face away from the concentrated acetic acid; wash skin immediately if any spills occur.) Swirl briefly, then allow the beaker to sit for 5 minutes. Observe the precipitate forming in the milk. Acid coagulates milk protein, forming the curd you are observing.
5. Weigh one fresh circle of filter paper. Record its mass. Support a short-stemmed funnel on a ring stand above a clean 150-mL beaker. Fold the filter paper and place it in the funnel.
6. Pour the coagulated milk into the funnel. Then add 2 mL of water to the empty beaker. With a spatula, attempt to remove as much of the white material clinging to the beaker walls as possible. Pour these loosened particles into the funnel. Repeat this beaker-cleaning step with a second 2-mL water sample.
7. The filtration will proceed slowly. Place your name on a paper towel next to your ring stand. In several hours, your teacher will remove the filter paper (with protein) from the funnel so the precipitate can air-dry overnight. The filtrate will also be saved for you to test tomorrow.
8. Wash your hands thoroughly before leaving the laboratory.

### Procedure, Day 2

**Part 2: Determining Percents of Milk Protein, Water, and Carbohydrate**

Students at each laboratory table should work in two groups. Group 1 will test and determine the percent protein in the milk. Group 2 will find the percent water in the milk. Each group's results will be shared with the other to provide complete information on the milk analysis. The total mass of nonfat milk is essentially made up of three parts—protein, water, and carbohydrate.

**Group 1:** Determining and testing milk protein

1. Weigh the filtered, dried protein and filter paper from yesterday's laboratory activity. Record the total mass.
2. Calculate the mass of protein collected.
3. Finally, calculate the percent protein in your milk sample. To verify that you actually extracted protein from your milk sample, complete steps 4–9.
4. Label four test tubes 1, 2, 3, and 4. Add 1 mL of milk filtrate (collected in yesterday's filtration) to Tubes 1 and 2. Add two chips of dried milk protein to Tube 3 and to Tube 4.
5. Add 2 mL of Molisch reagent to Tubes 1 and 3. Gently shake to mix.

6. Holding Tube 1 on a 45-degree angle with a test tube holder, slowly add—without stirring or shaking—about 15 drops of concentrated sulfuric acid, allowing the drops to run down the inside surface of the tube. Do not point the open end of the tube toward yourself or anyone else. (*Caution:* Concentrated sulfuric acid is highly corrosive to skin, clothing, books, and other materials. Avoid spilling. Immediately flood skin with tap water if any acid is touched.) Allow the sulfuric acid to form an undisturbed layer on the bottom of the tube.
7. Repeat Step 6, using Tube 3.
8. Observe and record any differences between Tubes 1 and 3. Molisch reagent, in the presence of carbohydrate and concentrated sulfuric acid, will produce a purple layer near the bottom of the test tube. Based on your observations, which tube contains a high level of carbohydrate?
9. Add 5 mL of Biuret reagent to Tubes 2 and 4. If the solution changes to a purple or purplish blue color, protein is present. If no color develops at room temperature, place the two tubes in a hot-water bath, assembled from a 250-mL beaker containing 150 mL of water, supported on a ring stand. Heat the water bath for 2 minutes, but do not allow the water to boil vigorously. Which tube gives a positive protein test?
10. Wash your hands thoroughly. Share your calculated results with Group 2, and obtain Group 2's calculated results. Then move on to Part 3.

**Group 2:** Determining percent of water in milk

1. Determine the combined masses of an empty evaporating dish and stirring rod. Record their combined mass.
2. Add 5 mL of fresh nonfat milk to the evaporating dish.
3. Weigh the evaporating dish, milk, and stirring rod together. Record their combined mass.
4. Calculate the mass of milk in the evaporating dish.
5. Set the evaporating dish and milk on top of a beaker half-full of water. (A 250-mL beaker probably will support the dish securely. Place a boiling chip in the beaker.)
6. Place the beaker–evaporating dish assembly on the wire gauze on a ring stand. Begin heating the water in the beaker to slowly evaporate the liquid from the dish. As a thin "skin" develops on the milk surface, break it with the stirring rod. Prevent the milk from burning by stirring it gently and continuously. Do not allow any milk to transfer from the evaporating dish. (Do not set the stirring rod down on the laboratory bench top, for example.)
7. As the milk loses its moisture, it may resemble liquid paste. When there is no further change in consistency, stop heating. Allow the evaporating dish to cool. Wipe condensed water from the outside of the dish.
8. Determine the combined mass of the evaporating dish, solid milk residue, and stirring rod. Record this value.
9. Calculate the mass of milk solids left in the evaporating dish.
10. Now calculate the mass of water in your original 5-mL milk sample.
11. Finally, calculate the percent water in your milk sample.
12. Wash your hands thoroughly.

### Part 3: Calculating the Energy Value of a Serving of Nonfat Milk

Proteins and carbohydrates deliver the food energy found in nonfat milk. You now have laboratory values for the percents of protein and carbohydrate in a nonfat milk sample. From these data, calculate the food energy delivered by a typical serving of this milk.

Assume we have a 1-cup serving of nonfat milk that has a mass of 244 g. First, calculate the mass of each major nutrient category in this milk sample, then convert these masses to the quantity of food energy each contains. Organize your results in charts similar to this one:

## Data Table

| Mass of Milk | × | % Expressed as Decimal Fraction | = | Mass of Nutrient | Cal/g | Cal from Each Major Nutrient |
|---|---|---|---|---|---|---|
| 244 g | × | ___Protein | = | ____g Protein | 4 Cal/g | ______Cal |
| 244 g | × | ___Protein | = | ____g Protein | 4 Cal/g | ______Cal |
| Total Cal/cup (244 g) of nonfat milk | | | | | | ______Cal |

## Questions

1. Compare your results with the values given below for fresh nonfat milk.

## Data Table

| | Your Laboratory Values | Average Values for Nonfat Milk |
|---|---|---|
| Percent protein | ________% | 4.9% |
| Percent carbohydrate | ________% | 3.3% |
| Percent water | ________% | 91% |
| Total Calories/cup | ________% | 80 Cal |

2. If you add all the "average values" given in the Question 1 data table, you will find they do not total 100%. Can you think of any reasons why? (No fair saying that it is due to poor laboratory results!)
3. Now add up your laboratory values shown in the Question 1 data table. They *should* total 100%. Why?
4. Young children sometimes discover at the breakfast table that they can make milk curdle if they pour orange juice into it. What is the composition of these milk curds? (*Hint:* Orange juice is rich in citric acid and vitamin C, ascorbic acid.)
5. A student analyzed the composition of whole milk and found that 1 cup (244 g) contained 8.0 g fat, 8.0 g protein, and 11.0 g carbohydrate.
   a. How many Calories of food energy are contained in this 1-cup sample?
   b. Compare your calculated Calories for whole milk with the comparable value for nonfat milk. What percent of total Calories are eliminated from whole milk when all the fat is removed?

## D.2 LABORATORY ACTIVITY: VITAMIN C

### Getting Ready

Vitamin C, also known as ascorbic acid, is a water-soluble vitamin. It is among the least stable vitamins, meaning that because it reacts readily with oxygen, its potency can be lost through exposure to light and heat. In this laboratory activity, you will investigate how much vitamin C is in a variety of popular beverages, including fruit juices, milk, and soft drinks.

This analysis for vitamin C is based on the chemical properties of ascorbic acid and iodine. Iodine ($I_2$) solution is capable of oxidizing ascorbic acid, forming the colorless products dehydroascorbic acid, hydrogen ions ($H^+$), and iodine ions ($I^-$):

| | | | | | | | | |
|---|---|---|---|---|---|---|---|---|
| $I_2(aq)$ | + | $C_6H_8O_6(aq)$ | → | $C_6H_6O_6(aq)$ | + | $2\ H^+(aq)$ | + | $2\ I^-(aq)$ |
| iodine | | Ascorbic acid (vitamin C) | | Dehydroascorbic acid | | Hydrogen ions | | Iodide ions |

You will perform a **titration**, a common procedure used for finding the concentrations or amounts of substances in solutions. You will add a known amount of one reactant slowly from a buret to another reactant until just enough for complete reaction has been added. You will recognize the complete reaction by a color change or highly visible change at the **endpoint** of the reaction. Knowing the reaction involved, you can calculate the unknown amount of the second reactant from the known amount of the first reactant.

The titration equipment will be demonstrated by your teacher.

In this analysis, the endpoint is signaled by the reaction of iodine with starch suspension, producing a blue-black product. Starch is added to the beverage to be tested, then an iodine solution is slowly added from the buret. As long as ascorbic acid is present, the iodine is quickly converted to iodide ion, and no blue-black iodine-starch product is observed.

However, when all the available ascorbic acid has been oxidized, the next drop of added iodine solution reacts with starch and you will see the blue-black color. The endpoint in the titration, then, is the first sign of permanent blue-black color in the beverage-containing flask.

You will first complete a titration with a known vitamin C solution, which will allow you to find a useful conversion factor for finding the mass of ascorbic acid that reacts with each milliliter of iodine solution. You can then calculate the mass of vitamin C present in a 25-mL sample of each beverage.

Prepare a data table in your laboratory notebook like the one below.

### Data Table

| Beverage | mL Iodine Solution Used (from Part 2) | × | Conversion Factor (from Part 1) | = | mg Vitamin C per 25 mL Beverage | Rank (Highest to Lowest in Vitamin C) |
|---|---|---|---|---|---|---|
| 1. | | | | | | |
| 2. | | | | | | |
| 3. | | | | | | |
| 4. | | | | | | |
| 5. | | | | | | |

## Procedure

### Part 1: Determining Conversion Factor

1. Measure 25 mL of vitamin C (ascorbic acid) solution into a 125-mL Erlenmeyer flask.
2. Add 10 drops of 1% starch suspension.
3. Fill a clean buret with iodine solution, and record the starting volume.
4. Slowly add iodine solution to the flask as you gently swirl it. Continue until the endpoint is reached (when the first sign of blue color remains after at least 20 seconds of swirling). A piece of white paper placed under the flask will help you recognize the color.
5. Record the final buret volume. Calculate the volume of iodine solution used in the titration.
6. Calculate the number of milligrams of vitamin C that corresponds to 1 mL of iodine solution. This can be found by dividing 25 mg of vitamin C by the volume (in milliliters) of iodine solution used in the titration.
7. Record your calculated value in the "conversion factor" column of your data table. (The factor has the units mg vitamin C per mL $I_2$ solution.)

### Part 2: Analyzing Beverages for Vitamin C

Follow the procedure below for each beverage assigned for analysis by your teacher.

1. Measure 25 mL of beverage into a 125-mL Erlenmeyer flask.
2. Follow Steps 2 through 5 in Part 1. Note that the color of a beverage may affect the endpoint color. For example, red beverage + blue starch–iodine color → purple endpoint.
3. Write the calculated volume of iodine solution used in the titration in your data table. Also, enter the conversion factor value from the data table in Part 1.
4. Use the formula to find the mass (mg) of vitamin C in the sample.
5. Finally, rank by number (1–5) the five beverages, from highest vitamin C level (no. 1) to lowest vitamin C level (no. 5).
6. Wash your hands thoroughly before leaving the laboratory.

## Questions

1. Among the beverages tested, were any vitamin C levels
   a. unexpectedly low, in your opinion? If so, explain.
   b. unexpectedly high, in your opinion? If so, explain.
2. What other common foods contain a high level of vitamin C?

# SUPPLEMENTAL READINGS

"The Pesticide Dilemma." *National Geographic,* February 1980, 145-183.

"Krill: Untapped Bounty." *National Geographic,* May 1984, 626-643. Future source of protein.

"The World's Urban Explosion." *National Geographic,* August 1984, 179-185. Population growth and the rural to urban shift.

"Potential New Crops." *Scientific American,* July 1986, 33-37. Plants that show promise as new sources of food and industrial materials.

"What's So Bad about Oxygen?" *CHEM MATTERS,* February 1986.

"Food Additives ... Who Needs Them?" Chemical Manufacturers Association. Free 11-page booklet which answers questions frequently asked about food additives.

"Polysaccharides." *CHEM MATTERS,* April 1986.

"Say Cheese." *CHEM MATTERS,* February 1995.

"Iron for Breakfast." *CHEM MATTERS,* October 1994.

"Hot Meals." *CHEM MATTERS,* February 1992.

"Your Breakfast (Luncheon, Dinner) Chemicals." Chemical Manufacturers Association. Three, free pamphlets which list the chemicals which occur naturally in typical meals--the point: all foods are made up of chemicals; the term "chemical" should not be indiscriminately associated with "harmful."

"Chocolate." *CHEM MATTERS,* April 1987.

"Chocolate-covered Cherries." *CHEM MATTERS,* April 1987.

"The Salmonella Search." *CHEM MATTERS,* April 1987.

"Egg Cookery." *CHEM MATTERS,* December 1984.

"The Effect of Acid on Eggs." *CHEM MATTERS,* December 1984.

"Poison That Heals." *CHEM MATTERS,* December 1994.

"Chinese Restaurant Syndrome." *CHEM MATTERS,* April 1995.

"Using Concepts of Exercise and Weight Control to Illustrate Biochemical Principles." *Journal of Chemical Education*, 1984, 882-885. Discusses such topics as forms of stored energy, caloric balance, exercise, weight control, physiological effects of endurance training, biochemistry of running a marathon, and exercise programs.

"Crazy Candies!" *CHEM MATTERS,* October 1993.

"Peanut Brittle." *CHEM MATTERS,* December 1991.

"Fast Fat." *CHEM MATTERS,* February 1990.

"Vanilla." *CHEM MATTERS,* February 1988.

"Nutrition (Diet) and Athletics (Chem I Supplement)." *Journal of Chemical Education,* 1984, 536-539. Discusses such topics as nutritional requirements, energy use, carbohydrate loading, and myths and fallacies regarding food and athletic performance.

"Future Food." *CHEM MATTERS,* April 1989.

"Biochemistry Off the Shelf." *Journal of Chemical Education,* 1985, 796-797. Provides sources of inexpensive, readily available, nonanimal biochemical materials for use in experiments.

"Oil Changes." *CHEM MATTERS,* December 1989.

"Making the Grade." *CHEM MATTERS,* December 1989.

"The Vitamin Pushers." *Consumer Reports,* March 1986, 170-175. Discusses the science and the hype behind vitamin supplements and associated advertising "myths."

"Fast Food." *CHEM MATTERS,* February 1990.

"How Good Is Your Breakfast?" *Consumer Reports,* October 1986, 628-637. Rates (and rank orders) 59 cereals in terms of fiber, protein, sugar, sodium, fat, calories, and cost with some surprising results relative to advertising images. The same issue contains two other articles of interest: "What Has All-Bran Wrought?" (638-639) exploring health claims in cereal advertising; and "The Fiber Furor" (640-642) on the state of research on fiber's role in colon cancer prevention.

*FDA Consumer.* Issued 10 times per year, this periodical contains colorful, plain-language articles and features on consumer-oriented FDA activities.

"Salt." *CHEM MATTERS,* December 1992.

"Christmas Tree." *CHEM MATTERS,* December 1989.

"Clarifying Chemical Bonding: Overcoming Our Misconceptions." *The Science Teacher,* March 1991, 24-27.

"Macro Learning on a Micro Scale." *The Science Teacher,* September 1990, 61-65.

"Choosing a Diet Analysis System for Classroom Use." *Journal of Nutrition Education,* June 1984, 73-75.

# TEACHING SCHEDULE

| | TIME BLOCK 1 | TIME BLOCK 2 | TIME BLOCK 3 | TIME BLOCK 4 | TIME BLOCK 5 |
|---|---|---|---|---|---|
| Class Work | Read Introduction<br>Discuss<br>Introduction | Discuss A<br>YD A.1 | Discuss A.2-A.3<br>CQ A.2<br>CQ A.3 | Discuss A.4<br>Prelab LA A.5<br>LA A.5 | YT A.7<br>Prelab A.8<br>LA A.8 |
| Homework | Nuclear Survey<br>Read A.1 | Read A.2-A.3 | Read A.4 | Read A.6-A.7 | Read A.9 |
| | **TIME BLOCK 6** | **TIME BLOCK 7** | **TIME BLOCK 8** | **TIME BLOCK 9** | **TIME BLOCK 10** |
| Class Work | YT A.9<br>SQ Part A<br>Prelab LA B.1<br>Getting Ready for<br>LA B.1 | LA B.1 | LA B.1 | Discuss B.2-B.3<br>YT B.2 | CQ B.5<br>YT B.5 |
| Homework | Read B.2 | | Read B.3 | Read B.4-B.5 | Read B.6 |
| | **TIME BLOCK 11** | **TIME BLOCK 12** | **TIME BLOCK 13** | **TIME BLOCK 14** | **TIME BLOCK 15** |
| Class Work | YT B.6<br>SQ Part B | Review Part<br>A&B<br>Quiz Part A&B | Prelab C.2<br>LA C.2 | YT C.2<br>Prelab C.5<br>LA C.5 | Read C.6<br>YT C.6<br>SQ Part C |
| Homework | Read C.1 | | | Read C.3 | Read D.1-D.3 |
| | **TIME BLOCK 16** | **TIME BLOCK 17** | **TIME BLOCK 18** | **TIME BLOCK 19** | **TIME BLOCK 20** |
| Class Work | Review Part C<br>Quiz Part C | YD D.4 | Discuss D.5-D.6<br>SQ Part D | YD E.1<br>CQ E.1 | YT E.2<br>SQ Part E |
| Homework | Read D.1-D.3 | Read D.5-D.6 | Read E.1-E.2 | YT E.1<br>Read E.3 | |
| | **TIME BLOCK 21** | **TIME BLOCK 22** | **TIME BLOCK 23** | **TIME BLOCK 24** | **TIME BLOCK 25** |
| Class Work | Review Parts D&E<br>Quiz Parts D&E | PIAT F.1-F.2 | Review Nuclear<br>unit | Exam Nuclear<br>unit | |
| Homework | | | | | |

**LA**=Laboratory Activity; **CQ**=ChemQuandary; **YT**= Your Turn; **YD**=You Decide; **PIAT**=Putting It All Together; **SQ**= Summary Questions; One **Time Block**=Approximately 50 minutes

# SUPPLY LIST

## Expendable Items and Unusual Materials

| Quanity | Unit | Item | Sections Used |
|---|---|---|---|
| 100 | mL | 2-propanol | C.5 |
| 1 | | alpha, beta, gamma radiation set | B.1 |
| | | calipers | B.1 |
| 12 | pieces | cardboard | B.1 |
| 12 | | cardboard boxes | C.2 |
| 12 | | cloud chamber kits | C.5 |
| 48 | pairs | disposable safety gloves | B.1, C.5 |
| 1 | lb | dry ice | C.5 |
| 24 | | glass plates, 5 cm x 5 cm x 0.1 cm | B.1 |
| 24 | sheets | graph paper | B.1, C.2 |
| 6-12 | | high-intensity light sources | C.5 |
| 12 | | lead plates, 5 cm x 5 cm x 0.1 cm | B.1 |
| 12 | | meter sticks | B.1 |
| 24 | | objects for boxes | A.5 |
| 960 | | pennies | C.2 |
| 120 | | pre- and post-1982 pennies | A.8 |
| 1-5 | | ratemeters | B.1 |
| 24 | | sealed coded boxes | A.5 |
| 12 | | sealed opaque containers | A.8 |

The Nuclear unit confronts one of the most emotionally charged issues in our society, the use of nuclear energy. The unit traces the history and development of nuclear energy, from the discovery of radioactivity to modern-day reactors and fusion, and explores the benefits and risks of nuclear technologies. The nuclear accidents at Chernobyl and Three Mile Island add an element of urgency to this unit, which probes the structure of the nucleus as well as the question of nuclear waste.

Our goal here is not to discourage debate on the use of nuclear energy by providing the "right" answers, but rather to provide a firm grounding in basic facts and principles related to nuclear science and technology. In this context, it is important to note that neither the technology nor the ethics of nuclear weapons (the most controversial of nuclear technologies) is addressed in this unit.

# LEARNING GOALS

Upon completion of this unit, the student will be able to

1. List the benefits and risks of preserving food by irradiation. [A.1]
2. Distinguish between ionizing and non-ionizing radiation and their differing biological effects. [A.2, B.3]
3. Discuss general properties of electromagnetic radiation, and the energy characteristics of radiation of various regions of the electromagnetic spectrum. [A.2]
4. Describe the experiments of Roentgen, Becquerel, the Curies, and Rutherford, and explain how they led to modifications in the atomic model. [A.3, A.4, A.6]
5. Describe the properties of the three major subatomic particles. [A.7]
6. Define the term isotope, and interpret nuclear isotope notation. [A.7, A.8]
7. Use molar masses and abundance data of isotopes to calculate average relative mass of elements. [A.9]
8. Compare and contrast the general properties of alpha, beta, and gamma radiation, including penetrating power, and discuss safety considerations in terms of shielding abilities of cardboard, glass, and lead. [B.2]
9. Balance nuclear equations, and use them to describe natural radioactive decay. [B.2]
10. Explain the concept of half-life and discuss the implications of half-life for natural radioactivity and nuclear waste disposal. [C.1, C.2]
11. Describe Geiger counters and their operating principles. [C.4, C.5]
12. Define nuclear transmutation, using a nuclear equation to illustrate the process. [C.6]
13. Distinguish nuclear fission from nuclear fusion. [D.1, D.6]
14. Use the equation $E = mc^2$ to compare the energies produced by nuclear fission and by typical exothermic chemical reactions. [D.2]
15. Explain the energy effects of a chain reaction, and compare controlled with uncontrolled reactions. [D.3, D.5]
16. Identify the main components of a nuclear power plant, and describe their functions. [D.5]
17. Assess relative risks and benefits of various nuclear technologies (such as power generation, medical applications, industrial tracing techniques). [B.4–B.6, C.3]
18. List and briefly explain some factors that determine the extent of biological radiation damage. [B.4]
19. Compare the ionizing radiation produced by various sources, including radon, that are encountered by a typical citizen. [B.5, B.6]
20. Discuss the problems and possible solutions associated with nuclear waste generation and disposal. [E.2, E.3]
21. Identify the future hazards associated with long-term storage of high-level nuclear waste. [F]

# A. RADIATION, ENERGY, AND ATOMS

This section begins with a discussion of food irradiation and the controversy that surrounds this process. It goes on to introduce the processes associated with changes in the nuclei of atoms, namely, nuclear radiation, radioactivity, and nuclear energy. Remind students of the general *ChemCom* theme that scientific understanding alone will not provide the answers to complex societal problems and technological issues, but it can provide a more rational basis for public debate and action.

One of the laboratory activities simulates the accidental circumstances that led to Becquerel's observations, later associated with radioactivity. This particular example of serendipity, as well as several others, are discussed in further detail in Section A.3 on pages 300–303. From this, students learn that certain scientific discoveries are accidental and that in searching for the answers to one question, scientists often stumble upon another question that is even more interesting.

Students learn to distinguish among the different kinds of radiation and to describe the damage they do. They trace the discoveries of X rays, as well as alpha, beta, and gamma rays, and they learn about the structure of atoms and about how Ernest Rutherford discovered the atomic nucleus. Isotopes and nuclear notation are introduced.

## A.1 YOU DECIDE: SHOULD YOU EAT IRRADIATED FOOD?

*(page 298)*

Allow 10 minutes for students to discuss and answer the questions in small groups. Then groups may share their results with the class.

1. Questions will vary, but may include some of the following: Will I be eating radioactive chicken? How long will the chicken keep? Why must the chicken carry the international logo if it is not dangerous? Is the chicken more or less expensive than that preserved in other ways? Will it hurt my health in any way?
2. Questions will vary, but may include some of the following: Will workers be exposed to radioactivity? Will radioactivity escape from the plant to the outside surroundings? Could workers carry home radioactivity on their bodies or clothing? Will there be danger to children or future children of workers? What type of health and safety precautions will be taken?
3. Answers will vary according to the type of community, but most communities will have medical facilities that use radioactive isotopes for diagnosis or treatment of illness. Other communities may have a nuclear power plant. Many homes use a radioactive isotope in smoke detectors. Radiation can also be used to gauge thickness of paper or metals. If students are interested in pursuing the uses of radiation in their communities, some might be assigned to interview local citizens who use radiation in their work.

## A.2 KINDS OF RADIATION *(pages 299–300)*

This section introduces students to radiation, its characteristics, and the electromagnetic spectrum. Students learn the difference between ionizing radiation and non-ionizing radiation, as well as some of the uses of each.

We have focused on the energies of the various types of radiation. The ideas of radiation and the electromagnetic spectrum are likely to be new to students and perhaps difficult for them to understand. Provide help by calling attention to the illustrations and analogies in the text.

## CHEMQUANDARY *Always Harmful?* *(page 300)*

"All radiation" includes the entire electromagnetic spectrum, both ionizing and non-ionizing. It is impossible to avoid all exposure to radiation, as that would mean isolation from radio waves, microwave equipment, all lighting sources, and infrared heaters. Most forms of radiation are helpful if they are correctly used but can be harmful if misused. For example, some exposure to sunlight is necessary and healthy, but overexposure can cause painful sunburn and tissue damage. Microwaves cook our food, but shielding is necessary to protect the cook from exposure. The source of nuclear radiation from changes in the atom's nucleus is discussed.

## A.3 THE GREAT DISCOVERY *(pages 300–302)*

In most cases, students are given scientific information with little background as to the hows or whys of the research. This section provides a historical background to the discovery of radioactivity, including the often startling revelation that Becquerel's crucial experiment, the exposure of photographic film by uranium, was something of an accident. However, Becquerel's knowledge and attitude led him to investigate the accident and make an important discovery.

The discovery of radioactivity also demonstrates how scientific inquiries are often interrelated. In this case, Roentgen's X rays led Becquerel to seek X rays from fluorescent materials such as uranium. Becquerel's plates were exposed by the uranium salts. Photographic plates are composed of silver halides (chloride, bromide, or iodide), salts that are sensitive to radiant energy. Any radiation more energetic than infrared radiation will cause a photochemical reaction, leading to silver ions being reduced to silver atoms during development. This reduction results in the darkening of a photographic plate. Thus, uranium salts must have given off some type of radiant energy, as hypothesized by Becquerel.

Some common objects, such as certain types of Frisbees and light sticks, fluoresce or glow in the dark when activated. These objects contain compounds that are excited by ultraviolet light and re-emit the absorbed energy as visible light. Diverse materials such as laundry additives that get clothes "whiter than white" and certain flowers on cloudy, overcast days also illustrate the principle of absorbing ultraviolet light and emitting light of a different energy.

TV screens and computer monitors (CRTs) also demonstrate fluorescence. Their use reinforces the idea of using a beam of electrons to excite certain substances. As the substances return to a more stable state, energy may be released in the form of X rays. TV picture tubes and X-ray machines are both CRTs that have been modified to decrease and increase the emission of X rays, respectively.

Focus attention on the medical benefits of X rays, using Figures 2 and 3 and the students' own experiences.

## CHEMQUANDARY *Scientific Discoveries* *(pages 302–303)*

Each discovery was similar to Becquerel's uranium work in that something unexpected happened. In all cases, the observed events were discrepant with current scientific explanations. Becquerel did not expect the plates to be exposed, nor did Goodyear expect sulfur to harden the rubber. Fleming did not expect a mold to kill bacteria, and Plunkett did not expect to find a polymer of tetrafluoroethene. The point to be made is that careful observation is needed to reveal deeper scientific principles from seemingly ordinary events or mistakes. Pasteur's statement, "Chance favors the prepared mind," is good advice for all scientists, present and future.

These cases could be used for student library research reports.

## A.4 NUCLEAR RADIATION *(pages 303–304)*

Rutherford's investigations into the nature and energy of radioactive particles are presented here. The differences among the three types of radiation are important background for subsequent sections, so make sure your students are familiar with the differences highlighted in this section.

## A.5 LABORATORY ACTIVITY: THE BLACK BOX *(pages 304–305)*

The purpose of this activity is to simulate how scientists use indirect evidence to infer properties that cannot be perceived by our senses. The focus here is on the reasoning processes used.

### Time

30–45 minutes

### Materials (for a class of 24 working in pairs)

Twenty-four sealed, coded boxes (cigar or shoe box), each containing three unconstrained objects such as small scissors, small rubber ball, small fork, plastic comb, large paper clip, or small key. Commercially available black boxes could also be used.

### Pre-Lab Discussion

Discuss how the study of atomic structure is analogous to a black box. A brief review of how Rutherford discovered alpha, beta, and gamma rays is useful here.

### Lab Tips

Using common materials such as test tube holders, flame spreaders, or rubber stoppers is advisable.

### Post-Lab Discussion

Focus attention on how the activity simulated the work of atomic scientists rather than on students who got right answers. Ask students to explain how they arrived at their answers.

### Answers to Questions

1. Hearing
2. The structure of the atom is like a black box, which scientists cannot directly open in order to examine its contents.
3. Indirect evidence is used to formulate explanations (theories) about the natural world whenever the events occur beyond the range of our perceptions. The time for events to occur may be too short (kinetic molecular theory) or too long (evolution), or the objects of interest may be too small (atomic theory) or too large (the universe).

### Demonstration Ideas

An interesting variation on the black box involves suspending a disposable object (key, paper clip, fork, comb, scissors, rubber ball, etc.) inside a sealed shoe box. Cut rigid construction paper strips (white) 50–55 cm long by 7 cm wide to fit through slits in the back of the box, to serve as an imaging screen. Spray paint through a hole in the front of the box to create a "shadow image" of the concealed object. The demonstration reinforces the notion of using a source and a target, and shooting something at the object in the atomic black box to acquire information about its contents.

Another method of identifying an unknown object is to embed the object in modeling clay and try to define its shape or identity by sticking pins into the clay. Commercial Rutherford simulations are also available. These include a piece of Styrofoam positioned under a board; the investigator rolls marbles under the board and notes the trajectory of the marble after it collides with the hidden object.

## A.6 GOLD FOIL EXPERIMENT *(pages 305–306)*

This section provides an example of how experimental observations are sometimes unexpected when they are based upon a current model. The discovery of the nucleus was not the purpose of the gold foil experiment; Rutherford was investigating the behavior of alpha particles with metals. The unusual deflections prompted him to propose a very massive, very small particle—the nucleus.

It is important that students acquire a picture of the physical arrangement of the experiment in order for the resulting deductions to make sense. Focus the discussion on Figures 5 and 6. Point out that not only is there a considerable space between atoms, but most of the volume occupied by an individual atom is essentially empty space. (The actual numbers illustrating these ideas appear in Section A.7.)

## A.7 ARCHITECTURE OF ATOMS *(pages 306–308)*

The components of the nucleus are identified, and isotopes are explained in terms of their relationship to mass numbers. Nuclear notation is used to help identify the number of neutrons in a given isotope. Analogies can help students gain a sense of atomic scale. There are several points to emphasize.

1. Atoms are very small. If people were the size of atoms, the entire population of the world (5.7 billion persons) would fit on the head of a pin.
2. Most of the volume of an atom is essentially empty space. If the nucleus were the size of a billiard ball, the average electron would be located about 1.6 km away.
3. Protons and neutrons account for most of the mass of the atom because their masses are approximately 2,000 times greater than that of electrons. Encourage students to come up with other appropriate analogies.
4. Students may need a review of scientific notation to deal with the real masses of the particles.

Reemphasize the idea that, in an uncharged atom, the number of electrons must equal the number of protons (or the atomic number). The number of neutrons may vary, and the mass of an atom of a given element varies with the number of neutrons it contains. When an ion is formed, the number of protons remains unchanged, but the number of electrons is increased when a negative ion is formed and is decreased when a positive ion is formed. The charge on the ion reflects the number of electrons lost or gained.

## YOUR TURN *Isotope Notation* *(pages 308–309)*

This activity provides practice in using nuclear chemistry notation and illustrates that atoms with higher atomic numbers have a larger neutron/proton ratio than do the atoms of elements with lower atomic numbers.

### Answers to Questions

1.

| Symbol | Name | Number of protons (atomic number) | Number of neutrons | Mass number | Number of electrons |
|---|---|---|---|---|---|
| a. C | carbon | 6 | 6 | 12 | 6 |
| b. N | nitrogen | 7 | 14 | 7 | 7 |
| c. O | oxygen | 8 | 8 | 16 | 8 |
| d. $Mg^{2+}$ | magnesium ion | 12 | 12 | 24 | 10 |
| e. Hg | mercury | 80 | 120 | 200 | 80 |
| f. U | uranium | 92 | 146 | 238 | 92 |

2. a. The lighter elements exhibit roughly a one-to-one ratio between the number of neutrons and protons.
   b. Heavier elements such as mercury (1.5:1 ratio) and uranium (1.6:1 ratio) exhibit higher neutron: proton ratios.

## A.8 LABORATORY ACTIVITY: ISOTOPIC PENNIES *(pages 309–310)*

This activity simulates the determination of isotopic composition of an element. The simulation involves determining the number of pre- or post-1982 pennies in a sealed container of 10 pennies without opening the container. The calculations are typical of the type found in the *Your Turn* section that follows.

### Time

50 minutes

## Materials (for a class of 24 working in pairs)

12 sealed opaque containers (coded) with different 10-coin mixtures of pre-1982 and post-1982 pennies ( 35-mm film canisters make ideal containers)
12 pre-1982 pennies
12 post-1982 pennies
balance

## Advance Preparation

Weigh one of the empty containers. Provide the mass for students to use in their calculations, as specified in the procedure. Fill the containers, using various combinations of pennies. Use a coding system to identify the containers.

## Pre-Lab Discussion

Emphasize that most elements are composed of two or more isotopes that differ in mass. The atomic weight is a weighted average of the mass of the individual isotopes. Explain that a penny minted before 1982 has a different mass from one minted after 1982. Thus, a mixture of pennies from different years simulates the mixture of isotopes in a sample of an element.

## Lab Tips

Avoid using 1982 pennies (unless you use all 1982 pennies), since some have the pre-1982 composition and others are similar to post-1982 pennies. There were seven different mintings of pennies in 1982.

If you want all the pennies to look equally "new," soak them briefly in dilute hydrochloric acid (not nitric acid), then rinse and dry.

Put the pennies in preweighed film containers packed with cotton. Or use a Seal-a-Meal machine to seal the pennies into little bags, after squashing all the pennies together so students cannot see the dates on the individual coins.

To minimize differences in individual pennies, students may weigh 10 pre-1982 pennies and 10 post-82 pennies and use the average masses in their calculations.

## Post-Lab Discussion

To provide a conceptual bridge between the simulation and "real" isotopes, point out again that atomic-weight values for various elements are weighted averages of the masses of each element's isotopes. The calculations are the same as those used in the simulation activity. Depending on the mathematical ability of your students, you may wish to work through a sample calculation.

## Answers to Questions

1. Mass
2. a. The pre- and post-1982 penny mixture represents different isotopes of the fictitious element "coinium." For practical purposes, they look and "behave" the same, yet they have different masses. Similarly, isotopes of the same element behave the same chemically, yet have different masses.
   b. This analogy is limited, in that
      (1) there are only two "isotopes", while actual elements often have more than two;
      (2) in the case of real isotopes, one isotope is sometimes more useful than another—this is especially true in the case of isotopes used in nuclear applications; and
      (3) one cannot weigh real individual atoms using a balance.
3. Eggs (small, medium, large), car tires, soft-drink containers, or any class of objects that come in different masses, but look alike and serve the same function, might be used.

## A.9 ISOTOPES IN NATURE *(page 310)*

This section points out that in nature, most elements occur as mixtures of isotopes; though the various isotopes exhibit the same chemical properties, their masses differ because of varying numbers of neutrons.

## YOUR TURN *Molar Mass and Isotope Abundance* *(pages 311–312)*

This activity is designed to provide students with practice in calculating the average molar mass of an element from isotopic molar masses, and in determining the percent abundances of isotopes from an element's molar mass and isotopic molar masses. The similarity between these calculations and those done in the laboratory activity should facilitate student work on these problems.

### Answers to Questions

1. Student data. Students should use the formula given in the text.
2. Student data
3. The answers to 1 and 2 should be the same.
4. a. 238; the final mass should be closest to the most prevalent isotope.
   b. Average mass = $(238.1 \times 0.9928) + (235.0 \times 0.0071) + (234.0 \times 0.000054) = 238.1$

## PART A: SUMMARY QUESTIONS *(page 313)*

1. Characteristics: energy with no mass, travels at the speed of light, travels through a vacuum, emitted by decaying or energized atoms, consists of packets of energy called photons, higher frequencies have higher energy. Examples: infrared, ultraviolet, visible, gamma rays, X rays, microwaves, radio waves.
2. Roentgen's discoveries of X rays and fluorescence prompted many others to study these phenomena. Becquerel was studying the fluorescence of minerals, including that of uranium, and his work led to the discovery of radioactivity. When Rutherford experimented with radioactivity, he was able to develop the model of the nuclear atom.
3. Fluorescence depends on the absorption of ultraviolet light that is subsequently re-emitted as visible light. Because the uranium compound was able to cause blackening of the photographic plate without having been exposed to UV radiation from sunlight, the radiation from the compound had nothing to do with fluorescence. The penetrating power of this new radiation was observed to be much greater than X rays generated by a cathode ray tube (see Section A.3, page 300).
4. See Section A.4 (page 303). Simply, two components of the beam emitted by the radioactive source were influenced by a magnetic field. This response is typical of charged particles but not of electromagnetic radiation.
5. See Section A.6 (page 305).
   a. The majority of alpha particles passed straight through gold foil.
   b. A small percent of the alpha particles were reflected back at large angles.
6. Because a proton ($1.67261 \times 10^{-24}$ g) is about 1,836 times more massive than an electron ($9.1095 \times 10^{-28}$ g), it takes approximately 2,000 electrons to equal the mass of one proton.
7.

| Symbol | Name | Atomic number | Number of protons | Mass number | Number of neutrons |
|---|---|---|---|---|---|
| C | carbon | 6 | 6 | 12 | 6 |
| Co | cobalt | 27 | 27 | 60 | 33 |
| Pb | lead | 82 | 82 | 207 | 125 |

8. a. If the average molar mass of the element potassium is 39.098 g/mol, then the most abundant isotope must have the molar mass of 38.964 g/mol.
   b. Its molar mass is closest to the average as compared with the other isotopes. The element name and mass number is potassium-39.
9. a. The average molar mass of the element neon has to be closer to 20 g/mol, because the natural abundance of the neon-20 isotope with a molar mass of 19.992 g/mol is over 90%.
   b. Average molar mass of Ne = $(0.9051 \times 19.992) + (0.0922 \times 21.991) = 29.179$ g/mol
10. B-11, because the molar mass is closer to 11 g/mol than it is to 10 g/mol
11. Atomic weight, and atomic number or chemical symbol

## EXTENDING YOUR KNOWLEDGE (OPTIONAL) *(page 313)*

Students can use references such as the *Readers' Guide to Periodical Literature* and the *McGraw-Hill Encyclopedia of Science and Technology* to find fairly recent information on these particles. Many students will be able to search the Internet, and most public libraries have computerized searching programs, such as InfoTrac. A clearer understanding of these particles might help to better explain questions ranging from the origin of the universe to ways of dealing with nuclear waste.

# B. RADIOACTIVE DECAY

In Part B, students learn the nature of the three most common types of nuclear radiation: alpha, beta, and gamma. The effects of radiation on humans and the presence of radon in homes are discussed. The rem is introduced as a unit used to measure radiation effects on humans.

## B.1 LABORATORY ACTIVITY: α, β, AND γ RAYS *(pages 314–318)*

This activity is designed to acquaint students with emissions from radioactive materials and with the ability of materials to serve as radiation shields. In Part 1, students compare the penetrating power of alpha, beta, and gamma radiation through cardboard, glass, and lead. In Part 2 they determine the effect of distance on radiation intensity, and in Part 3 they compare the radiation-shielding effects of lead and glass. As they account for background radiation in their observations, students also discover that their everyday environments contain natural radiation.

If equipment is in short supply, this activity can be presented as a teacher-led demonstration. Because the activity covers a number of important concepts, however, students should at least witness a demonstration.

### Time

50–90 minutes

### Materials (for a class of 24 students)

1–5 ratemeters
Alpha, beta, gamma radiation source set
12 meter sticks
12 lead plates (5 cm × 5 cm × 0.1 cm)
24 glass plates (5 cm × 5 cm × 0.1 cm)
12 pieces of cardboard (5 cm × 5 cm × 0.1 cm)
calipers (optional)
12 sheets of graph paper
12 forceps or crucible tongs (for isotope handling)
12 pairs disposable safety gloves

### Advance Preparation

Check the background radiation before students begin the activity. If student readings are questionable, have them take more readings.

### Lab Tips

Radioactive sources sealed in plastic discs are available from most suppliers of school science materials. The American Nuclear Society, Civil Defense Authority, or college or university physics departments may have Geiger counters available for loan.

The thickness of the lead plates is easily determined by standing the plate on a centimeter scale. Most plates available for this purpose are 1 mm thick.

Because even a few centimeters of air will block alpha particles, the thick glass on the probe of the ratemeter will not allow alpha particles to be detected. You can order a very thin glass plate or totally remove the glass plate on the end of the ratemeter probe so that students can discover how a few centimeters of air act as a barrier to alpha particles.

Great care must be taken to ensure that the radioactive materials you and your students will be handling constitute no danger. Although the commercially available radioactive materials will be encased in plastic when they are purchased, share the following precautions with the students.

## Safety Precautions

1. Always wear rubber or plastic gloves and use tongs when handling radioactive materials.
2. Do not bring food into the room when working with radioactive materials (or at any other time). Food can be easily contaminated during handling and could result in the ingestion of radioactive materials.
3. Be sure that no extraneous radioactive materials come in contact with your radiation counter. The counter can become contaminated with radiation and record higher than normal background readings.
4. Collect all materials used in the activity in an appropriate storage container and check for radiation before disposal.
5. Check hands with a radiation monitor before leaving the laboratory. If the monitor reading is high, the hands should be rewashed.

## Part 1 Answers to Questions

1. Lead greatly reduces the level of alpha, beta, and gamma radiation. Cardboard greatly reduces the alpha radiation. Glass reduces the alpha radiation, but is ineffective against beta and gamma radiation.
2. Gamma radiation is the most penetrating; alpha radiation is the least penetrating. Gamma radiation is without mass and is classified as electromagnetic radiation. Alpha particles are the most massive and can be retarded easily by collisions with other particles.
3. a. Lead
   b. Cardboard
4. Density and thickness

## Part 2 Answers to Questions

1. Four
2. No
3. The intensity varies inversely with the square of the distance ($I = k/d^2$).

## Part 3 Answers to Questions

1. a. Doubling the shield made little difference.
   b. Doubling the shield reduced the intensity of the radiation.
2. Lead, because it is most effective at blocking radiation, including electromagnetic radiation such as X rays
3. Gamma radiation is the most dangerous to the body because it has the greatest penetrating power and the greatest energy per unit; thus it can cause the most cellular and molecular disruption. Alpha radiation poses a serious problem if it is taken into the body through air passages or breaks in the skin. Although it has less penetrating power, it has greater ionizing potential.

## Post-Lab Discussion

Focus on the data collected in the activity and the answers to the questions.

## B.2 NATURAL RADIOACTIVE DECAY *(pages 319–321)*

The characteristics of alpha, beta, and gamma radiation are explained. Given that radioactive decay is a dynamic process occurring below the limits of human vision, computer and film simulations can be effective aids to enhance student understanding (see the listing provided in this guide).

In addition, students are introduced to the concept of balanced nuclear equations. In the Resources unit, students learned that mass is conserved in chemical equations. Emphasize that mass numbers are conserved in nuclear equations, and while elements may change in a nuclear reaction, the number of protons is also conserved. After working through examples of nuclear equations, draw attention to Table 6 on page 320, which summarizes radioactive decay and the nuclear notation used.

**YOUR TURN** ***Nuclear Balancing Act*** *(pages 321–322)*

This activity provides the students with practice in balancing nuclear equations. Explain that during beta decay, a neutron is transformed into a proton and an electron. Since the new proton remains in the nucleus, the atomic number increases by 1 while the mass number remains the same. (The mass number remains constant because the number of neutrons is decreased by 1 and the number of protons is increased by 1. The consequence is the formation of a different element.)

**Answers to Questions** *(page 322)*

1. a. ${}^{0}_{-1}e$

   b. ${}^{4}_{2}He$

2. a. ${}^{4}_{2}He$

   b. ${}^{228}_{89}Ac$

   c. ${}^{228}_{89}Ac, {}^{228}_{90}Th$

   d. ${}^{224}_{88}Ra$

   e. ${}^{224}_{88}Ra$

## B.3 IONIZING RADIATION—HOW MUCH IS SAFE? *(page 323)*

This section defines the rem, a unit for measuring the ability of radiation to cause ionization in human tissue, and discusses how radiation can cause tissue damage.

## B.4 RADIATION DAMAGE: NOW AND LATER *(pages 323–325)*

This section considers the biological effects of ionizing radiation on human tissue, as well as the link to cancer. Focus student attention on Tables 7 and 8 on page 324. The text outlines the uncertainties involved in estimating the long-term effects of exposure to low dose levels (below 25 mrem). This is related to the similar discussion concerning the use of food additives in the Food unit.

One problem with assessing radiation damage is that radiation does not produce unique effects that can be unequivocally distinguished from the results of other injuries or illnesses. Cancer is all too common: one in five persons will die of cancer from a variety of causes, and experts estimate that only about 2% of cancer deaths are due to radiation. Contrast this with 30% of all cancer deaths attributed to voluntary cigarette smoking, 3% to alcohol consumption, and 2% to asbestos exposure.

## B.5 EXPOSURE TO RADIATION *(page 325)*

This section clarifies sources of radiation and presently accepted levels of exposure. This material and Section B.6 provide the background needed for the *Your Turn* activity. The point is made that we are all regularly exposed to natural low-level background radiation of about 135 mrem/yr and that, through various conscious or unconscious decisions (where we choose to live, what medical treatment we receive, etc.), we are exposed to varying amounts of additional radiation generated as a result of human activities. Note that just because human additions to natural background radiation seem small (with the exception of medical X rays—see Figure 16 on page 326) does not mean they are unimportant in terms of potentially negative health effects. Since each individual is different, the effect of equal exposure will vary for different persons.

## CHEMQUANDARY *Radiation Exposure Standards* *(page 326)*

Certain individuals who have more rapidly dividing cells, such as children and pregnant women, are more at risk from a given dose level than the general population. On the other hand, for some people needing the benefits of radiation treatment, higher doses may be worth the risks. Also, for practical (including economic) reasons, occupationally exposed individuals (such as medical X-ray technicians and workers handling radiopharmaceuticals, defense program workers, and civilian workers in nuclear power plants) are allowed to receive higher, though apparently still safe, doses.

## YOUR TURN *Radiation from Foods* *(pages 326–327)*

This *Your Turn* gives students the opportunity to calculate the quantity of radiation derived from just one isotope, potassium-40, present in food. They also consider the risks involved in eliminating foods high in potassium from their diet. The main point of this *Your Turn* is for students to realize that radioactivity is all around them, even in the food they eat.

### Answers to Questions *(pages 326–327)*

1. Student data
2. Calculate by multiplying the number of servings of each food by the d/s (disintegrations per second) per serving.
3. a. $\frac{\text{60 days}}{\text{servings/kg}}$ student weight in lb $\times \frac{2.2\text{ kg}}{\text{lb}} = \frac{\text{d}}{\text{s}}$

   b. Some potassium-40 is excreted. Values given for servings are approximate.
4. Potassium is necessary for proper functioning of the body. It is particularly important in maintaining heart rhythms.

## B.6 RADON IN HOMES *(pages 327–328)*

The radon issue is a particularly interesting one to explore with students for the following reasons: (1) It is a recent problem that has only received widespread attention since the mid-1980s. (2) It is a case where nature, not humans, is the source of the pollution. (3) It illustrates how solving one problem (high home energy bills) can sometimes help create a new one. (4) It provides a good context for discussing risk-benefit analysis and the concept of comparative risks.

## YOUR TURN *Your Annual Radiation Dosage* *(pages 328–329)*

This activity heightens student awareness of the sources and quantities of radiation in their everyday environment. They are asked to estimate the dosage of radiation received per year, analyze the sources, and consider ways to reduce the quantity of their annual radiation exposure. "Personal Radiation Dose Charts" are available in quantity at special prices from the American Nuclear Society, Public Communications Department, 555 North Kensington Avenue, La Grange Park, IL 60525, (708) 579-8265.

### Answers to Questions

1. a. Individual student answers

   b. Individual student answers, comparing their values with the standard 135 mrem.
2. a. Yes; some ways are practical, such as avoiding unnecessary X rays by keeping records of medical and dental X rays, inquiring about the necessity of each, and being sure that technicians use lead to shield body parts not being photographed. Other ways to reduce your exposure are harder to implement—flying less (which may involve switching to more accident-prone automobile transportation), moving to lower elevations, or switching to a wood home, etc.

   b. Individual answers; in many cases, such as choice of location or type of home, students have no say in the choice.

## PART B: SUMMARY QUESTIONS *(page 330)*

1. a. Beta
   b. Gamma
   c. Alpha
2. a. $^{60}_{27}Co \rightarrow \, ^{60}_{28}Ni + \, ^{0}_{-1}Ni$
   b. $^{40}_{19}K \rightarrow \, ^{40}_{20}Ca + \, ^{0}_{-1}e$
   c. $^{226}_{88}Ra \rightarrow \, ^{222}_{86}Rn + \, ^{4}_{2}He$
3. Ionizing radiation such as alpha, beta, or higher-energy electromagnetic radiation (high UV and above on the spectrum) causes disruption of molecular bonds in cellular materials. Non-ionizing radiation (visible, infrared, and the lower end of the spectrum) is not sufficiently energetic to effect such changes.
4. Rem (roentgen equivalent man) is a measure of the radiation's power to ionize human tissue.
5. Alpha particles have little penetrating power. If the source is outside of the body, the skin blocks their entry. However, if the source is internal, the alpha particles can travel short distances and cause damage to surrounding tissue. If an alpha emitter is ingested or inhaled and if its chemical behavior causes it to be concentrated by the body, more severe damage can result.
6. (1) Length of time of exposure; (2) area of body and type of tissues exposed; (3) density of ionization; (4) dose, or quantity of radiation received.
7. a. Benefit—radiation treatment for cancer
   b. Risk—undesirable radiation exposure for fetuses and infants
8. a. Problems—lack of good experimental data, complication of long-term effects, and the difficulty of isolating radiation-induced effects from other environmental factors
   b. Without better data, regulations may err on the side of safety and not assume a threshold between no effect and a hazardous effect.
9. a. Minimum dose for immediate observable effects: 25,000 mrem (25 rem)
   b. Maximum allowable dose for radiation workers: 5,000 mrem (5 rem)
   c. Maximum allowable dose for individual: 170 mrem (0.17 rem). Both the occupational and individual allowances are substantially less than the minimum short-term effect dose to allow for the possibility of delayed, long-term effects. Radiation workers are allowed higher levels, at least partially, because of the practical difficulties of protecting them at the 170-mrem level.

# C. RADIOACTIVITY: NATURAL AND ARTIFICIAL

This section begins with discussion and a laboratory activity on the concept of half-life. It continues by introducing Geiger counters as a type of radiation detector. Medical benefits of radioactive tracers are discussed, and the section concludes with transmutation of elements and artificial radioactivity.

## C.1 HALF-LIFE: A RADIOACTIVE CLOCK *(pages 331–332)*

The concept of half-life is introduced. Students may have heard the term but may not thoroughly understand it. Some students may think that one half of the material changes in one half-life and the other half in the next half-life. They may also believe that the atom "disappears" rather than being changed into another isotope. Begin any explanation using whole numbers of half-lives so that the concept is not obscured by attention to difficult mathematics. The concept of half-life is introduced through a simulation. Another activity demonstrates the use of the cloud chamber. Beneficial uses of radioisotopes are discussed.

## C.2 LABORATORY ACTIVITY: UNDERSTANDING HALF-LIFE *(pages 332–333)*

This activity models the exponential decay curve of radioactive samples through several half-lives. The "decaying" atoms are pennies, and decay is represented when a coin flips from heads to tails. The simulation makes the point that because atomic decay is a statistical matter, we can never be sure which particular atom will survive and which will decay at any given time.

### Time

30 minutes

### Materials (for a class of 24 working in pairs)

960 pennies
12 cardboard boxes
12 sheets of graph paper

### Advance Preparation

Set up 12 containers, each holding 80 pennies.

### Pre-Lab Discussion

Avoid discussing half-life until after the completion of the activity.

### Post-Lab Discussion

Now is the time to discuss the concept of half-life. The graphs made by the students represent a radioactive decay curve. Make sure that students understand the random, statistical nature of radioactive decay. One cannot know when a given atom will decay but can predict the rate of decay for a large number of atoms. Once again, various computer and film simulations can be helpful.

### Answers to Questions

1. a. The graph slopes downward as a curved line.
   b. The pooled class data should produce a smoother curve and be more reliable, representing a larger number of decayed atoms.

2.

| Number of half-lives | Headsium nuclei remaining |
|---|---|
| 0 | 600 |
| 1 | 300 |
| 2 | 150 |
| 3 | 75 |

3.

| Number of half-lives | Headsium nuclei remaining |
|---|---|
| 0 | 2,800 |
| 1 | 1,400 |
| 2 | 700 |
| 3 | 350 |
| 4 | 175 |

4. The simulation is similar to the actual radioactive decay, in that the graph results model the half-life relationship.
   This simulation, however, involves only 80 pennies, and the half-life of any radioisotope is based upon a much larger number of particles. The decay of nuclei is a chance occurrence, as observed in this simulation. The averaging of several results from the simulation helps adjust for this fact. Also, identification and separation of real radioactive atoms are not as simple as this simulation might suggest.
5. To model the half-lives of different isotopes, use an object with more than two sides, such as dice or sugar cubes.
6. a. Four half-lives
   b. (1) Yes (2) Yes
7. a. No
   b. A half-life value only predicts how long it will take for half of a sample of atoms to decay, not which ones.
8. If you have $1,000 and spend half of the money remaining every 10 days, the half-life of the money is 10 days. Students will suggest other ways.

## YOUR TURN *Half-Lives* *(pages 334–335)*

This activity provides practice in calculating half-lives.

## Answers to Questions

1. a. Fifth year (first year $500, second year $250, third year $125, fourth year $62.50, fifth year $31.25)
   b. $0.98 (sixth year $15.62, seventh year $7.81, eighth year $3.91, ninth year $1.95, tenth year $0.98)
2. a. 750 mg cobalt-60. No; the rate of decomposition is independent of usage and depends only on time.
   b. (1) Four (50%–one half-life, 25%–two half-lives, 12.5%–three half-lives, 6.25%–four half-lives)
   (2) Seven (3.125%–five half-lives, 1.563%–six half-lives, 0.7813%–seven half-lives)
   (3) Ten (0.3906%–eight half-lives, 0.1953%–nine half-lives, 0.09766%–ten half-lives)
3. a. Theoretically, the person never reaches the telephone booth because one half of any distance leaves a distance yet to be traveled to the destination, even though that distance becomes vanishingly small.
   b. See the following table:

| Distance(cm) | Time(min) | Percent remaining |
|---|---|---|
| 51,200 | 0 | 100 |
| 25,600 | 1 | 50 |
| 12,800 | 2 | 25 |
| 6,400 | 3 | 13 |
| 3,200 | 4 | 6.3 |
| 1,600 | 5 | 3.1 |
| 800 | 6 | 1.6 |
| 400 | 7 | 0.78 |
| 200 | 8 | 0.39 |
| 100 | 9 | 0.20 |
| 50 | 10 | 0.05 |

4. a. (1) Zero–1963; one–1992; two–2021; three–2050; four–2079; five–2108
   (2) 1963–100%; 1992–50%; 2021–25%; 2050–12.5%; 2079–6.25%, 2108–3.13%
   (3) Graph
   (4) Graph
   b. Values can be read from graph; 50% in 1992
   c. More than 3%

## C.3 BENEFITS OF RADIOISOTOPES *(pages 335–336)*

The medical benefits of radioisotopes—including diagnostic tracers, high-energy emitters for cancer treatment, and sterilization of equipment—are discussed in this section.

## C.4 RADIATION DETECTORS *(pages 337–339)*

This section discusses two types of radiation counters: ionization and scintillation. A Geiger counter is a kind of ionization counter that can detect alpha, beta, and gamma particles. Scintillation counters emit light when radiation strikes them.

## CHEMISTRY AT WORK/High-tech Soldiers in the War Against Cancer *(page 338)*

A radiation oncologist uses tools from chemistry, as well as many other fields, to work with patients who have cancer. Students can find out more about this career by contacting hospitals, which often have in-house training programs, community colleges, or their own high school career counseling center.

## C.5 LABORATORY ACTIVITY: CLOUD CHAMBERS *(pages 339–340)*

In this activity, students construct a simple device for indirect observation of ionizing radiation. The objective is to use a simple cloud chamber with supersaturated 2-propanol (isopropyl alcohol) vapor and observe the condensation trails generated by the radiation source.

### Time

45 minutes

### Materials (for a class of 24 working in pairs)

12 cloud chamber kits (see note below)
100 mL of 2-propanol
1 lb of Dry Ice (thin slabs)
6–12 high-intensity light sources (20 W or greater)
12 squeeze bottles for 2-propanol
24 pairs of disposable gloves
12 crucible tongs

## Advance Preparation

You will need to obtain cloud chambers and Dry Ice. Cloud chamber kits, including the radiation source, may be purchased or may be made from transparent plastic disposable boxes (the kind used for salads in grocery stores or fast-food restaurants). Sometimes Dry Ice can be purchased from ice cream stores. Check the Yellow Pages to find a local source.

Place the radioactive source in the cloud chamber with forceps. Do not allow students to handle it.

## Pre-Lab Discussion

Remind students of the Black Box Activity (A.5), which also used indirect methods of detection.

## Lab Tips

Students need only to dampen the felt band (on the top cover) with alcohol, avoiding direct contact with the radiation source. If available, a small piece of old lantern mantle (newer ones do not contain thorium) can be used as a source. The Dry Ice should be crushed unless you have a smooth block. The objective is to get good contact between the Dry Ice and the bottom of the cloud chamber. The light source should be directed at an oblique angle rather than straight down.

Practice is necessary for this activity. Rubbing the outside of the chamber with a silk cloth to promote a charged atmosphere may also help.

## Answers to Questions

1. The length, direction, and intensity of the tracks will differ owing to variations in temperature.
2. Alpha particles will produce more visible tracks owing to their large mass and size and their relatively slow speed.
3. Since the Dry Ice sublimes directly from the solid to the gaseous state, it absorbs heat from the surroundings, including the gases inside the closed chamber. Thus, the Dry Ice forms a heat sink, reducing the temperature of the gases inside the closed chamber. This allows the already vaporized alcohol to supersaturate the air. The supersaturated gaseous mixture is unstable and will condense to a liquid with the slightest disturbance, such as the introduction of an ion speeding through the chamber.

## Post-Lab Discussion

Discuss the conditions that support the direct observation of ion vapor trails (supersaturation) and the importance of cloud chambers even in current research. Discuss the answers to the questions.

## C.6 ARTIFICIAL RADIOACTIVITY *(pages 340–342)*

A brief history of the development of nuclear chemistry is given, beginning with Rutherford's experiments with alpha particles and transmutation, which had been the alchemists' goal. The phosphorus-30 produced by the Joliot-Curies had a half-life of 3.5 minutes, changing into silicon-30 by the elimination of a positron (an electron with a positive charge). The transuranic elements have been produced by bombarding heavy nuclei with particles such as helium nuclei (*Your Turn* question 3). Most recently, a few atoms of elements 110 and 111 have been produced by bombarding bismuth-209 nuclei with relatively heavy particles such as nickel-64.

## YOUR TURN *Bombardment Reactions* *(pages 342–343)*

1. ${}^{59}_{27}\text{Co} + {}^{1}_{0}\text{n} \rightarrow {}_{27}\text{Co}$

2. ${}^{96}_{42}\text{Mo} + {}^{1}_{1}\text{H} \rightarrow {}^{97}_{43}\text{Tc} + {}^{0}_{1}\text{n}$

3. $^{209}_{83}Bi + ^{58}_{26}Fe \rightarrow ^{266}_{109}Mt + ^{1}_{0}n$

## CHEMQUANDARY *Transmutation of Elements* *(page 344)*

Lead and iron could be changed into gold; however, the method would probably be far too expensive to be profitable. Reactions of the type used to produce elements 109–114 could possibly be used.

## PART C: SUMMARY QUESTIONS *(page 344)*

1. 15 minutes (1/4 hour)
2. 6.3 μg
3. a. Graph
   b. (1) 12.6 mg, (2) 0.75 mg, (3) less than 0.5 mg
   c. The jewelry would be radioactive, and it would no longer be made of gold because the gold atoms change to atoms of another element when they decay.
4. An isotope used to detect abnormal functioning in the body
5. a. Diagnostic purposes and cancer treatment
   b. Isotopes that have short half-lives decay rapidly and leave the body of the patient more quickly. Those used externally can have longer half-lives because the time the patient is in contact with the radiation can be monitored.
6. Alpha particles do not have great penetrating power. The relatively long half-life means fewer particles are given off over a given interval, and the detector does not have to be changed as often.
7. Scintillation counter, solid-state detector, cloud chamber
8. Alpha, beta, gamma
9. a. Target is cerium-242; projectile is helium-4; product is californium-245; ejected particle is a neutron.
   b. Target is uranium-238; projectile is a neutron; product is neptunium; ejected particle is an electron.
10. The uranium-238; those with shorter half-lives would decay too quickly. Space probes are generally in use for a longer time.

## EXTENDING YOUR KNOWLEDGE (OPTIONAL) *(page 345)*

- A cloud chamber would not be expected to show neutrino trails. Whenever a charged fast-moving particle passes through a gas, ionization is produced. The neutrino is not charged.
- Many students will be able to search the Internet, and most public libraries have computerized searching programs, such as InfoTrac. Students will also find information about these topics in college chemistry and physics texts that are found in many classrooms.
- $^{235}_{92}U \rightarrow ^{207}_{82}Pb + 7\,^{4}_{2}He + 4\,^{0}_{-1}e$ All of the mass loss (28 units) must be accounted for by adding alpha particles requiring seven. The charge is four units too high, which requires four beta particles.

# D. NUCLEAR ENERGY: POWERING THE UNIVERSE

Part D introduces students to the huge amount of energy that is locked inside the atom. They learn about the strong force (the force that holds together the nucleus) and see how nuclear energy is released via chain reactions in nuclear power plants and in nuclear weapons. Major components of a nuclear plant are pictured and discussed. Students see the enormous difference in the magnitude of nuclear energy compared with chemical energy. Fusion—the nuclear reaction that powers the sun and the stars, and a potentially potent source of energy on Earth—is also introduced.

## D.1 SPLITTING THE ATOM *(pages 346–347)*

The brief historical sketch about the first nuclear fission reaction introduces Lise Meitner, an Austrian Jew who worked with Hahn and Strassman but was forced to flee from Nazi Germany. Hahn and Strassman found what appeared to be barium when uranium was bombarded with neutrons. They were unable to explain the presence of barium and sent a letter to Lise Meitner, who was then residing in Sweden. Meitner and her nephew, Otto Frisch, proposed that barium was one of the products formed when the uranium atom split into two parts of similar size. In addition, they predicted that krypton would also be formed, a prediction that was later verified by Hahn and Strassman.

## D.2 THE STRONG FORCE *(page 347)*

Students are introduced to the strong force that holds the protons and neutrons together in the nucleus. Energy changes in nuclear reactions are compared and contrasted with those of chemical reactions. You may want to emphasize that a small quantity of mass, if converted into energy, yields an enormous quantity of energy, according to the equation $E = mc^2$.

## D.3 CHAIN REACTIONS *(pages 347–348)*

The terminology of nuclear reactions is introduced and explained. Use Figure 20 and the following *You Decide* activity as the focal points for your discussion. You may wish to point out that the term "critical mass" refers not only to nuclear bombs and explosions, but also to controlled chain reactions.

## D.4 YOU DECIDE: THE DOMINO EFFECT *(pages 348–349)*

This activity demonstrates the concept of expanding versus limited chain reactions. You will need 2–3 sets of dominoes for a demonstration, or as many as 30–45 sets for a student-centered activity.

### Answers to Questions

1. (No answer required)
2. a. (No answer required)
   b. This is an expanding (out of control) chain reaction because the number of falling dominoes increases with each row.
3. (No answer required)
4. This is a limited (controlled) chain reaction because the number of falling dominoes does not increase as it did above.
5. a. The expanding chain reaction models the explosion of a nuclear weapon, in which a tremendous amount of energy is released in a very short time.
   b. The limited chain reaction models a nuclear power plant. In this case the energy is released over a longer period of time and is kept under control.

6. a. The limited reaction may be stopped by removing one or two of the dominoes in the sequence.
   b. It is nearly impossible to stop the expanding reaction, since you would have to remove a whole row of dominoes at one time, instead of one or two. The farther the expanding reaction proceeds, the lower the probability of stopping.
7. a. This simulation is like a nuclear chain reaction, in that the falling dominoes (representing fissioning atoms) release energy, which triggers other dominos to fall.
   b. The simulation is limited, in that it does not produce new substances or produce significant quantities of energy.
   c. Other modeling alternatives include loaded mouse traps and Ping-Pong balls in a closed container.

## D.5 NUCLEAR POWER PLANTS *(pages 349–353)*

This section deals with the basic components and operation of nuclear power plants. The functions of plants are explained, with diagrams and text. Several different types of reactors are in use: pressurized water, boiling-water, high-temperature gas-cooled, and European breeder reactors. The text explains pressurized and boiling-water reactors.

## D.6 NUCLEAR FUSION *(pages 353–354)*

This section explains nuclear fusion and introduces positrons. The tremendous energy potential, as well as difficulties with fusion plants, are explained. The promise of generating electricity via nuclear fusion is that fusion reactors will produce neither the troublesome air pollutants of fossil fuel plants, nor the long-lived fission fragments and transuranic elements resulting from fission reactors. However, fusion reactors will contain a large inventory of radioactive tritium, and during operation, structural components will become radioactive. Mid-1970s forecasts of full-scale fusion power plants by the year 2020 are now considered unrealistically optimistic by most experts. Students may prepare reports on the history and present state of fusion research for extra credit.

## CHEMISTRY AT WORK/Nuclear Chemistry at Full Speed Ahead *(page 355)*

A naval officer, in charge of the water circulating system for a nuclear reactor, tests the water using methods that are familiar to students.

## PART D: SUMMARY QUESTIONS *(page 356)*

1. a. Uranium-235, uranium-233, plutonium-239, and californium-252
   b. Radioactive decay is a naturally occurring process, resulting in isotopes that are more stable than the original one. Nuclear fission can also be artificially induced by neutron bombardment.
2. $^{235}_{92}U + ^{1}_{0}n \rightarrow ^{87}_{35}Br + ^{146}_{57}La + 3^{1}_{0}n$
3. a. A chain reaction is a self-propagating reaction in which the products of one reaction initiate the next. When fissionable material such as uranium-235 is bombarded with neutrons, its atoms are split into lighter isotopes, producing more neutrons. These neutrons split other uranium-235 nuclei, thus setting off a chain reaction.
   b. A chain reaction will occur when the fissionable mass captures neutrons, creating the chain reaction. The particular mass in this quantity of fissionable material is termed the *critical mass.*
   c. An expanding chain reaction cannot be stopped and will lead to an explosion, such as in a nuclear weapon. In a controlled reaction the process is slowed down to a manageable speed, mainly by absorbing neutrons and lowering the percentage of fissionable material. Therefore, a nuclear power plant cannot explode like a nuclear weapon.

4. Nuclear reactor fuel pellets are only 3% uranium-235. Weapon-grade uranium is enriched to 90% uranium-235, enough to cause a chain reaction leading to a nuclear explosion. The uranium used in nuclear power plants is sufficient to allow the chain reaction to occur, but not enough to cause a nuclear explosion.
5. a. Both nuclear power plants and fossil-fuel-fired plants heat water to drive turbines that generate electricity.
   b. Nuclear power plants generate energy by bombarding fissionable materials with neutrons, while fossil-fuel-fired plants burn materials such as coal to release thermal energy.
   c. The advantage of using fossil fuels is that they are readily available. Disadvantages include the large amounts of carbon dioxide and other gases, such as sulfur dioxide, dumped in the atmosphere. Burning fossil fuels also releases radioactive materials into the atmosphere.

   One advantage of nuclear fuel is that relatively small volumes are needed; a disadvantage is that large volumes of long-term radioactive waste will accumulate over the years.
6. The control rods limit the chain reaction in nuclear power plants. (See Section D.5 on page 352.)
7. Nuclear fission is the splitting of radioisotopes, by neutron bombardment, into lighter isotopes. Nuclear fusion is the combining of lighter isotopes to form heavier isotopes. In effect, fission is the breakdown of larger nuclei into smaller pieces. Fusion is the joining together of small nuclei to make a larger one.
8. a. Nuclear fusion in the laboratory is problematic and is not yet available for use as an economically sound source of energy.
   b. Because energy from the sun drives the physical and chemical processes of the Earth, and because this energy is derived from nuclear fusion reactions, it may be said that nuclear fusion is already the "number one" energy source on Earth.

## EXTENDING YOUR KNOWLEDGE (OPTIONAL) *(page 356)*

In a very real sense, we are "star dust." Astrochemists believe that all the heavier elements found in the universe were produced via fusion reactions from the lighter elements (hydrogen and helium) that make up the bulk of the mass of stars. Check the *Readers' Guide to Periodical Literature* and the Supplemental Readings list provided at the end of this unit.

# E. LIVING WITH RISKS AND BENEFITS

What risks are justified by the benefits of nuclear technology? Before delving into the intricacies of this question as it relates to nuclear technology, students perform risk–benefit comparisons of several means of transportation. Next, they explore the benefits of radioisotopes used widely in medicine, industry, and biological research. Then they consider disposal problems presented by high-level and low-level nuclear waste products. Finally, they consider the risk of a catastrophic nuclear accident.

## E.1 YOU DECIDE: THE SAFEST JOURNEY *(pages 358–359)*

This activity is designed to point out that risks and benefits enter into every part of our lives. How we deal with risks may be a function of information coming from the news media and shaping our perception. Controversy arises when the benefits do not seem to outweigh the risks, or when society (government) makes decisions in opposition to the wishes of certain individuals or subgroups.

### Answers to Questions

1. a. Bicycle: $500 \times (0.000001/10) = 0.00005$, or $5 \times 10^{-5}$
   Auto: $500 \times (0.000001/100) = 0.000005$, or $5 \times 10^{-6}$
   Scheduled airline: $500 \times (0.000001/1{,}000) = 0.0000005$, or $5 \times 10^{-7}$
   Train: $500 \times (0.000001/1{,}200) = 0.00000042$, or $4.2 \times 10^{-7}$
   Bus: $500 \times (0.000001/2{,}800) = 0.00000018$, or $1.8 \times 10^{-7}$
   b. Bus travel is the safest.
   c. Bicycling is the least safe.
2. Students may have the false impression that air travel is the most dangerous form of transportation, owing to media attention given to air disasters; rarely do we read about nonlocal bicycle fatalities. Some people, because of their familiarity with driving, have the false impression that automobiles are the safest.
3. a. Traveling by bicycle is the cheapest; autos, trains, and buses are faster than the bicycle, and transportation by scheduled airlines is the fastest and most worry-free in terms of responsibility for guiding the vehicle. Traveling by train or bus takes longer than scheduled airlines, but offers the lowest risks.
   b. Increased time, cost, personal responsibility, and increased risk of accidents would be seen by most as negative factors. In this case the cost saved in traveling 500 miles by bicycle probably does not seem worth the risk of death and the extended travel time. Other factors (see Question 5), such as the desire for personal mobility and luggage capacity provided by the auto, may influence decision making.
4. These statistics will probably not hold true for 25 years, given an increase in the population and increased automobile transportation and air travel. On the other hand, new technologies, such as proposed "smart highways", might make transportation safer or even provide totally new options.
5. Cost (including meals and overnight expenses), time, responsibility for vehicle, space, payload (luggage, load), freedom to stop vehicle or detour, exercise, pleasure, lack of worry regarding mechanical breakdown
6. a. (Individual answers)
   b. A risk-benefit analysis could never yield the same result for every individual because each person places a different priority and value on each mode of transportation.

## CHEMQUANDARY *Risk-Free Travel?* *(page 359)*

There is no risk-free way to visit a friend. Even deciding not to visit a friend may pose unknown risks. A conscious decision not to take a risk may put one in a position of greater risk because of future events unknown at the time of the decision. The idea here is to minimize total risk by taking a more knowledgeable approach toward known risks.

The use of irradiation in many contexts, including food irradiation, is discussed. There is no simple answer to questions concerning food irradiation. Rather, the risks and benefits must be considered, including the hazards of eating spoiled food. Students may be surprised to learn that traffic fatalities per passenger mile (30 deaths per 108 passenger miles on horses) were approximately 30 times higher in the pre-automobile days than today, or roughly equivalent to the fatality rate for motorcycles. There is no such thing as a risk-free ride.

### YOUR TURN *Assessing Risks and Benefits* *(page 361)*

1. Refer to student questions raised in Section A.1.
2. Student answers. Evaluate answers not in terms of which side of the controversy is chosen, but as to how well the student supports the position taken.

## E.2 NUCLEAR WASTE: PANDORA'S BOX *(pages 361–366)*

This section introduces students to a highly controversial and pressing issue. The nature of high-level nuclear waste disposal, with the proposed solution, is presented and clarified. Even if all high-level nuclear waste generation were to end tomorrow, it would still be an issue requiring continued attention.

Encourage students to watch newspapers for articles pertaining to this issue. Given the nature of radioactive decay (particularly the concept of half-life), the problem will not go away in the near future.

### YOUR TURN *High- and Low-Level Waste Disposal* *(page 362)*

1. a. 29.1% + 57.1% = 86.2%
   b. 8.3% + 0.03% + 5.3% = 13.6%
2. Defense
3. Commercial and defense
4. Evaluate student answers according to how well they defend their answers with evidence.

## E.3 CATASTROPHIC RISK: A PLANT ACCIDENT *(pages 366–369)*

The two major accidents in nuclear power plants, at Three Mile Island (TMI) and Chernobyl, are discussed, including the reasons for the accidents. It is important to put the two "accidents" in perspective. The Chernobyl incident resulted in significantly more release of radiation than TMI. There was significant loss of life and radiation sickness following Chernobyl, which was not the case following the TMI incident. The section concludes by posing the central question found in many parts of *ChemCom:* Are the benefits worth the risks?

## PART E: SUMMARY QUESTIONS *(page 369)*

1. Radioactive materials decay at a rate (half-life) that is unalterable by any known means. We must be able to isolate most of our nuclear wastes for thousands of years from both manmade (terrorism and war) and natural (water seepage and geological shifts) intrusions. The shielding material must block beta and gamma radiation and withstand thermal, mechanical, and chemical stress.
2. a. low level
   b. high level
   c. low level
   d. low level
3. a. Burial in a geologically stable location
   b. Problems include concentrating and safely shipping the wastes to the site and the political reality that nobody seems to want such a site near their home (or even in their state).

4. Are the containers safe from breaking or cracking? Is it possible for the radiation to become part of the water supply? Are adequate precautions taken to deal with heat produced? (Many other answers are possible.)
5. The difficulty in disposing of the waste, the cost, the fears people have about using nuclear power, and building new nuclear power plants

# F. PUTTING IT ALL TOGETHER: LONG-TERM NUCLEAR WASTE STORAGE

Long-term nuclear waste disposal must be effective for about 10,000 years. During such a long period of time, civilization as we know it can change drastically, and people in the future might unwittingly expose themselves to radiation from the waste we leave. There would be serious problems with communicating danger over that long a period.

## F.1 A SIGN FOR THE TIMES *(page 370)*

Students are challenged to design markers that will last for 10,000 years and will communicate with humans who may be around at that time. They will need to call on the knowledge they have accumulated by studying this unit as well as knowledge gained in previous units, particularly Resources.

## F.2 LOOKING BACK *(page 371)*

In this unit we have tried to give students an adequate background for making responsible decisions about nuclear energy policies. Insights on the problems associated with the use of nuclear energy continue to change regularly, and so it is impossible, even if it were desirable, to teach "right" answers or solutions for nuclear issues.

# ENRICHMENT MATERIALS

## TEACHER BACKGROUND

### Section C.5

Scientists continue to add elements to the periodic table, creating new elements by bombarding a relatively large target atom with a high-energy beam of smaller particles or atoms. In late 1994, three atoms of element 111 were created in Darmstadt, Germany, by bombarding bismuth-209 with a high-energy beam of nickel-64 atoms for several days.

In 1994 and 1995 a controversy erupted over the naming of elements 101–109. Actually, the controversy over the naming of these elements is decades old and had its roots in disagreement over who had discovered several elements, among them 104 and 105. (Traditionally, the discoverer of an element has the right to propose its name.) For a long time, German and U.S. scientists have used "rutherfordium" for 104 and "hahnium" for 105, while the Russians use "kurchatovium" for 104 and "nielsbohrium" for 105. The ACS Committee on Nomenclature proposed "nielsbohrium" for element 107. Further, the committee proposed for element 106 the name "seaborgium" after Nobel Laureate Glenn T. Seaborg, who has made major contributions to the discovery of many transuranium elements.

An International Union of Pure and Applied Chemistry (IUPAC) panel rejected all the proposed names and chose a slate of its own. Seaborgium was rejected because Seaborg is still alive. This decision and the rejection of hassium for element 108 provoked international outrage in Germany and the United States. IUPAC responded to the criticism by agreeing to a five-month review process beginning on January 1, 1996. ACS Executive Director John K Crum says that U.S. chemists should continue to use the names recommended by the ACS Committee on Nomenclature.

**Different element names favored by ACS and IUPAC**

| Atomic Number | ACS Slate | IUPAC Slate |
|---|---|---|
| 104 | Rutherfordium | Dubnium |
| 105 | Hahnium | Joliotium |
| 106 | Seaborgium | Rutherfordium |
| 107 | Nielsbohrium | Bohrium |
| 108 | Hassium | Hafnium |
| 109 | Meitnerium | Meitnerium |

Source: *Chemical and Engineering News*, p. 7, May 8, 1995.

## DEMONSTRATION IDEA

### Section D.4

This demonstration is an alternative model of the concept of chain reaction and critical mass. Each student is given two corks (or other light objects such as Styrofoam balls or crumpled paper) and instructed to shut his/her eyes. The teacher begins the chain reaction by tossing a cork over his/her head in the direction of the student desks. If hit, a student must toss his/her corks in a random direction. Within the confines of an average classroom, the reaction usually takes a few tries to get started and dies out before all the corks are thrown. Repeat the process by having students take one step closer together. Students will probably be amazed at how close they have to get to reach "critical mass."

(From *Reactions*, September 1986, a free newsletter of the American Nuclear Society.)

## CHEMQUANDARIES

### Section D.5

In arguing for the use of nuclear power, some claim that the industry is held to unrealistically strict standards. They argue that its environmental and health impacts are less than those of coal plants, which are estimated to kill thousands of people each year by polluting the air. In assessing the relative risks of nuclear and coal plants, what considerations might lead to especially strict standards being imposed on nuclear power plants?

**Answer:** Although the probability of a catastrophic nuclear accident may be small, the potential effects in terms of massive loss of life and property damage far exceed the destruction possible from the worst coal plant accident. It is important to point out that a nuclear plant cannot explode like a nuclear bomb. One might compare the regulation on construction, maintenance, and traffic control within the airline industry with that in the automotive industry. Also, the impact on future generations (in terms of genetic disorders and long-term environmental contamination) is greater than in nonnuclear cases. This is not to say that coal plants are the best answer to our energy needs.

### Section E.2

1. One proposal for solving the nuclear waste problem is to use rockets or the space shuttle to shoot the wastes out into space. What are some problems with this idea?
   **Answer:** High costs, possibility of a launch accident spreading radioactive wastes across a wide area
2. Even if all nuclear power plants were phased out in the next five years, nuclear waste disposal would remain a controversial issue needing attention. Why is this so?
   **Answer:** We would still need to find an effective solution for the high-level nuclear waste accumulated at the power plants and the much larger volumes produced by military weapons production. Additionally, some small, nonelectricity-generating reactors would need to remain in operation to supply radioisotopes for medical and other essential applications.

# SUPPLEMENTAL READINGS


"Chemiluminescence." *CHEM MATTERS,* October 1995.

"Worlds within the Atom." *National Geographic,* May 1985, 634-663. [Includes historical milestones and a discussion of accelerators, quarks, etc.]

"Fine Art Fraud." *CHEM MATTERS,* December 1994.

"Hot Shirts." *CHEM MATTERS,* October 1992.

"John Dalton's First Paper and Last Experiment." *CHEM MATTERS,* April 1984.

"Nuclear Diagnosis." *CHEM MATTERS,* December 1985, 4-7.

"Radioactive Dating: A Method for Geochronology." *Journal of Chemical Education,* 1985, 580-584. [Gives historical background on the discovery of natural radiation and discusses the use of K-40/Ar-40 in geochronology.]

"Starborn–The Origin of the Elements." *CHEM MATTERS,* October 1984.

"Nuclear Synthesis and Identification of New Elements." *Journal of Chemical Education,* 1985, 392-395. [Reviews nuclear terms, reactions, and experimental methods.]

"Radioactivity in the Service of Man." *Journal of Chemical Education,* 1982, 753-738.

"Radiation Chemistry: State of the Art Symposium." Special Issue of *Journal of Chemical Education,* February 1981. [This contains many articles including "Principles and Techniques of Radiation Chemistry" (84), "Radiation Chemistry and the Preservation of Food" (162-167), "Radiation Processing: Industrial Applications of Radiation Chemistry" (168-173), "The Effects of Ionizing Radiation on Mammalian Cells" (144-156), and "Radiolytic Damage to Genetic Material" (135-139).]

"Biological Effects of Low Level Radiation." *Scientific American,* February 1982, 41-49.

"Radon Gas: A Deadly Threat." *Newsweek,* August 18, 1986, 60-61.

"Carbon-14 Dating." *CHEM MATTERS,* February 1989.

*A Citizen's Guide to Radon: What It Is and What to Do about It* (OPA-86-004/August 1986) and *Radon Reduction Methods: A Homeowners Guide* (OPA-86-005/August 1986). [Two publications of the Environmental Protection Agency.]

"Safety of Fission Reactors." *Scientific American,* 242 (3), 33, 1980.

"Special Report on Energy." *National Geographic,* February 1981. [Whole issue].

"Fusion Power." *Chemical & Engineering News,* April 2, 1979, 32-47. [An overview of theory and research.]

"Fission in the Fusion Camp." *Discover,* December 1989.

"Light Your Candy." *CHEM MATTERS,* October 1990.

"Microwave Chemistry." *CHEM MATTERS,* December 1993.

*Overview—Nuclear Waste Policy Act. The Illustrated Mechanics of Nuclear Waste Isolation. What Will a Nuclear Waste Repository Look Like? What Is Nuclear Waste? What Is Spent Fuel? Can Nuclear Waste be Transported Safely? Radiation and Nuclear Waste—How Are They Related? How Much High-level Nuclear Waste Is There? What Rock Types Are Being Considered for Nuclear Repositories and Why?* [A series of publications of the Office of Civilian Radioactive Waste Management, U.S. Department of Energy, Mail Stop RW-40, Washington, DC 20585.]

"The Promise and Peril of Nuclear Energy." *National Geographic,* April 1979, 459-493.

"The Disposal of Radioactive Wastes from Fission Reactors." *Scientific American,* June 1977, 21-31.

"The Reprocessing of Nuclear Fuels." *Scientific American,* December 1976, 30-41.

"New Era of Inherently Safe Nuclear Reactor Technology Nears." *Chemical & Engineering News,* June 30, 1986, 18-22. [Describes work that has occurred since the Three Mile Island accident and contrasts U.S. plant design to Chernobyl.]

"Rethinking Nuclear Power." *Scientific American,* March 1986, 31-39. [Discusses how a new generation of lower power, centrally fabricated reactors could avoid the safety and political impasse facing nuclear power; figures show the growth of the industry both here and abroad from 1964 to present; and details the new designs.]

*The Three Mile Island Accident: Diagnosis and Prognosis.* American Chemical Society, 1986. [The story of the TMI accident told by scientists and engineers who operated the reactor or worked in the national laboratories that analyzed the accident.]

"The Three Mile Island Story." *Time,* April 9, 1979.

"Decisions, Decisions." *Discover,* June 1985, 22-31. [Research shows how irrational considerations play a large part in human decisions.]

"Advanced Light-Water Reactors." *Scientific American,* 262 (4), 82, 1990.

# TEACHING SCHEDULE

| | TIME BLOCK 1 | TIME BLOCK 2 | TIME BLOCK 3 | TIME BLOCK 4 | TIME BLOCK 5 |
|---|---|---|---|---|---|
| Class Work | Read Introduction<br>Discuss Introduction<br>YD A.1 | LD A.2 | SQ Part A<br>Prelab LA B.1 | Review Part A<br>Quiz Part A<br>LA B.1 | LA B.1 (cont.) |
| Homework | Read A.3 | YT A.3 | Getting Ready for LA B.1 | | Read B.2 |
| | **TIME BLOCK 6** | **TIME BLOCK 7** | **TIME BLOCK 8** | **TIME BLOCK 9** | **TIME BLOCK 10** |
| Class Work | YTs B.2 (2) | Discuss B.4<br>CQ B.4 | Discuss B.5<br>CQ B.5<br>YT B.5<br>Prelab LA B.6 | LA B.6 | CQ B.6<br>YT B.6 |
| Homework | YD B.3<br>Read B.4 | Read B.5 | Read B.7 | | |
| | **TIME BLOCK 11** | **TIME BLOCK 12** | **TIME BLOCK 13** | **TIME BLOCK 14** | **TIME BLOCK 15** |
| Class Work | YTs B.7 (2) | Review Part B<br>Quiz Part B<br>YT C.2<br>CQ C.2 | Discuss C.3-C.4<br>YT C.3 | Prelab C.6<br>LA C.6 | YD C.7 |
| Homework | SQ Part B<br>Read C.1-C.2 | Read C.3-C.4 | YT C.5 | | Read C.8 |
| | **TIME BLOCK 16** | **TIME BLOCK 17** | **TIME BLOCK 18** | **TIME BLOCK 19** | **TIME BLOCK 20** |
| Class Work | SQ Part C | Review Part C<br>Quiz Part C<br>Discuss D.1-D.4 | Discuss D.5-D.6<br>D.6<br>LA D.5<br>CQ D.5 | Discuss D.8<br>YD D.7<br>CQ D.8 | Discuss D.9<br>Prelab LA D.10<br>LA D.10 |
| Homework | Read D.1-D.4 | YD D.2<br>Read D.5-D.6 | Read D.8 | Read D.9 | Read D.11 |
| | **TIME BLOCK 21** | **TIME BLOCK 22** | **TIME BLOCK 23** | **TIME BLOCK 24** | **TIME BLOCK 25** |
| Class Work | YD D.11<br>SQ Part D | Review Part D<br>Quiz Part D<br>PIAT E.1-E.3 | PIAT E.3-E.4<br>YT E.1 | Review Air Unit | Air unit Exam |
| Homework | Read E.1-E.3 | | | | |

**LA**=Laboratory Activity; **CQ**=ChemQuandary; **YT**= Your Turn; **YD**=You Decide; **PIAT**=Putting It All Together; **SQ**=Summary Questions; One **Time Block**=Approximately 50 minutes

# SUPPLY LIST

## Expendable Supplies and Unusual Materials

| Quantity | Unit | Item | Sections used |
|---|---|---|---|
| 125 | mL | ammonia, conc | A.2, D.5 |
| 2 | | apples, red | D.10 |
| 12 | | aspirators (for faucets) | C.6, D.5 |
| 10 | large | balloons | A.2 |
| 1 | | bottle with lid, clean, plastic | A.2 |
| 25 | mL | bromthymol solution | C.6 |
| 1 | pkg | candles, birthday | A.2 |
| 12 | large | candles | C.6 |
| 1 | cylinder | carbon dioxide | A.2 |
| 1 | | container, glass or plastic | A.2 |
| 18 | gauge | copper wire | D.5 |
| 300 | mL | corn oil | B.6 |
| 1 | pkg | cotton balls | A.2 |
| 2 | | drinking glasses | A.2 |
| 1 | 500-mL | filter flask | D.5 |
| 12 | 500-mL | filter flasks | C.6, D.5 |
| 1 | 500-mL | flasks, Florence | D.5 |
| 12 | 7-cm-diameter | funnels, glass | D.10 |
| 1 | | glass tube open on both ends | A.2 |
| 12 | 30-cm | glass tubing | B.1 |
| 12 | 15-cm | glass tubing | B.1 |
| 1 | cylinder | helium | A.2 |
| 1 | pkg | Hydrion paper (pH paper) | D.5, D.10 |
| 25 | mL | hydrochloric acid, 6M | D.10 |
| 25 | mL | hydrochloric acid, conc | A.2, D.5 |
| 1 | cylinder | hydrogen | A.2 |
| 1.5 | L | hydrogen peroxide, 3% | B.1 |
| | | iodine crystals | A.2 |
| 12 | g | iron(III) chloride hexahydrate | B.1 |
| 22 | mL | limewater | B.1 |
| 1 | pkg | litmus paper, blue | D.10 |
| 1 | pkg | litmus paper, red | D.10 |
| 27 | cm | magnesium ribbon | B.1, D.10 |
| 24 | | marble chips | D.10 |
| 12 | | melting point tubes, capillary | B.6 |
| | | meter stick | A.2 |
| | | methane (burner fuel) | A.2 |
| 1 | | paramecium culture | D.10 |
| 2 | | paring knives | D.10 |
| 1 | bottle | perfume | A.2 |
| 5 | mL | phenolphthalein solution | A.2 |
| 12 | pint size | plastic bags, self sealing | D.10 |
| 24 | small | rubber bands | B.6 |
| 12 | 60-cm | rubber tubing | B.1 |

## Expendables, continued

| Quantity | Unit | Item | Sections used |
|---|---|---|---|
| 12 | g | sodium sulfite | D.10 |
| 1 | | soft drink bottle, empty | A.2 |
| 1 | | soft drink can, empty | A.2 |
| 1 | g | steel wool | B.1 |
| 12 | | straws | C.6 |
| 1 | cylinder | sulfur hexafluoride | A.2 |
| 1 | | Tesla coil | D.5 |
| 350 | mL | universal indicator | B.1, D.5 |
| 1 | | wire cutters | D.5 |
| 12 | | wood splints | B.1 |

Returning to the theme of conserving resources, in this unit we examine the role of chemistry in establishing and maintaining air quality. The physical properties of gases, including the relationship of volume, pressure, and temperature, are explored, and the chemical properties of the air's major constituents are tested in laboratory activities. Technological issues explored include global warming, ozone depletion, and acid rain. Finally, the importance of air quality—indoor as well as outdoor—is discussed.

# LEARNING GOALS

Upon completion of this unit, the student will be able to:

1. Describe common physical and chemical properties of gases. [A.1, A.2]
2. Describe the changes in air that occur as a result of its being inhaled and exhaled. [A.3]
3. Compare the chemical properties of nitrogen, oxygen, and carbon dioxide. [B.1]
4. Identify the major components of the troposphere, and indicate their relative concentrations. [B.2]
5. Show how Avogadro's law and the concept of molar volume clarify the interpretation of chemical equations involving gases. [B.2]
6. Describe with words and mathematical equations the relationships among temperature, volume, and pressure of a gas (Charles' and Boyle's laws), and list one practical application of each law. [B.2, B.4-B.6]
7. Define, and apply in appropriate situations, the terms *molar volume, pressure, kelvin temperature scale, and absolute zero.* [B.2, B.5, B.6]
8. Plot and interpret graphs that describe the relationship between altitude and air pressure. [B.3]
9. Discuss air pressure and explain how to measure it. [B.4]
10. Account for the gas laws in terms of the kinetic molecular theory of gases. [B.7]
11. Compare the energies of the various types of electromagnetic radiation from the sun. [C.1]
12. Describe how reflection, absorption, and reradiation of solar radiation account for Earth's energy balance. [C.2]
13. Explain how differing heat capacities and reflectivities of various land covers and water can influence local climates. [C.3]
14. Describe the greenhouse effect, its natural incidence and causes, and the significance of industrial contributions. [C.4, C.5]
15. Use graphical extrapolation to predict future $CO_2$ concentrations in the air, and outline assumptions and problems associated with such predictions. [C.5]
16. Compare the quantity of $CO_2$ produced by combustion with that provided by respiration. [C.6]
17. Describe the function of the ozone layer and how human activities may be affecting it. [C.8]
18. Identify some of the consequences of the Montreal Protocol. [C.8]
19. List the major categories of air pollutants, and discuss the relative contributions of various human and natural factors to each category. [D.1-D.3]
20. Describe general strategies for combating pollution, and specific strategies for particulates [D.4, D.5]
21. Describe chemical reactions and geographic and meteorological factors that contribute to photochemical smog. [D.6]
22. Interpret graphs and tables related to automotive-induced air pollution. [D.7, D.8]
23. Explain the role of activation energy in a chemical reaction, and give an example of how a catalyst affects it. [D.8]
24. Describe the role of catalytic converters in reducing automotive emissions of unburned hydrocarbons, carbon monoxide, and nitrogen oxides. [D.8]
25. Describe sources and consequences of acid rain. [D.9-D.11]
26. Define acidic and basic solutions in terms of pH. [D.11]
27. Relate hydrogen ion concentration to the pH scale. [D.11]
28. Describe the major indoor air pollutants, and identify their sources. [E.1, E.2]
29. Collect detailed information about a single indoor air pollutant, and use it to present a report to the class. [E.3]

# A. AIR—THIS MARVELOUS CANOPY

The major issues to be considered in this unit are introduced by having students complete a true-false questionnaire designed to assess their understanding of gases. Students identify and explain some physical properties of gases by observing a series of teacher demonstrations.

Local newspapers and TV weather news are useful resources for increasing student awareness of local air quality issues. Student teams can be assigned to keep records of data reported by these sources and make graphs and tables from them. Newspaper articles about local air quality issues can be used by students to create bulletin boards or other visual representations.

## A.1 YOU DECIDE: THE FLUID WE LIVE IN *(pages 375–376)*

This activity focuses students' attention on air—a key resource they probably take for granted. The questions go farther than normal true-false tests by asking students to correct their false answers, and to apply true answers to additional situations. After the students complete the "test" give them true–false answers to the questions and remind them that each item will be discussed in this unit. The reasons for the true–false answers are given for the teacher's benefit and are not meant to be shared with students at this time.

### Answers to Questions

1. True. An air supply is needed for scuba diving, flying at high altitudes, or space travel.
2. True. Volume is a function of temperature and pressure.
3. False. All forms of matter have mass.
4. True. The human body is designed to handle external pressures of approximately 15 $lb/in^2$.
5. False. If this were true, living organisms would be much more limited in the habitats they could occupy.
6. True. Most of the more harmful, high-energy ultraviolet radiation is absorbed by ozone and other triatomic species ($CO_2$, $H_2O$).
7. False. Air temperature decreases until the altitude is about 12 km, and increases until the altitude is 50 km. Then it again decreases.
8. True. Atmospheric water is a part of the hydrologic cycle. Also, together with carbon dioxide, it plays a crucial role in helping to maintain the heat balance on our planet.
9. True. Oxygen is number five (35 billion lb/yr in the United States) on the "top 10" substance list. This atmospheric gas is used in welding, medicine, propellants, and steel production. Nitrogen is number three (37 billion lb/yr in the United States) on the top 10 substance list. This atmospheric gas is used as an inert atmosphere for metals and electronics, as a freezing agent for foods and cryogenics, and in the production of ammonia.
10. True. In the lower atmosphere, ozone is a product of photochemical smog that corrodes metals and plastics and is a major health hazard. Ozone also interacts with other smog components such as hydrocarbons to produce such pollutants as peroxyacetyl nitrate (PAN). In the stratosphere, ozone acts to filter out harmful ultraviolet radiation.
11. False. Air is a mixture that contains a variety of substances ($N_2$, $O_2$, Ar, $CO_2$, $H_2O$) of both biological and industrial importance.
12. False. Air pollution is at least as old as civilization. Early fires from cooking and smelting operations polluted the air in ancient times. Today the planet supports a much larger number of people. Since the Industrial Revolution, the amount of pollution/person has increased along with our increased use of resources.

13. False. Many large cities experience conditions that require warnings suggesting that citizens, particularly those with health problems, remain indoors. While it is difficult to attribute an individual death to air pollution, medical authorities attribute the higher rate of deaths due to lung diseases in cities and industrial centers at least partially to polluted air.
14. True. The eruption of the Mt. St. Helens and Mt. Pinatubo volcanoes, and the fires that ravage dry western states, vividly demonstrate sources of natural air pollution.
15. True. Estimates of economic loss due to air pollution run as high as $16 billion annually in the United States.
16. False. In Section D.2 the students discover that in terms of total mass of pollutants emitted, industrial activity falls behind transportation, space heating, and electricity generation. The important implication is that each of us, by our lifestyle and personal habits, contributes to air pollution. It is not just a problem of industry.
17. True. The greenhouse effect is one of Earth's natural thermal control mechanisms. Part C discusses this in some detail.
18. False. The rain in industrialized nations has become more acidic, causing both environmental and political problems.
19. True. Major combustion products are $CO_2$ and $H_2O$ and a variety of other substances such as unburned hydrocarbons, $NO_x$, $SO_x$, and particulates.
20. False. Throughout the unit, students will find evidence for improving air quality.

## A.2 LABORATORY DEMONSTRATION: GASES *(pages 376–377)*

These demonstrations are a very effective way to promote interest in and understanding of the properties of gases. Since atmospheric gases are colorless, odorless, and tasteless, students tend to think of them as nothing. Of course, gases, like solids and liquids, have definite physical and chemical properties.

Students could record their observations in a table. If you use a table like the one suggested below, be sure to explain what you are going to do (but do not explain what will happen) prior to the demonstration.

| Descriptions | Predictions | Observations | Gas Property | Explanations |
|---|---|---|---|---|

### Time

50 minutes

### Materials

10 large balloons
meter stick
1-L beaker
3 50- or 100-mL beakers
250-mL Erlenmeyer flask
ring stand, clamp, wire gauze, ring
Bunsen burner
empty soft drink bottle
2 drinking glasses
piece of cardboard or wax paper
clean plastic bottle with lid (detergent bottle)
empty aluminum soft drink can
bottle of perfume
100 mL conc. ammonia, $NH_3$(aq)
5 mL phenolphthalein
large plastic or glass container
long glass tube, open on both ends, with stoppers
pkg of cotton balls
10 mL concentrated hydrochloric acid, HCl(aq)

cylinder of helium (optional)
cylinder of carbon dioxide (optional)
cylinder of sulfur hexafluoride (optional)
natural gas, methane or propane (burner gas)
small funnel
soap-glycerin mixture
Dry Ice
pkg of birthday candles
iodine crystals

## Pre-Demonstration Discussion

As you do the demonstrations, ask students to:

1. try to predict the outcome of the demonstration;
2. carefully observe the demonstration and, if necessary, account for differences between their prediction and the actual outcome;
3. try to identify the specific gas property that is being demonstrated, and why gases exhibit this property.

## Demonstration Tips

The following demonstrations are effective for illustrating that gases (even colorless, odorless ones) are "real" substances. They use readily available materials and, if properly performed, they are quite safe. They do not require elaborate setups or much advance setup time.

When performing the demonstrations, be sure to capitalize on the element of surprise. Avoid forecasting expected outcomes to your students. You do not need to try to "teach" concepts on this day; all will be developed more fully in subsequent text and activities. The main purpose at this time is to raise questions and motivate your students to want to learn more about the "sea" of gases in which we live.

## Demonstrations

### Part 1. Is air really matter?

**Demonstration 1:** Air has weight. Balance two filled balloons by hanging them from a meter-stick balance. Pop one of the balloons and note the direction of the balance's tilt. Its tilt toward the filled balloon demonstrates that "invisible" gases do possess weight, like other forms of matter.

**Demonstration 2:** Having a student observer close to the glass is essential for this demonstration. Air occupies space. Lower an "empty" glass, open end downward, into a larger beaker (or pot) of water. Does water fill up the "empty" glass? Tilt the glass. What happens? The observation that the water does not enter the glass shows that air does take up space.

**Demonstration 3:** Not all gases are colorless. Prior to the demonstration, place several crystals of iodine in a sealed flask. A faint purple color will soon be observed in the flask above the crystals. (This is best displayed with white paper behind the flask.) Contrast this "colored gas"–iodine vapor, $I_2(g)$–to air or other colorless gases.

### Part 2. What is air pressure all about?

**Demonstration 4:** Air has pressure. Select a balloon that is easy to blow up. Insert the bottom of the balloon partway into an empty soft drink bottle, stretching its neck over the bottle mouth. Ask a talkative student to try blowing up the balloon so it fills the container. Ask the class to explain the difficulty of this task.

**Demonstration 5:** Temperature changes the air pressure. Place 10 mL of water in a 250-mL Erlenmeyer flask. Bring the water to boiling. Remove the flask from the source of heat and quickly and carefully place an empty balloon over the mouth of the flask. (It helps to blow up the balloons once before they are used.) As the flask cools, the pressure differential between the outside and inside of the flask will cause the balloon to invert and line the inside walls of the flask. In class, show your students the flask. Ask them to explain how the balloon got into the flask and how they could remove the balloon from the flask. (Heating will cause the liquid water to vaporize and push out the balloon.)

**Demonstration 6:** Air exerts pressure upward and in all directions. Fill a glass to the rim with water, then cover it with either wax paper or a piece of plastic or cardboard (as cut from a coffee can lid or cereal box). Press down along the edges to make a tight seal, turn the glass upside down over a sink, and let go of the cover. Air pressure (acting upward against the covering) will support the weight of the water and prevent it from spilling out. Repeat the process without filling the glass completely. Ask your students to account for the difference in the two cases.

**Demonstration 7:** Air exerts pressure sideways and in all directions. Punch a small hole in the side of an empty, clean, plastic bottle (such as a liquid-detergent bottle). Holding your finger over the hole, completely fill the bottle with water. Replace the cap. Remove your finger from the hole. Explain your observation. Unscrew the lid and note what happens. Ask students to account for the difference. A related question is, "Why is it difficult to pour juice from a can if only one hole is punched in it?"

**Demonstration 8:** Air exerts pressure in all directions. Place about 20 mL of water in a clean, empty aluminum soft drink can. Bring the water to a rapid boil and quickly invert the can in a container of water. It is not necessary to submerge the can, but only to have its top touch the water's surface. Be careful, as the can may be quite hot. The boiling process will force out much of the air that is inside the can and replace it with water vapor. Upon cooling, the water vapor will condense, leaving a partial vacuum inside the can. Atmospheric pressure outside the can will crush it

**Part 3. Why does air sometimes have an odor?**

**Demonstration 9:** Diffusion of gases. Standing at the front of the classroom, open a bottle of perfume or other safe, odoriferous substance with high vapor pressure. Have each student raise a hand as soon as the odor is detected. This demonstrates that gas molecules are in motion.

**Demonstration 10:** Diffusion rates of gases. Place two small beakers (one containing ammonia, the other, water and phenolphthalein) under a large sealed plastic or glass container. Do not allow ammonia fumes to escape. Ask your students to account for any changes they note. (You will see a pink ring forming nearing the water surface.) Placing the setup on an overhead projector makes it more visible, and the heat causes the diffusion to occur faster. A similar demonstration using hydrochloric acid and ammonia (to produce ammonium chloride "smoke" particles) can also be performed, either with this setup or with the traditional long glass tube setup.

**Part 4. Is air heavy? Will it burn?**

**Demonstration 11:** Different gases have different densities. Contrast the behavior of equal-sized balloons filled with available gases such as hydrogen (0.09 g/L), helium (0.18 g/L), natural gas from a burner jet which is mainly methane (0.55 g/L), "air" (1.0 g/L), carbon dioxide (2.0 g/L), or sulfur hexafluoride ($SF_6$ 6.6 g/L) by attempting to throw the balloons. Alternative and especially effective demonstrations include: (a) "blowing bubbles" using a liquid soap-glycerin mixture and a funnel connected via rubber tubing to a gas source and (b) pouring carbon dioxide down a tilted tray of burning candles. Observing a container of Dry Ice is also effective. Exercise extreme care when using Dry Ice.

**Demonstration 12:** Flammability. Contrast the flammability of hydrogen (or methane) to that of helium, using either balloons or soap bubbles. If using the soap bubble approach, be sure to run enough gas through the tubing to clear out any air. This precaution will ensure even, rapid, nonexplosive burning. Use a candle attached to a meter stick to attempt igniting the bubbles after they detach themselves from the bubble generator and begin to rise. Exercise extreme care when performing this demonstration.

## At-Home Activities

1. When one sucks on a single straw, air pressure is decreased in the straw. The atmospheric pressure on the surface of the fluid remains the same. It is this difference in pressure that allows the fluid to flow upward. If two straws are used as described in the student text, no pressure is built up, and therefore the fluid does not rise. If resourceful students use their tongues to close off the end of the second straw, they can accomplish the task.
2. As above, the action of a straw depends on a pressure differential. If the container is filled and sealed, atmospheric pressure cannot act on the surface of the fluid, and the fluid will not rise.
3. It is possible to drink liquid through a straw because of the difference in pressure inside and outside the straw.

## A.3 AIR: THE BREATH OF LIFE *(pages 377–379)*

This section focuses on the importance of atmospheric oxygen to human beings and the changes it undergoes as it is used. The complementarity of photosynthesis and respiration is emphasized. The *Your Turn* exercise provides practice in using chemical equations for calculations. A demonstration is found in the Enrichment Materials section at the end of this unit.

## YOUR TURN *Breath Composition and Glucose "Burning"*

*(pages 377–378)*

This activity is designed to compare quantitatively the composition of inhaled and exhaled air, and to reemphasize the role of oxygen in the metabolic conversion of glucose to water and carbon dioxide. An enjoyable extension of the *Your Turn* questions is to have students measure their own respiration rate and use it in the calculations.

### Answers to Questions

1. a. The percent of oxygen decreases, carbon dioxide increases, and nitrogen plus other gases increase.

   b. Human respiration converts carbohydrates (or fats and proteins) and oxygen into carbon dioxide, water vapor (accounting for the slight increase in percent of nitrogen plus other gases), and energy.

   c. $\frac{14 \text{ breaths}}{1 \text{ min}} \times \frac{60 \text{ min}}{\text{h}} \times \frac{24 \text{ h}}{\text{day}} = \frac{20{,}000 \text{ breaths}}{\text{day}}$

   d. Factors that could change this include level of activity, personal metabolism (age, sex, weight, body build), and food consumption. In general, larger, more active people require more oxygen for respiration.

   e. (1) $\frac{500 \text{ mL}}{\text{breath}} \times \frac{1 \text{ L}}{1{,}000 \text{ mL}} \times \frac{14 \text{ breaths}}{\text{min}} = \frac{7 \text{ L}}{\text{min}}$

   (2) $\frac{7 \text{ L}}{\text{min}} \times \frac{60 \text{ min}}{\text{h}} \times \frac{24 \text{ h}}{\text{day}} = \frac{10{,}000 \text{ L}}{\text{day}}$

2. The oxygen required for functioning of a healthy body is less than that obtained by respiration. In cases of illness, more oxygen is sometimes given.
3. $C_6H_{12}O_6 + 6\ O_2 \rightarrow 6\ CO_2 + 6\ H_2O$

   a. Given an equal number of molecules, oxygen would be the limiting reactant because six molecules are needed for every molecule of glucose.

   b. Given equal masses, oxygen would be the limiting reactant because for each 180 g (1 mol) of glucose, 192 g (6 mol) of oxygen gas are needed.

   c. $\frac{20 \text{ mol } O_2}{1 \text{ day}} \times \frac{1 \text{ mol glucose}}{6 \text{ mol } O_2} = \frac{3.3 \text{ mol glucose}}{1 \text{ day}}$

   d. $\frac{3.3 \text{ mol glucose}}{1 \text{ day}} \times \frac{180 \text{ g glucose}}{1 \text{ mol glucose}} = \frac{600 \text{ g glucose}}{1 \text{ day}}$

e. $\frac{600 \text{ g glucose}}{1 \text{ day}} \times \frac{17 \text{ kJ}}{1 \text{ g glucose}} = \frac{10{,}000 \text{ kJ}}{1 \text{ day}}$

4. a. Water vapor is a by-product of animal respiration.
   b. Exhale into cold air or onto a cold surface to confirm the presence of water.

## PART A: SUMMARY QUESTIONS *(page 379)*

1. a. Air is similar to other resources, in that our planet has fixed quantities of air and other resources, and its behavior can be explained in terms of atoms and molecules.
   b. Air is different, in that it is uniformly distributed across all geographic locations and therefore is not usually mined, stored, or exported; it is absolutely essential for life; and it is a gas at room temperature and pressure.
2. a. True. Empty bottles seldom contain a vacuum, but contain air molecules.
   b. False. The can was crushed from all directions.
   c. True. The odor of perfume was detected in all areas of the classroom after the bottle was opened.
   d. False. Helium, air, carbon dioxide, and hydrogen had different densities. Hydrogen burned; helium and carbon dioxide did not.
3. The density of air is less at higher altitudes. Therefore the same volume of air contains less oxygen at higher altitudes. In order to maintain metabolism, the rate of respiration must increase.
4. a. $\frac{0.15 \text{ mol } O_2}{1 \text{ min}} \times \frac{1 \text{ mol glucose}}{6 \text{ mol } O_2} \times \frac{180 \text{ g glucose}}{1 \text{ mol glucose}} = \frac{4.5 \text{ g glucose}}{1 \text{ min}}$

   b. $\frac{4.5 \text{ g glucose}}{1 \text{ min}} \times \frac{17 \text{ kJ}}{1 \text{ g glucose}} = \frac{77 \text{ kJ}}{1 \text{ min}}$

   c. $\frac{77 \text{ kJ}}{1 \text{ min}} \times \frac{1 \text{ Cal}}{4.2 \text{ kJ}} = \frac{18 \text{ Cal}}{1 \text{ min}}$

# B. INVESTIGATING THE ATMOSPHERE

Part B explores the chemistry of the individual gases contained in the atmosphere, and then turns to the gas laws (Avogadro's law, Boyle's law, and Charles' law). Some mathematical manipulation of the gas laws is included. The properties of nitrogen, carbon dioxide, and oxygen, and the relationships between volume and temperature are covered in laboratory activities. Absolute zero, standard temperature and pressure, and the kelvin temperature scale are introduced. This section concludes with applications of the kinetic molecular theory to explain the behavior of gases.

## B.1 LABORATORY ACTIVITY: THE ATMOSPHERE *(pages 380–384)*

This activity provides students with a basic understanding of chemical properties of the gaseous components of air. Specifically, students test nitrogen, carbon dioxide, and oxygen for flammability, acidity, and reactivity with limewater. Setting up the apparatus for collection of carbon dioxide and oxygen by water displacement may be challenging for some students.

### Time

100 minutes

### Materials (for a class of 24 working in pairs)

12 ring stands
12 wire screens (with ceramic center)
24 iron rings
12 500-mL Florence flasks
12 10-mL graduated cylinders
12 1-hole rubber stoppers to fit flasks
12 30-cm lengths of glass tubing (right-angle bend)
12 60-cm lengths of rubber tubing
12 pneumatic troughs or trays
36 test tubes (18 × 150 mm)
36 corks or rubber stoppers to fit test tubes
12 spatulas
12 tongs
12 Bunsen burners
balances
12 grease pencils
12 dropper bottles (for universal indicator solution)
5 cm magnesium ribbon
300 mL universal indicator solution (25 mL/bottle)
70 mL limewater, saturated solution of calcium hydroxide, $Ca(OH)_2$
1 g steel wool
12 wood splints
1.5 L 3% hydrogen peroxide ($H_2O_2$) solution
12 g iron(III) chloride ($FeCl_3$)

### Safety

Before class begins, insert the glass bends into the one-hole stoppers to prevent the risk of student accidents. Follow conventional precautions, lubricate glass tubing with glycerin and insert the tubing gently with a twisting motion. A handy technique for reducing the chances of breakage involves choosing a brass cork borer just large enough to go around the glass tubing. Insert the cork borer into the stopper, pass the tubing through the borer until it is in the right position, then remove the cork borer, leaving the glass entrapped within the stopper. Use heavy towels or gloves to protect your hands.

Students should wear safety goggles and aprons throughout the laboratory activity. **Caution students not to look directly at the flame produced when magnesium is burned.** It is bright enough to do eye damage.

## Advance Preparation

Bend glass for generator tubes: a bend at each end will prevent the pinching of the tube (see Figure 2 on page 383). The hydrogen peroxide solution should be purchased just before use.

## Pre-Lab Discussion

Be sure your students understand that the substances being generated in this activity are normal components of the gaseous mixture called air. The gases will be prepared in the activity, since this is easier than trying to separate them directly from air.

Review the general technique of collecting gases by water displacement (you may wish to have your students explore what factors limit the general use of this technique, such as the extent to which a given gas is water soluble or reacts with water).

## Lab Tips

Begin the laboratory session by producing nitrogen gas by either Option A or Option B (see below). Generating nitrogen by this reaction is a convenient way to model the setup for student generation of gases.

Perform all of the tests outlined in Part 2.

Assign half of the student groups to collect and test oxygen and the other half to collect and test carbon dioxide. Check the laboratory setups before students begin work. Provide a place on the board or on an overhead transparency sheet for student groups to display and share their observations.

*Nitrogen gas—Option A*

Use shielding to protect students and yourself. To produce nitrogen gas, place 5 g ammonium chloride, $NH_4Cl$, 7 g sodium nitrite, $NaNO_2$, and 50 mL water in the Florence flask used in the gas-generating apparatus. Heating is necessary only to begin this reaction. After bubbling starts, remove the heat source. **Do not overheat!** Remind students that they do not need to use heat for the reactions that produce carbon dioxide and oxygen. Demonstrate the tests outlined in Part 2. Ask a student volunteer to record the observations in the space provided.

*Nitrogen gas—Option B*

Use the gas collecting trough, but dispense the nitrogen gas from a compressed nitrogen gas cylinder. Be certain to properly secure the gas cylinder. Demonstrate the tests outlined in Part 2. Ask a student volunteer to record the observations in the space provided.

After the students have collected the gas samples and performed the splint, limewater, and indicator tests, demonstrate the magnesium and steel wool tests. Remind students to record their observations in their data tables.

## Expected Results

| Test | Nitrogen | Oxygen | Carbon dioxide |
|---|---|---|---|
| Magnesium ribbon | Burns | Burns | Burns |
| Steel wool | No reaction | Burns | No reaction |
| Wood splint | Extinguished | Burns | Extinguished |
| Limewater | Clear | Clear | Cloudy |
| Universal indicator | Yellow | Yellow | Red |

The pH testing of oxygen and nitrogen gases may give inconsistent results. Suggest that students compare their data with other groups.

## Answers to Questions

1. If the first bubbles were saved, air would contaminate the sample of the desired gas. Test results would be inconclusive.

2. a. Oxygen is the most reactive.
   b. Nitrogen is the least reactive.
3. a. Oxygen supports rusting and burning.
   b. Air contains 20.7% oxygen molecules.
   c. A higher oxygen concentration in the atmosphere would lead to more fires and possibly to a lower general respiration rate in animals.
4. Carbon dioxide from the fire extinguisher is denser than air and smothers the fire by preventing the less dense oxygen gas from reaching the fuel. Carbon dioxide does not burn.
5. Blow through a straw into limewater. If the solution turns cloudy, carbon dioxide is present. Secondly, carbon dioxide may be detected by an increase of acidity of tap water when exhaled gas is bubbled through water.

### Post-Lab Discussion

Reemphasize the points made during the Pre-Lab discussion. Prepare a class summary table of the results and discuss the questions.

As a review, ask students to write balanced equations for the reactions of this activity.

$2\ Mg + O_2 \rightarrow 2\ MgO$ (white solid)

$3\ Mg + N_2 \rightarrow Mg_3N_2$ (magnesium nitride)

$2\ Mg + CO_2 \rightarrow 2\ MgO + C$ (some may see a black residue)

$CO_2 + Ca(OH)_2 \rightarrow CaCO_3 + H_2O$

A demonstration is found in the Enrichment Materials section at the end of this unit.

### Additional Post-Lab Questions

1. Would carbon dioxide fire extinguishers be useful in putting out a magnesium fire? [No, because carbon dioxide and magnesium react with each other.]
2. Would combustion reactions that produce carbon dioxide contribute to acid rain? [Yes, carbon dioxide produces an acidic solution when dissolved in water.]
3. Why should limewater test solution be freshly prepared and/or kept tightly stoppered? [Limewater can react with carbon dioxide in the air.]

## B.2 A CLOSER LOOK AT THE ATMOSPHERE *(pages 385–386)*

After focusing on individual gases in the atmosphere, this section leads to information about the general behavior of gases, Avogadro's law, and molar volume. The point is made that the physical properties of all gases are nearly the same, provided the temperature is not too low and the pressure not too high. Have students sketch the results of the four observations. This activity will make Avogadro's law more concrete for them and allow you to assess their current thinking about moles and gases.

## YOUR TURN *Avogadro's Law* *(pages 386–387)*

Construct a box having the dimensions of 28.1 cm × 28.1 cm × 28.1 cm and label it "1 mole of any gas at STP = 22.4 L." In the *Your Turn,* students have an opportunity to use Avogadro's law to explain observations.

### Answers to Questions

a. If all three balloons have equal numbers of molecules, they will expand the same amount under the same conditions (a 10 °C temperature increase at constant pressure).
b. If all three balloons begin with the same number of molecules, they will have the same number of molecules when half of the molecules are removed from each balloon. All balloons will have the same volume.

## YOUR TURN *Molar Volumes and Reactions of Gases* *(pages 387–388)*

This activity helps reinforce an important practical consequence of Avogadro's law. If all gases (at the same temperature and pressure) have the same molar volume, then in a chemical reaction involving gases, the coefficients indicate the relative volumes of reactants and products as well as the relative number of moles.

### Answers to Questions

1. $3.0 \text{ mol } CO_2 \times \frac{22.4 \text{ L } CO_2}{1 \text{ mol } CO_2} = 67 \text{ L } CO_2$
2. The ratio of moles NO to moles $O_2$ is 2:1; therefore, the ratio of liters NO: liters $O_2$ is also 2:1. Two liters $O_2$ would react with 4 L of NO.
3. a. $2\ CO(g) + O_2(g) \rightarrow 2\ CO_2\ (g)$

   b. $50 \text{ mol CO} \times \frac{1 \text{ mol } O_2}{2 \text{ mol CO}} = 25 \text{ mol } O_2$

   c. $1{,}120 \text{ L CO} \times \frac{1 \text{ L } O_2}{2 \text{ L CO}} = 560 \text{ L } O_2$

## B.3 YOU DECIDE: ATMOSPHERIC ALTITUDE *(pages 388–390)*

Students explore the changes in the atmosphere during an imaginary flight into the stratosphere. The spacecraft is designed to collect data on the composition of the atmosphere at various altitudes. Students will plot the data, revealing a direct relationship between altitude and pressure; the relationship between altitude and temperature is inconsistent. During the discussion of this activity, consider inviting two students to try to separate a pair of plungers as discussed in Question 3.

### Answers to Questions

1. Temperature decreases linearly to 12 km, increases linearly to 50 km, decreases again linearly to 80 km. Pressure decreases in a more consistent and regular pattern. The graph is a curve.
   a. Pressure
   b. Pressure is due to the collisions of air molecules against a surface. With increasing altitude, there is a steady decrease in the number of molecules in a given volume of air.
2. a. Fall
   b. Rise
   c. In Part A, an increase in altitude means fewer molecules in a given volume of air, which will reduce the pressure. In Part B, there are more molecules at the lower altitude, and therefore more pressure.
3. a. The plungers are hard to separate because there is little pressure inside them, and 14.7 lb/in$^2$ (assuming 1 atm) pushing them together from the outside. [Two students could demonstrate this problem.]
   b. Easier
   c. On top of a mountain, the pressure pushing them together would be less.
4. a. (1) The mass decreases.
   (2) The number of molecules decreases with increasing altitude.
   b. A straight line
   c. Mass depends only on the number of molecules. Each additional molecule contributes the same mass to the sample.
5. a. The temperature-change graph shows three thermal layers.
   b. Layer boundaries appear to be at the 12 km, 50 km, and 80 km altitude levels. The fourth, the thermosphere, begins around 90 km.

## B.4 AIR PRESSURE *(pages 390–391)*

This section introduces the scientific meaning of "pressure" with several common examples. SI units for both force and pressure are used. Placing a brick in different positions illustrates the meaning of pressure. This section also describes a mercury barometer and explains how it works.

## CHEMQUANDARY *Pressure Puzzles* *(pages 391–393)*

1. The force is the same in both cases, but the area over which the constant force is exerted is different. More pressure is exerted on the floor by the shoes with stiletto heels because they have much less surface area in contact with the floor.
2. When walking on the thin ice, the force is confined to the area of the soles of the shoes. Lying down on the ice, or using plywood, spreads the force over a larger area, which may prevent the ice from breaking.
3. The tires should be wide. Thin tires would exert more pressure and tend to sink in the sand, slowing the vehicle's movement. Because of greater surface area, the wider tires exert less pressure and tend to ride on the sand.

## B.5 BOYLE'S LAW: PUTTING ON THE SQUEEZE *(page 393)*

This section is an introduction to the pressure-volume relationship of gases. The mathematical statement of Boyle's law is used to carry out calculations. Students sometimes have difficulties doing calculations for inverse relationships. Help them think through the problems and predict whether the answer should be larger or smaller before they begin calculations. Two methods of calculating are included.

Consider allowing your students to see Boyle's law in action before they are asked to work quantitative problems. Apparatus to demonstrate Boyle's law are available from suppliers of school science materials.

## CHEMQUANDARY A *Volume Discount?* *(pages 393–395)*

The volume of a gas is not a good measure of the quantity of gas unless temperature and pressure are specified. Even though Company B's cylinder costs almost twice as much as the same size cylinder from Company A, the pressure in Company B's cylinder could be substantially higher, leading to more hydrogen molecules per liter. A better way is to buy in terms of cost per unit mass (gram or kilogram) of gas.

## YOUR TURN P-V *Relationships* *(page 395)*

This activity provides qualitative and quantitative practice in applying Boyle's law to predict the pressure and volume of gas at constant temperature.

### Answers to Questions

1. a. (Hint: What will gases in the human body do if the body is exposed to a sudden drop in air pressure?) Gases in the lungs, stomach, and sinuses, and those dissolved in blood, will expand under conditions of reduced pressure, causing an uncomfortable feeling.
   b. Carbon dioxide inside a sealed bottle or can of soda is under high pressure. When the container is opened, the gas expands and rushes out due to lower external pressure. Some dissolved carbon dioxide also comes out of solution as pressure drops.
   c. (Hint: Each tennis ball contains gases at elevated pressure to give a good bounce.) If the tennis ball cans are not pressurized, the pressurized gas in the balls themselves will leak out at a faster rate during storage. Tiny pores in the structural material of the ball will leak gas, just as helium leaks from an inflated balloon.
   d. Air is normally found in the middle ear, its volume dictated by the pressure outside the ear. A sudden decrease in this external pressure caused by a swift elevator ascent causes the air in the middle ear to expand rapidly and rush out into the throat cavity and ear canal. (The "pop" is the releasing of the skin flap that covers the opening into the throat.)
2. $V_1 = 7{,}100$ mL, $P_1 = 1$ atm, $V_2 = 492$ mL, $P_2 = x$

   $$P_2 = 1 \text{ atm} \times \frac{7{,}100 \text{ mL}}{492 \text{ mL}} = 14.4 \text{ atm}$$
3. a. $V_1 = 430 \text{ m}^3$, $P_1 = 760$ mmHg, $P_2 = 596$ mmHg, $V_2 = x$

   b. $$V_2 = 430 \text{ m}^3 \times \frac{760 \text{ mmHg}}{596 \text{ mmHg}} = 548 \text{ m}^3$$

4. a. $V_1 = x$, $P_1 = 14\ lb/in^2$, $V_2 = 40.0\ L$, $P_2 = 44\ lb/in^2$

$$40.0\ L \times \frac{44\ lb/in^2}{14\ lb/in^2} = 126\ L$$

b. Pumping by hand takes such a long time because it takes many strokes of the handle to pump such a large volume of air (126 L).

## B.6 LABORATORY ACTIVITY: T-V RELATIONSHIPS *(pages 396–398)*

This laboratory activity introduces students to the relationship between temperature and volume of a gas sample at constant pressure. Students observe the volume of a trapped air column at different temperatures. After making a graph of their temperature/volume data, the students are asked to predict the temperature that corresponds to zero volume. Students are introduced to the kelvin temperature scale. (273 K is read 273 kelvin.)

### Time

50 minutes

### Materials (for a class of 24 working in pairs)

600-mL beaker and a hot plate or a small deep-fry appliance
300 mL corn oil (to fill heating container deep enough to allow capillary tube immersion)
24 small rubber bands, orthodontist's type, or rings cut from rubber tubing (2 bands/team)
12 capillary melting point tubes
12 non-mercury thermometers (capable of reading to 150 °C)
paper towels
12 centimeter rulers

### Advance Preparation

Obtain the rubber bands, or prepare from rubber tubing.

### Safety

Students should wear safety goggles and aprons throughout the laboratory activity. Be especially careful with the hot oil, as it can cause serious burns. Do not heat the oil with a burner. Maintain the temperature below 130 °C.

### Pre-Lab Discussion

This laboratory activity investigates the relationship between volume and temperature at constant pressure. The gas of interest is that trapped below the oil droplet in the capillary tube. Point out that measuring the length of trapped air in the tube is valid because length is proportional to volume.

### Lab Tips

Heat corn oil to 130 °C. As hot oil can be very hazardous, use one oil bath (under your personal control) for the entire class. Heat the oil in a small deep-fry appliance or in a beaker on a hot plate. Maintain the oil temperature as close to 130 °C as possible. Remind students to read and follow the directions carefully to avoid burns from hot oil.

Corn oil is specified because it has given consistently satisfactory results. If you use some other type of oil, be sure you test it in this activity first. Results have been less satisfactory, for example, with mineral oil.

The capillary tubes should be fully immersed in oil for good results.

If the procedure is followed as written, students will record temperature-length data over about at least a 90 °C to 100 °C range. This is necessary to provide a reasonable number of data points for extrapolation to zero volume. The first data point is at the hot-oil temperature and is the full length of the capillary tube. Measure the distance from the closed end of the tube to the oil plug, and record the temperature.

## Expected Results

The measured temperature-length data will yield reasonably linear plots. Experimental data collected over five years of teaching this activity (involving several hundred results) have led to a median extrapolated temperature of –280 °C at zero volume, compared to the "expected" value of –273 °C.

## Answers to Questions

1. (Student answers will vary but will be close to –273 °C.)
2. Zero milliliters
3. The answer was obtained by extrapolation. We only assume the gas behaves the same way as the temperature decreases. Real gases condense before absolute zero is reached.
4. (No answer required)
5. 273 K, 373 K (student answers will vary but will be close to these numbers.)

## Post-Lab Activity

Focus discussion on student data.

## CHEMQUANDARY *Behavior of Gases* *(page 398)*

1. The carbon dioxide produced by the yeast creates pockets of gas within the dough structure that expand during the high-temperature baking process.
2. The new value for the volume is the old value multiplied by 236/293 (0.80).
3. The anesthetic gas will increase in volume when it enters the warmer environment of a patient's body.
4. The thermostat should be installed on the first floor. If it were installed on the second floor, it would give higher readings (and therefore cause the furnace to be turned on less) than if the thermostat were on the first floor because the hot air is less dense than cold air, and rises. Placing it on the first floor would result in a warmer house overall (and more costly heating bills). If the house is in a warm climate where air conditioning is the issue, and the thermostat were installed on the first floor, the second floor of the house might get very warm before the air conditioning was turned on.

## YOUR TURN *T-V-P Relationships* *(page 399)*

1. a. Pressure will increase by three times.
   b. Examples: pressure cooker, incineration of aerosol cans, etc.
2. a. The absolute temperature (in kelvins) will decrease by a factor of four; the absolute temperature will be 1/4 its original value.
   b. A balloon placed in a freezer
3. Car owners add air to their tires in winter because the decreased temperatures decrease the pressure of the enclosed volume of air. More air molecules are needed to ensure proper tire pressure. Air is released when summer arrives because the pressure from the enclosed volume of air increases as the temperature increases. Both underinflation and overinflation cause excessive wear on the tires.
4. a. Changes in pressure and temperature change the volume.
   b. The volumes of liquids and solids remain nearly constant over a small range of temperatures.
5. a. A given mass of warm moist air will occupy more volume than the same mass of cold air (Charles' law) and will rise above the more dense, cold air.
   b. When the balloon rises, the pressure decreases, which should cause the volume to increase. Temperature also decreases, which should decrease the volume. The data in Table 2 show that the pressure changes are larger than the temperature changes. For example, from 0 km altitude to 10 km, the temperature changes 65 °C, which is a 22% change, while the pressure changes from 760 mmHg to 218 mmHg, a change of 71%. Therefore, Boyle's law dominates, and the volume of the balloon will increase.

## B.7 IDEALLY, GASES BEHAVE SIMPLY *(pages 399–400)*

This section introduces the kinetic molecular theory as a way to understand the regularities in the behavior of gases. Analogies are introduced to help students better comprehend the unseen world of molecules in a gas. The concept of ideal gases is also introduced.

## YOUR TURN *Kinetic Molecular Theory: An Analogy* *(pages 400–401)*

Students answer questions comparing a group of dancers to gas molecules.

### Answers to Questions

1. a. Number of molecules
   b. Volume
   c. Temperature
   d. Pressure
2. a. Boyle's law
   b. Other (P/T = k)
   c. Avogadro's law

## YOUR TURN *Gas Molecules in Motion* *(pages 401–402)*

### Answers to Questions

1. Decreasing the volume means that there is less room for the molecules to move around in. This will cause molecules to hit the walls more often per unit time, causing an increase in pressure.
2. At higher temperatures, the kinetic energy of the molecules is greater because of their higher average velocities. This increased kinetic energy will cause more energetic and frequent collisions with the walls, and thus higher pressure.
3. External body pressure is caused by collisions of air molecules against the skin. Since our body volume remains relatively constant, the internal body pressure must be nearly equal to the external pressure.
4. Given their high rate of motion and small size, gas molecules will eventually pass through the molecular structure of the balloon wall.
5. Helium molecules are smaller in size and exhibit a higher average velocity than molecules of air ($N_2$, $O_2$, $CO_2$). Helium molecules collide more frequently with the balloon walls and fit more easily between the molecular-scale openings.

## PART B: SUMMARY QUESTIONS *(page 403)*

1. A higher atmospheric concentration of oxygen would cause more rapid corrosion (oxidation) of metals. Fires would be more common and difficult to control. Living organisms could probably gradually adapt to the metabolic effects of a higher oxygen concentration.
2. a.

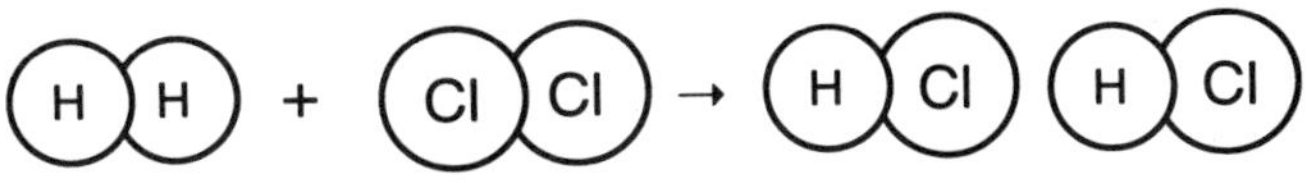


   b. $H_2(g) + Cl_2(g) \rightarrow 2\ HCl(g)$
3. a. II
   b. II
   c. II All answers are the same because gases fill all the space available.
4. As the suction cup is pressed inward, air is forced out from beneath. The atmospheric pressure acting on the outer surface is greater than the pressure beneath the cup, causing the device to remain pressed against the wall.
5. The extreme temperature, pressure, and oxygen concentration above the troposphere make life there (to understate the case) inhospitable.

6. a. constant temperature

$V_1 = x$, $P_1 = 760$ mmHg, $V_2 = 0.210$ L, $P_2 = 1.30$ mmHg

$$V_1 = 0.210 \text{ L} \times \frac{1.30 \text{ mmHg}}{760 \text{ mmHg}} = 0.00036 \text{ L } (3.6 \times 10^{-4} \text{ L})$$

b. constant volume

(1) $T_1 = 300$ K, $P_1 = 1.92$ atm, $T_2 = x$, $P_2 = 5.76$ atm

$$T_2 = 300 \text{ K} \times \frac{5.76 \text{ atm}}{1.92 \text{ atm}} = 900 \text{ K} = 627 \text{ °C}$$

(2) $T_1 = 300$ K, $P_1 = 1.92$ atm, $T_2 = 1{,}275$ K, $P_2 = x$

$$P_2 = 1.92 \text{ atm} \times \frac{1{,}275 \text{ K}}{300 \text{ K}} = 8.16 \text{ atm}$$

Yes, the tank will explode at 1,275 K (1,000 °C).

c. $V_2 = 2.5 \text{ L} \times \frac{280 \text{ K}}{298 \text{ K}} = 2.3 \text{ L}$

7. Comparisons of equal volumes of warm and cold air show the warmer air to be less dense (fewer number of molecules per unit volume), with a greater average molecular velocity. In short, the less dense air rises above the cooler, denser air, causing convection currents. Just as objects that are less dense than water rise to its surface, warm air rises and cold air falls.

## EXTENDING YOUR KNOWLEDGE (Optional) *(page 404)*

- Different environments are encountered at different heights. Pressure and temperature change with depth and height. In some ways the atmosphere is more complex than the hydrosphere: (1) pressure changes exponentially, not linearly; (2) the temperature profile creates fairly permanent strata; and (3) its chemical composition varies with height and depth.
- During inhaling, the diaphragm moves down and out, allowing the volume of the lungs to increase. This results in a pressure decrease within the lungs, and the greater atmospheric pressure forces air into the lungs. Exhaling is the reverse process.
- Boyle's law and diving
  a. Pressure at 10.3 m = 1 atm (from air) + 1 atm equivalent (10.3 m water), or 2 atm pressure.
  b. Pressurized tanks are needed for two purposes: to withstand the increased external pressure without being crushed, and to enable sufficient oxygen gas to be transported in a reasonable volume.
  c. It would be decreased (crushed).
  d. If the diver does not exhale and rise slowly, the volume of the various gases trapped in the lungs and sinuses will rapidly expand, causing extreme pain. In addition, at decreased pressure the gases dissolved in the blood, particularly nitrogen, are released, causing "bends" or even rupturing of blood vessels.
  e. A diver experiences increased pressures. A pilot experiences reduced pressures at high altitudes. In both cases, Boyle's law applies; the body must be protected with a pressurized suit or cabin.
- Mass of a roomful of air:
  a. $\frac{1.28 \text{ g}}{\text{L}} \times 2.00 \times 10^5 \text{ L} = 2.56 \times 10^5 \text{ g}$
  b. (i) If the room is completely sealed, the mass of air in the room will remain constant.(ii) If the room is not sealed, the increased molecular motion would cause some of the air molecules to spread out to adjacent rooms.
- A Celsius temperature can be converted to the corresponding kelvin temperature by adding 273: °C + 273 = K
- Kelvin temperatures
  a. Scientists have cooled cesium atoms to one-millionth of a degree of absolute zero.
  b. Students should use electronic and paper search techniques that are available in their library media center

# C. ATMOSPHERE AND CLIMATE

Part C focuses on complex issues associated with the chemistry of the atmosphere. The interaction of the energy from the sun with Earth's atmosphere and surface is explored. The impact of human activity on the planet's climate, the greenhouse effect, and global warming are discussed. Students study carbon dioxide levels in a *Your Turn* and compare the amounts of $CO_2$ in air samples in a laboratory activity. A discussion of the importance of the ozone layer and damage to it by human activity, particularly the use of CFCs, concludes the section. While the data given in the student text are valid for the time of publication, students may be able to find more current information in newspapers and periodicals.

An excellent resource for this section is *World's Resources: 1994-5 Guide to the Global Environment*, World Resources Institute, Oxford University Press, NY ISBN: 0-19-521044-1.

## C.1 THE SUNSHINE STORY *(pages 406–407)*

This section provides information on radiation that showers Earth during daylight hours. Properties of electromagnetic radiation and its interaction with molecular structure are briefly presented. Only energy and wavelength relationships are used, not frequency and energy.

## C.2 EARTH'S ENERGY BALANCE *(pages 407–409)*

Earth's energy balance is discussed, along with the roles of molecules such as $CO_2$ and $H_2O$.

## YOUR TURN *Solar Radiation* *(page 408)*

These questions are designed to focus attention on the radiation from the sun and the factors that determine the temperature of the atmosphere.

### Answers to Questions

1. Ultraviolet radiation has a shorter wavelength than infrared radiation. Because the shorter the wavelength, the higher the energy of the photon, ultraviolet radiation has enough energy to disrupt chemical bonds and harm biological organisms.
2. Visible light energizes the electrons in some chemical bonds, making sight possible. It also provides energy for photosynthesis reactions.
3. Increased concentrations of carbon dioxide may result in a warming trend with subsequent side effects such as the melting of polar ice caps and the shifting of sites suitable for agriculture.
4. The average daytime temperature would be higher and the average nighttime temperature would be lower. If the atmosphere contained fewer molecules to absorb and reradiate thermal energy to Earth, it would reach thermal balance at a lower temperature.

## CHEMQUANDARY *Grab Another Blanket!* *(page 409)*

On a clear night there are fewer water molecules in the air to absorb and reradiate thermal energy to Earth's surface.

## C.3 AT THE EARTH'S SURFACE *(pages 409–410)*

This section explains the properties of reflectivity and heat capacity and how they affect the quantities of energy absorbed. Point out that materials that quickly become hot or cold have a lower heat capacity than those that become hot or cold slowly; as an example, contrast how much hotter a concrete sidewalk feels than the grass nearby. The unique role of water in moderating temperatures and influencing wind patterns is pointed out.

To show the different reflectivities of black and white, paint one metal can black and another white. Fill with equal quantities of water, place a thermometer in each, and heat with a heat lamp. Within an hour the water in the black can will have a higher temperature than that in the white can.

## YOUR TURN *Thermal Properties of Materials* *(page 410)*

Students apply what they have learned about heat capacity and reflectivity to analyze situations they encounter in life.

### Answers to Questions

1. a. The white convertible fender reflects more sunlight than the dark seat covers.
   b. The concrete is nearly white and reflects more light than the dark colored asphalt.
2. The average temperature would be lower in the winter with large quantities of snow, because the snow would reflect more solar energy.
3. A given mass of water can store more heat energy, and release it over a longer period of time, than alcohol or other fluids having lower heat capacities.
4. Because the higher reflectivity of sand would tend to make it cooler than the grass, heat capacity must be responsible for the higher temperature of sand.
5. The molecules of water in clouds absorb and reradiate thermal energy more efficiently than do air molecules.
6. Water has a moderating effect, and so the city located away from the water would be hotter. Water temperature rises more slowly than that of air, soil, or asphalt. It also cools more slowly. During the day there is a sea breeze and at night there is a land breeze. Therefore, near the water the daytime temperatures do not get as hot, and the nighttime temperatures do not drop as much as they do away from the water.

## C.4 CHANGES ON THE EARTH'S SURFACE? *(pages 410–414)*

Several factors have led to a 30% increase in levels of $CO_2$ since 1800. This section points out that although not all of the variables controlling climate are well understood, almost all climate specialists agree that the average temperature of Earth will increase in the next 100 years. Probable consequences of this global warming trend are discussed.

## CHEMISTRY AT WORK/Keeping Cities Green is a Nonstop Job
*(page 413)*

An arboriculturist combines his knowledge of botany, biology, and soil chemistry to help Seattle plan ahead so that trees and power lines can peacefully coexist.

## C.5 YOU DECIDE: TRENDS IN $CO_2$ LEVELS *(pages 414–415)*

This activity illustrates using trends to predict future behavior. Students plot known carbon dioxide concentrations by years and extrapolate to predict the carbon dioxide concentration in the year 2050. Call attention to the marginal note that emphasizes the tentative nature of extrapolations.

If your students do not yet have a good grasp of the concept of parts per million (ppm), you may wish to have them consider the following analogies: One ppm is equivalent to one penny in 10,000 dollars or one millimeter in one kilometer.

### Answers to Questions

**Part 1**

(Answers are based on student graphs.)

**Part 2**

1. Carbon dioxide levels increase.
2. a. Depends on the year
   b. 390 ppm
   c. 450 ppm
3. No
4. The predictions nearest the last measured data are more accurate than the one for 2050. The rate of carbon dioxide buildup in the atmosphere may change from

previous rates. The further into the future we extrapolate, the greater the change in rate can become, and the more uncertain our prediction.

5. Some possible factors are change in the burning of fossil fuels, the amount of land we clear annually, forest fires, and a change in climatic temperature due to the beginning of a hypothesized ice age.
6. The assumption is that there will be no drastic changes in the processes that have produced the known data.

## C.6 LABORATORY ACTIVITY: $CO_2$ LEVELS *(pages 415–417)*

Students estimate the amount of carbon dioxide present in normal, combusted, and exhaled air by monitoring the change in acidity caused by $CO_2$ absorption in a solution. The indicator is bromthymol blue.

### Time

50 minutes

### Materials (for a class of 24 working in pairs)

12 small glass (not plastic) funnels (7 cm diameter)
12 500-mL filter flasks, wrapped with duct tape or towels
12 1-hole stoppers to fit filter flasks
12 aspirator adapters for faucets
12 small (30-cm) pieces of rubber aspirator hose
24 straight glass tubes
24 glass bends
24 ring stands and clamps
36 rubber connectors
36 test tubes (18 × 150 mm)
12 large candles
12 straws
bromthymol blue solution
12 250-mL Erlenmeyer flasks

### Safety

Students should wear safety goggles and aprons throughout the laboratory activity. Insert the glass tubes into the rubber stoppers yourself to prevent student injury. If you decide to allow students to do this, emphasize proper technique, using glycerin, protective towels, and a twisting motion.

### Advance Preparation

Bromthymol blue can be purchased as either a 0.04% solution or a powder to be made up to a solution that indicates pH in the range from 6.0 (yellow) through green to 7.6 (blue).

If the number of aspirators presents a problem for your classroom, Part 2 can be done as a demonstration.

### Pre-Lab Discussion

Draw student attention to Figure 14 and display the proper setup. For health reasons, only one student from each group should use the straw for the testing of exhaled air.

### Lab Tips

Heavy-walled rubber tubing works better with aspirators. Thin-walled tubing will collapse.

### Answers to Questions

1. The bromthymol blue changed to yellow in the sample from the burning candle. The yellow color indicates that $CO_2$ has dissolved. The burning candle produced the most $CO_2$.
2. Because combustion of paraffin (a hydrocarbon) produces $CO_2$, less air had to be drawn through the apparatus to effect a color change.

3. Exhaled air
4. a. Green plants use $CO_2$—time should increase.
   b. People produce $CO_2$—time should decrease.
   c. Better ventilation reduces the concentration of exhaled $CO_2$—time should increase.
   d. Smoking produces $CO_2$—time should decrease.

## Post-Lab Discussion

Combustion has fascinated humans throughout recorded history and probably earlier. The French chemist Antoine Lavoisier was the first to connect combustion and human respiration; and the British chemist and physicist Michael Faraday delivered a whole lecture series on the history of the candle. Some chemistry texts have an entire laboratory activity patterned around observations of a burning candle. Present-day research in combustion processes is focused on ways of extracting more energy and less pollution from fossil fuels. Both the history and present-day research make interesting topics for library research.

One focus of the post-laboratory discussion can be on the inevitable production of carbon dioxide when fuels, including foods, are combined with oxygen. Ask students how this type of pollution can be decreased. Some of their answers may include using alternative energy sources and reducing use of energy.

Some classes may wish to discuss the analytical techniques used in the activity. Chemists routinely make quantitative analyses, using automated, sensitive instrumental techniques, on the order of ppm. Even parts per billion, ppb, and parts per trillion, ppt, are possible. Especially interested students may be challenged to design an experiment that would allow quantitative measurements of carbon dioxide. One possibility could be based on the reaction of carbon dioxide with limewater and the subsequent increase in mass. An additional *ChemQuandary* is found in the Enrichment Materials at the end of this unit.

## C.7 YOU DECIDE: REVERSING THE TREND *(page 417)*

This activity asks students to consider what can be done to reverse a warming trend from increased carbon dioxide concentration levels. The objective is to list possible courses of actions and decide on those that would gain the strongest support, given economic, political, and social considerations.

This activity can be assigned as homework. On the following day, you may want to divide students into groups to discuss their answers to the questions. Bring the students back together as a class; list the various options on the chalkboard or overhead projector. Analyze each option in terms of economic, political, and social problems.

### Answers to Questions

1. Answers will vary. Some possible answers include: reduce the need for fossil fuel combustion by relying more on conservation (to decrease energy demand) and alternative energy sources (solar, wind, nuclear, etc., to replace the energy presently supplied via combustion), find better ways to prevent and control forest fires, halt the rapid elimination of forest and land clearing, and plant more green plants.
2. a. A number of practical and political problems would probably limit cooperation. Some of these include: (1) Not all scientists agree on the seriousness of the problem. (2) Less technologically advanced nations may be less capable of switching to alternative technologies and may feel that the real burden should be borne by the industrial nations that contribute most heavily to the problem. (3) A rapid reduction in the consumption of fossil fuels would cause major economic problems for both individuals and nations, especially those with economies driven by the use of petroleum. (4) The demand for increased food production may make it difficult to implement some of the suggested solutions.
   b. Students should realize that public support is most likely for measures that do not interfere with safety or comfort, such as lowering the number of forest fires or providing high-quality mass transit for commuters.

## C.8 OFF IN THE OZONE *(pages 417–421)*

The formation of ozone and its decomposition are explained. It is important to distinguish between the ozone that is a pollutant near the planet's surface, and the ozone layer that is necessary to reduce the ultraviolet radiation that reaches Earth. The work of Rowland and Molina on CFCs was recognized by a Nobel Prize in Chemistry in 1995. The Montreal Protocol requiring phase-out of CFCs is discussed, and the complexity of atmospheric chemistry is acknowledged.

## CHEMISTRY AT WORK/A Lean, Green, Dry-Cleaning Machine *(page 422)*

An interview format introduces students to a company that finds new consumer uses for products developed for the defense industry.

## PART C: SUMMARY QUESTIONS *(page 421–423)*

1. a. The order of relative energy per photon is infrared (least), visible (intermediate), and ultraviolet (highest).
   b. Infrared is important in heating our atmosphere, visible enables us to see, and a minimal exposure to ultraviolet is essential to vitamin D production (see discussion in Food unit).
2. a. High-energy solar radiation penetrates the atmosphere, but low-energy infrared radiation is temporarily trapped.
   b. This helps maintain a comfortable average temperature.
3. The stratosphere contains the ultraviolet-absorbing ozone, protecting organisms. It also serves as a barrier that prevents the escape of molecules from the troposphere. The latter is accomplished by a temperature inversion between the troposphere and the stratosphere, which prevents molecules from passing between the layers.
4. The color white reflects all frequencies of visible light, whereas colored materials absorb some of the frequencies. Absorbed visible light is converted into molecular motion (heat). Thus, wearing white will help keep one cooler.
5. a. Beach sand heats up faster during the day and cools off faster at night than water does.
   b. The heat capacity of water is much greater than that of beach sand. Also, while sand is stationary, water can dissipate some of the heat by convection and evaporation.
6. Carbon dioxide and water trap lower-energy infrared radiation that is re-emitted by Earth. As a result, Earth's average temperature is maintained at a higher level (15 °C) than would be the case without this insulating blanket. This is called the greenhouse effect.
7. Burning fossil fuels, and removing green plants from the surface of Earth, increase the amount of carbon dioxide in the atmosphere. Burning appears to be the major factor.
8. a. Increased concentrations of carbon dioxide and water.
   b. Decreased concentrations of carbon dioxide and water.
9. Ozone molecules in the stratosphere absorb high-energy ultraviolet photons that would be capable of breaking the covalent bonds in biomolecules if they were allowed to reach Earth's surface. If the ozone layer were depleted, sunburn, cancer, and other biological problems would increase.
10. CFCs contain chlorine atoms, which in the stratosphere are split off by high-energy ultraviolet photons. These chorine atoms, in a series of reactions, break down ozone to oxygen.
11. The Montreal Protocol established a timetable for phasing out all CFC production in developed nations by 1996.
12. The HCFCs decompose in the troposphere.

## EXTENDING YOUR KNOWLEDGE *(page 423)*

The supplemental reading list provides resources for students to use in investigating the topics suggested in this list. Students can also use electronic search techniques and the Internet to find up-to-date information.

# D. HUMAN IMPACT ON AIR WE BREATHE

Part D explores human impact on air quality by comparing natural and human sources of common pollutants. In particular, the sources of photochemical smog and acid rain and their impact are discussed, with students making acid rain in the laboratory. Different methods of controlling and preventing pollution are considered. This Part includes the study of catalysis and pH as they relate to air pollution.

## D.1 TO EXIST IS TO POLLUTE *(pages 424–427)*

Primary air pollutants are identified, and their natural and human-generated sources are compared. The effects of air pollution will be considered as Part D continues.

## D.2 YOU DECIDE: WHAT ARE THE MAJOR POLLUTANTS? *(pages 427–429)*

Students analyze data from Table 4 (comparing quantities of pollutants from human and natural sources) and from Table 5 (comparing quantities of pollutants from various activities).

### Answers to Questions

**Table 4**

1. a. The pollutants are listed in decreasing order of annual emissions from human sources.
   b. It would be difficult to change the quantity of pollutants from natural sources.
   c. No, this should not be the only factor considered.
   d. Some substances are much more toxic than others to humans and the environment. The figures are for global emissions; the need for specific control measures varies by location. For example, pollutants from automobiles are of greater concern in regions with temperature inversions and a lot of sunlight.
2. a. No, it does not. Natural pollution occurs over a wide area and mechanisms exist in nature for dealing with such substances. Pollutants produced by human activities, which are generally localized, overwhelm the ability of nature to deal with them.
   b. Sulfur dioxide from human sources exceeds (by more than a factor of 10) that from natural sources.
   c. This suggests that modern society is capable of significantly affecting the amount of trace components in the atmosphere.
3. $CO_2$, $NH_3$, $CH_4$
4. Combustion, especially of fossil fuels
5. Nature may not have mechanisms for converting new pollutants into harmless by-products. The chlorofluorocarbons, CFCs, are inert in the lower atmosphere and are not decomposed until they reach the stratosphere, where they pose a threat to the ozone layer.

**Table 5**

1. a. No
   b. Transportation is the main source.
2. a. TSP, HC
   b. $NO_x$, CO
   c. $SO_x$, $NO_x$

Students should realize that all of us—by our lifestyles based on high energy consumption and our use of many industrial products—contribute to the pollution problem.

## D.3 SMOG: HAZARDOUS TO YOUR HEALTH *(pages 429–430)*

This section introduces the idea that the type of smog produced depends on the type of activity, climate, and geography.

## D.4 POLLUTION CONTROL AND PREVENTION *(pages 430–433)*

Section D.4 provides a brief overview of air pollution control options. Some proposals for controlling pollution are offered, while the two questions concerning cost and benefit are raised. An additional *ChemQuandary* is found in the Enrichment Materials at the end of this unit.

## D.5 DEMONSTRATION: CLEANSING AIR *(page 433)*

**Part 1: The Electrostatic Precipitator**

### Materials

1 100-mL glass graduated cylinder
18 gauge bare copper wire (exact gauge is not critical)
1 1-hole rubber stopper to fit the graduated cylinder
wire cutters
transparent tape
Tesla coil
source of smoke

### Safety

Teacher and students should wear goggles. Be aware of electric hazard due to high voltage. Keep the Tesla coil unplugged until ready for use and unplug immediately after use.

### Procedure

1. Assemble the apparatus as shown in the diagram.

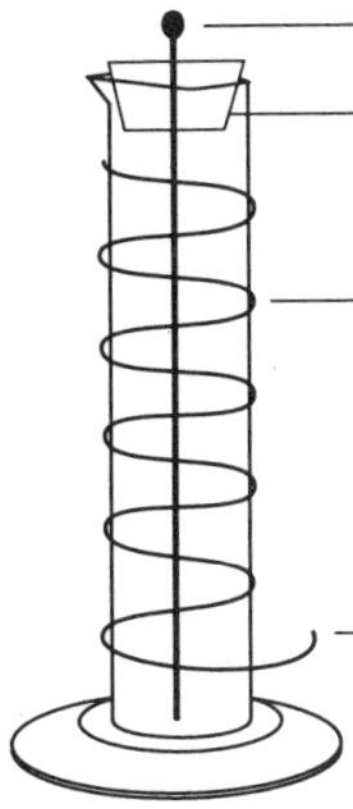


2. Insert the rubber stopper in a clean, dry graduated cylinder.
3. Wrap several turns of wire around the outside of the cylinder.
   Make a loop at the bottom end of the wire. Attach a longer piece of copper wire that can be wrapped around a water faucet to create a ground. Be sure that the water supply line is metallic.
4. Cut a piece of wire about 4 cm longer than the depth of the cylinder and make a small loop on one end. Insert the wire into the graduated cylinder through the stopper with the loop at the top. This wire should not touch the sides of the cylinder.
5. Remove the wire-stopper assembly, and ignite a small piece of heavy cardboard. Drop the burning cardboard into the cylinder and replace the wire-stopper assembly. (Again, be careful not to touch the sides of the cylinder.)
6. Plug in and turn up the Tesla coil.
7. Hold the Tesla coil as close as possible to the central wire in the cylinder, and allow a spark to jump to the top of the central wire. Exercise extreme care with the Tesla

coil. Smoke should clear almost immediately. If this demonstration is performed against a dark background, it will be more visible.

8. Disassemble the apparatus and clean. The cylinder must be clean and dry before it can be used again.

## Answers to Questions

1. The smoke disappears.
2. The copper rod produces a greenish color.

**Part 2: Wet-scrubbing**

Wet-scrubbing can also be done as a student activity using the instructions found in the Enrichment Materials section at the end of this unit. The photograph in the text shows the flasks or bottles hooked up to a centralized vacuum system. The vacuum created using a water aspirator also works if a centralized vacuum system is not available.

## Materials

1 pkg Hydrion (pH) paper
50 mL universal indicator solution
10 mL 12.0 M hydrochloric acid (concentrated HCl)
10 mL 14.8 M ammonium hydroxide (concentrated $NH_3$(aq))
2 500-mL flasks
500-mL filter flask (500-mL flask can be used with 2-hole stopper).
If using filter flask: 1-hole stopper to fit; 2 2-hole stoppers for other flasks
thistle tube
2 straight pieces of glass tubing (to fit in first two flasks)
2 pieces rubber tubing; for first to second flask (10-cm); and for final flask to aspirator (40-cm)
aspirator adapter for sink faucet

## Safety

Students should wear goggles throughout this activity. Be sure there is adequate ventilation when using conc. HCl and conc. $NH_3$(aq). Wrap the flasks with duct tape, leaving space for viewing the contents. Exercise caution when inserting the thistle tube into the stopper.

## Procedure

1. Add 5–10 drops of universal indicator to each collecting bottle (or flask). Then fill each with water up to the neck, allowing just a small space for the inlet tube.
2. Place a moistened piece of pH paper between the stopper and glass rim so it hangs in the space between the stopper and the water level. (This will indicate the pH of the gases hovering above the water level.)
3. Add 10 mL of concentrated hydrochloric acid to the smoke-generating flask. Immediately stopper it.
4. Pour 10 mL concentrated ammonium hydroxide through the thistle tube and immediately turn on the aspirator.
5. Direct students' attention to the color change in the water and pH paper for each flask.

## Answers to Questions

1. $HCl + NH_3 \rightarrow NH_4Cl$
2. Universal indicator gives the pH of the solution of smoke in water as the ammonium chloride enters the flasks.
3. The pH paper indicates the pH of the gases hovering above the water layer just before being exhausted.
4. The pH of the water solution and gas above the water layer approaches neutrality as the reaction proceeds to the second flask and on to the aspirator. Scrubbing can effectively remove acidic or basic compounds from exhaust gases prior to their emission.
5. Wet-scrubbing can remove particulate matter and lower the acidity (or basicity) of exhaust gases.

6. a. Electrostatic precipitators yield a solid waste product, whereas wet-scrubbers result in a water solution that needs to be concentrated.
   b. Precipitators require expensive electricity to operate and do not remove gases such as sulfur dioxide and nitrogen dioxide that can be removed by wet-scrubbers.

### CHEMQUANDARY *Steps toward Clean Air* *(page 433)*

(1) Visible particulate matter in the air is both an eyesore and a potential health problem. Therefore its removal is desirable.

(2) Just because a power plant no longer has a visible, dirty plume does not necessarily mean that colorless, invisible gases are also being removed. Thus, the general public could gain a false sense of security.

## D.6 PHOTOCHEMICAL SMOG *(pages 434–435)*

The discovery of the causes of smog is outlined. Equations for typical reactions that produce photochemical smog are given. Remind students again that the ozone produced in photochemical smog remains in the troposphere.

## D.7 YOU DECIDE: AUTOS AND SMOG *(page 436)*

In this activity, students analyze the graph in Figure 20 to determine why the concentration of certain air pollutants varies over a 24-hour period.

### Answers to Questions

1. a. Nitrogen oxides and hydrocarbons peak between about 6 A.M. and 10 A.M.
   b. These times correspond to the morning rush hours when traffic is at a peak.
2. A given pollutant may decrease because it is being physically dispersed, adsorbed onto surfaces, or chemically converted to another substance.
3. Nitrogen monoxide is the source of nitrogen dioxide. Most of the $NO_x$ emissions from automobiles is in the form of nitrogen monoxide. Once released into the atmosphere, nitrogen monoxide is further oxidized to form nitrogen dioxide.
4. a. Ozone is at a maximum when NO is at a minimum.
   b. This might suggest that NO is involved in ozone formation.

## D.8 CONTROLLING AUTOMOBILE EMISSIONS *(pages 437–438)*

The Environmental Protection Agency has set emissions standards for automobiles, and the automobile industry has developed several techniques to reduce emissions. The main technique, catalytic conversion, is familiar to most students in name only. The chemistry behind the converter leads to an introduction of catalysts and their role in reducing activation energies in chemical reactions.

### CHEMQUANDARY *Controlling Air Pollution* *(page 438)*

It is easier and cheaper to install, maintain, and update control technologies when relatively few sites and relatively few people are involved, but automobiles, buses, and trucks are located at many sites and controlled primarily by individuals. It is also more difficult to ensure compliance with standards when many individuals are involved. However, many states require inspection of motor vehicles before licenses are issued. An additional ChemQuandary can be found in the Enrichment Materials section at the end of this unit.

## D.9 ACID RAIN *(pages 439–440)*

The history, causes, and consequences of acid rain are introduced. Equations illustrate that oxides of nonmetals react with water to produce acids.

## D.10 LABORATORY ACTIVITY: MAKING ACID RAIN *(pages 440–441)*

## Time

50 minutes

## Materials (for a class of 24 working in pairs)

1-2 red apples
12 cm magnesium ribbon
12 pint plastic bags with "sealing strips" (Baggies)
12 stirring rods
12 10-mL graduated cylinders
12 test tubes (18 × 150 mm)
12 paring knives
36 Beral pipets
24 marble chips, calcium carbonate ($CaCO_3$)
12 g sodium sulfite ($Na_2SO_3$)
25 mL 6.0 M HCl (hydrochloric acid)(50 mL 12 M HCl/100 mL solution)
1 pkg Hydrion (pH) paper
1 pkg red litmus paper
1 pkg blue litmus paper
paper towels
distilled water
Paramecium (or mixed) culture

## Safety

Students should wear safety goggles and aprons throughout the laboratory activity. Exercise caution when handling 6 M HCl. Acetone is highly flammable. Use only with proper ventilation.

## Advance Preparation

The red apple skins should be wiped with acetone to dissolve any protective wax, then rinsed and dried with water.

## Lab Tips

Caution students to make sure the plastic bag is quickly sealed after adding required materials. The 6 M HCl is corrosive.

Orange peel may be substituted for apple peel.

As marble chips sometimes give poor results, small pieces of chalk may be substituted.

## Demonstration

Place a drop of Paramecium (or mixed) culture on a microscope slide. (Pond water, if available, is an interesting alternative.) Allow student groups to examine the drop under the microscope and record their observations. Add three drops of the acid rain solution prepared in the Baggie.

## Answers to Questions

1. $SO_2$, sulfur dioxide
2. The dissolved gas increased the acidity of the water.
3. Because the marble in statues reacts with acid, and because the iron in the steel girders reacts with acid, the statue and bridge would deteriorate.

## Post-Lab Discussion

Focus on the experimental results and the answers to the questions.

## D.11 pH AND ACIDITY *(pages 441–442)*

The pH concept, originally part of the Riverwood fish kill information in the first *ChemCom* unit, is explored more thoroughly in this section. It is not necessary to teach students to solve problems involving pH, but they should recognize that the smaller the pH number the greater the acidity. They should also know the range of values on the scale corresponding to acidic, basic, and neutral solutions, and know that one unit change in pH represents a ten-fold change in acidity.

## YOUR TURN *pH* *(page 443)*

1. d. A solution of household lye, pH 13 (most basic)
   g. Household ammonia, pH 11 (basic)
   b. A solution of baking soda, pH 9 (basic)
   f. Sugar dissolved in pure water, pH 7 (neutral)
   e. Milk, drinking water, pH 6 (acidic)
   c. A cola drink, pH 3 (acidic)
   a. Stomach fluid, pH 1 (most acidic)
2. According to the data presented in Question 1, a cola drink is 1,000 ($10^3$) times more acidic than milk.

## PART D: SUMMARY QUESTIONS *(page 443)*

1. Oxides of carbon (CO and $CO_2$), nitrogen (NO and $NO_2$), and sulfur ($SO_2$ and $SO_3$), particulates, hydrocarbon derivatives, and ozone
2. The combustion reaction for the burning of a hydrocarbon fuel, even under the best circumstances, produces carbon dioxide and water. Also, in reality, high-temperature combustion will always result in the production of nitrogen oxides, and catalytic conversions are not 100% efficient. Additionally, combustion is rarely complete, which produces particulates and CO.
3. Photochemical smog is produced by the interaction between sunlight, auto exhaust components (HC, CO, and $NO_x$), and naturally occurring oxidizing agents. This interaction produces a mixture of ozone, nitrogen oxides, complex organics, carbon dioxide, and water. Smog originating with other combustion processes (such as from industries, electric utilities, smelting operations) usually contains particulates and sulfur dioxide.
4. Synergism refers to the phenomenon whereby the environmental and health effects of two or more pollutants or drugs are greater than their simple additive effect. An example related to air pollution is the health impact of sulfur oxides and particulate matter. This phenomenon works against simple pollution control strategies that aim at reducing only one pollutant.
5. Because of wind patterns, acid rain is the most prevalent in the northeastern United States. The burning of high-sulfur coal in industrial states such as Ohio is believed to be the principal source of the acidic pollutants.
6. a. Nitrogen oxides and sulfur oxides are the main contributors to acid rain.
   b. Human-generated $SO_x$ is a result of the burning of sulfur containing fuels and smelting processes, whereas $NO_x$ is a result of high-temperature combustion processes.
7. Given the law of conservation of matter, technologies that capture air pollutants before emission into the atmosphere generate solid matter or aqueous solutions that must be recycled or "thrown away." If the material is discarded improperly, it can become a source of solid waste or water pollution.
8. Ozone is a resource in the stratosphere and a serious pollutant in the troposphere. Also, sulfur is a resource when extracted from the ground in pure form and a pollutant in fuels such as coal and oil.
9. $2\ NO_2\ (g) + H_2O\ (liq) \rightarrow 2H^+\ (aq) + NO_3^-\ (aq) + NO_2^-\ (aq)$
10. Bases: NaOH; Acids: $HNO_3$, $H_2SO_3$; Neutral: $CH_4$, $C_{12}H_{22}O_{11}$
11. Lowest pH = (b) stomach fluid; highest pH = (c) drain cleaner.
12. Photochemical smog is produced by the interaction of automobile emissions with sunlight, while London-style smog is produced by industrial activity and power-generating stations. The ingredients of photochemical smog include nitrogen oxides, hydrocarbons, ozone, and carbon monoxide; London-style smog consists mainly of sulfur oxides and particulates.
13. Automobile emissions produce the substances that are converted to the ingredients

of smog. Unburned hydrocarbons and carbon monoxide, also components of smog, are present in automobile exhaust.

## EXTENDING YOUR KNOWLEDGE *(page 444)*

• There is a strong positive correlation between carbon monoxide levels and traffic accidents. In heavy traffic, sustained levels of 100 or more ppm are common. A few hours of breathing such carbon monoxide levels will result in 17% or more of the body's hemoglobin molecules being tied up with carbon monoxide. The subsequent cellular oxygen deprivation results in a lowering of body efficiency and probably contributes to some loss of function.

| Engine | Advantages | Disadvantages |
|---|---|---|
| Electric | Clean, quiet. | Limited range of driving distance between charges, limited speed, requires more electric power generation. |
| Gas turbine | Lightweight, compact volume, long life. | Poor efficiency, poor acceleration, high production costs. |
| Stratified charge | Cleaner, more energy efficient, already in use. | |
| Wankel | High power/weight ratio, lower vibration-noise, uses variety of fuels, fewer parts. | Higher pollution, lower mileage, maintenance problems. |
| Stirling | External combustion, cleaner, less vibration and noise, little oil consumption, can burn a variety of liquid and gaseous fuels. | Needs larger radiator, more appropriate to larger vehicles. |
| Diesel | Operates at a higher temperature, more energy efficient, produces less unburned hydrocarbons, does not need spark plugs, can burn a wide variety of petrofuels. | Produces more particulates and $NO_x$, noisier, heavier, more costly, requires larger battery, more expensive to build. |
| Expanded mass transport | Cleaner, more space and energy efficient. | Less convenient for some drivers, high initial development cost and time. |

• The April 1983 issue of *Chem Matters* focuses on acids and bases in everyday living and includes (on page 7, "Experimenter's Corner") a procedure for making a red-cabbage juice indicator. Remove several large leaves from the red cabbage head. Tear them in small pieces, put in a pan, and cover with water. Heat, watching constantly (burned cabbage leaves smell very unpleasant!). Remove the leaves with tongs and discard in the trash. Stored in the refrigerator, the juice will be useable for two weeks. Extra red cabbage juice can be frozen in ice trays and thawed for use later. At higher temperatures it deteriorates rapidly. Cabbage test paper can be made by dipping filter paper strips or strips of white paper towels in the juice and allowing them to dry.

• The Toxics Release Inventory (TRI) is available from a variety of sources. Many EPA, Federal Depository, county, public, university, and medical libraries have online access to the National Library of Medicine that makes TRI accessible electronically. NLM can be reached 7 days/week; 24 hours/day.

National Library of Medicine (NLM)

Specialized Information Services
8600 Rockville Pike
Bethesda, MD 20894
Tel: 301-496-6531

Toxics Release Inventory User Support Service (TRI-US) can be reached from 8:00 AM - 4:30 PM (Eastern time).

Toxics Release Inventory User Support Service
U.S. EPA
401 M Street, SW
Washington, DC 20460
Tel: 202-260-1531
Fax: 202-260-4659

# E. PUTTING IT ALL TOGETHER: INDOOR AIR QUALITY

Part E focuses on indoor air quality. This is a complex but important subject because of how much indoor air each of us breathes daily. Indoor air quality in workplaces and homes is getting increased attention as we learn more about health issues and airborne pollutants. In the final activity, students will do library research to learn more about individual indoor air pollutants.

## E.1 INDOOR AIR POLLUTANTS *(pages 445–446)*

Different categories of indoor air pollutants are described. In the *Your Turn,* sources of air pollution in the home are described.

## YOUR TURN *Air in the Home* *(pages 446–447)*

1.

| Number | Room | Source of Pollutant |
|---|---|---|
| 1 | Furnace room | Combustion by-products |
| 2 | Family room | Smoking |
| 3 | Laundry room | Pipe insulation, cleaning supplies |
| 4 | Garage | Car exhaust, paint solvents, pesticides |
| 5 | Den | Smoking, carpet fibers |
| 6 | Bathroom | Shower, hair spray |
| 7 | Living room | Furniture materials, carpet fibers |
| 8 | Kitchen | Cleaning supplies, natural gas, spoiled food |
| 9 | Bedrooms | Heater fumes, mothballs, dry cleaning fluids |
| 10 | Grounds | Rocks |

2. Garage, furnace room, places where members of the family might smoke
3. a. Biological contaminants—molds, fungi, bacteria, insects, cleaning solvents, combustion products
   b. Solvents, combustion products, pesticides, insects, paints
   c. Radon
   d. Environmental tobacco smoke, ozone from office equipment, volatile organic compounds from carpets and furnishings, and plywood
4. Ozone, carbon monoxide, nitrogen oxides, sulfur oxides
5. (Students' answers will vary.)

## E.2 A CLOSER LOOK *(pages 448–451)*

This section gives more detailed information about four important indoor air pollutants—radon, formaldehyde, environmental tobacco smoke, and asbestos.

## E.3 EVALUATING A POLLUTANT *(page 451)*

Assign students to groups of three or four and assign each group one of the four indoor pollutants described in the previous section. Each group will do its own research and then compare results with other groups assigned the same pollutant to prepare one report to give to the class. All members of the group should be involved in all phases of the assignment. Students can get information from the following list of sources as well as other books and periodicals in the classroom and from sources they may locate by computer search in the library media center.

Student groups do not need to be able to perform the tests, but they should come up with the following testing circumstances: the need for a control, the test should be specific for the pollutant, the minimum threshold level compared to the threshold for physiological damage in humans.

### Sources


1. *Merck Index*
2. EPA's *Toxics Release Inventory* (referenced at the end of Part D)
3. EPA's *Indoor Air Clearinghouse*, phone 202-260-2080
4. *Indoor Air Pollution: A Health Perspective*, Samet, J. and Spengler, J. D., Eds.; Johns Hopkins University, Baltimore, MD, 1991.
5. Godist, T., *Indoor Air Pollution Control*, Lewis, Chelsea, ME, 1989.


## E.4 LOOKING BACK *(page 451)*

The final section is a synopsis of the topics covered, leaving students with a sense of the importance of their future decisions as to the quality of air we all must have.

# ENRICHMENT MATERIALS

## CHEMQUANDARIES

**Section C.6**

1. Of the various gases that contribute to the greenhouse effect ($CO_2$, $NO_2$, $CH_4$, etc.), carbon dioxide accounts for about 50%. List and briefly discuss some ways of limiting this impact. Is there a single right or best answer to the problem?

**Answer** (a) Increase energy efficiency, so that less fuel would be needed, and less carbon dioxide produced, to accomplish a given task

(b) Promote conservation measures that result in an overall lowered demand for energy

(c) Switch to solar, wind, geothermal, and nuclear energy

(d) Halt the destruction of forest and begin a reforestation plan, etc. Given our high dependence on carbon-based fuels and given the variability of local natural and technological resources, a variety of "solutions" probably are necessary.

2. How does changing the global climate via the greenhouse effect relate to issues raised in the previous *ChemCom* units?

**Answer** *Water:* Increased temperatures would affect both the demand and supply of water on a regional basis. *Resources:* Much of mineral production and processing (mining, roasting of ores, electrolysis, etc.) depends on the combustion of fossil fuels. *Petroleum:* The cleanest, most efficient burning of petroleum inevitably produces carbon dioxide and water—if less petroleum were burned, more would be available for building purposes well into the future. *Food:* Climate changes would trigger shifts in agricultural productivity—some presently productive areas would become deserts.

**Section C.8**

In what respects are the problems of controlling the greenhouse effect qualitatively and quantitatively different from limiting the depletion of the ozone layer?

**Answer** The greenhouse effect is created by a wide variety of pollutants including carbon dioxide, nitrogen oxides, methane, and chlorofluorocarbons (CFCs). The single, major contributor, carbon dioxide, is produced whenever a carbon-based fuel is burned. Short of switching to nuclear, solar, wind, or geothermal energy, people cannot avoid producing carbon dioxide. And, at present, economic, political, and technical issues block any significant shift in our energy consumption patterns away from carbon-based fuels. On the other hand, ozone depletion is believed to be governed almost exclusively by CFCs that are emitted in much smaller quantities from a much smaller number of sites. Substitutes for CFCs either already exist or can be developed for most applications (witness the successful 1978 ban on using CFCs as propellants in aerosol cans). Also, in terms of economic development of developing countries, use of CFCs is far less critical than use of carbon-based fuels.

**Section D.4**

Previous *ChemCom* units have discussed chemical contamination of water and food. In what respects is the problem of air pollution different from these other types of pollution?

**Answer** (1) At the local level, individuals typically have more choices about whether or not to drink water or eat food from a specific source than they do with respect to breathing the air.

(2) A much larger quantity of air (approximately 20,000 L/day or 14 kg/day) passes through the body on a daily basis than water or food. Thus, even modest contamination of the air can result in appreciable doses of a given pollutant.

**Section D.8**

Electric utilities are responsible for about 80% of the sulfur dioxide in the United States. Would use of low-sulfur coal from the western states be a viable solution? Why or why not?

**Answer** Probably not, since (a) Such coal typically has a lower energy value than coal from the East and more would have to be burned to obtain the same quantity of energy. (b) The increased transportation (and the pollution generated in transit) costs would probably be prohibitive. (c) Mining requires water, and increasing mining in western states could exacerbate the water shortage problem there.

## Demonstrations

**Section A.3**

Set up the following classroom display. Place an Elodea plant in a large test tube containing water and an indicator that changes color near pH 6, such as bromthymol blue. Set up another test tube without the plant as a control. Breathe into both test tubes and note the color change due to the presence of carbon dioxide in your exhaled breath. Stopper both test tubes and place them in the light. Note the change in color in the first test tube as the carbon dioxide is used by the plant.

**Section B.1**

To demonstrate that air is composed of at least two components, approximately 1/5th by volume of a gas that supports oxidation (oxygen) and 4/5th of a gas (nitrogen) that does not, wedge a piece of steel wool moistened with vinegar in the bottom of a graduated cylinder. Place the cylinder upside down in a beaker of water so that the water levels inside and outside the cylinder are equal. Mark the level and let the apparatus sit overnight. As the steel wool rusts, oxygen gas is used up, and the water rises in the cylinder. Water will replace 1/5 of the air in the cylinder. A wood splint test on the remaining gas shows that it does not support combustion.

Studies on the physical and chemical properties of gases gave birth to the modern science of chemistry (Cavendish, Black, Priestley, Lavoisier, etc.) and make an interesting library research project for students.

**Section D.5**

The wet-scrubbing demonstration can be done as a student activity by using the microscale setup, which follows. Provide a beaker for disposal of the acidic contents of the pipets.

## A Microscale Wet-scrubber

### Getting Ready

This activity demonstrates the wet-scrubbing process. The pollutant, sulfur dioxide, is produced by mixing $Na_2SO_3$ and $H_2SO_4$.

### Materials (for a class of 24 working in pairs)

60 Beral pipets
12 24-well plates
24 50-mL beakers
12 burners
60 cm 16-18 gauge wire, cut in 5-cm pieces
12 tongs
12 scissors
10 mL universal indicator
20 mL 0.5 M $Na_2SO_3$ (7.96 g $Na_2SO_3$/100 mL solution)
20 mL 2.0 M $H_2SO_4$ (11.2 mL conc. $H_2SO_4$/100 mL solution)

### Safety

Students should wear goggles and aprons throughout the laboratory activity.

### Procedure

1. Hold the wire in a flame with tongs and use the hot wire to make a small hole in the base of the stems of four of the five Beral pipets, as shown in the illustration. The

diameter of the hole should be slightly smaller than that of the stem of the pipet.

2. Make a diagonal cut near the ends of the stems of three of the four pipets with holes. Remove one-half of the stem of the pipet without a hole, using a diagonal cut.
3. Label the pipets 1–5, and fill as follows:
   No. 5 (uncut with hole), no. 4, and no. 3 (cut with holes)—half full of distilled water plus two drops of universal indicator. The color of the liquid in all three pipets should be green. No. 2 (cut with hole) is empty.
4. Connect the pipets as shown below. The inserted stems should reach the bottom of the filled pipet bulbs.
5. Add 2 drops 0.50 M $Na_2SO_3$ and 4 drops 2.0 M $H_2SO_4$ through the hole in pipet no. 2. *Without releasing the bulb, remove the pump.* (If you allow the pipet to expand before removing it, all the reagents will be drawn back into the pump.) Gas should bubble through the distilled water, carrying the pollutants with it.
6. Reinsert the expanded pipet into the generator and pump gently four times. Record any color changes.
7. Estimate the pH of the contents of the pipets.
8. Dispose of the contents of the pipets into a beaker provided by your teacher.
9. Wash your hands before leaving the laboratory.

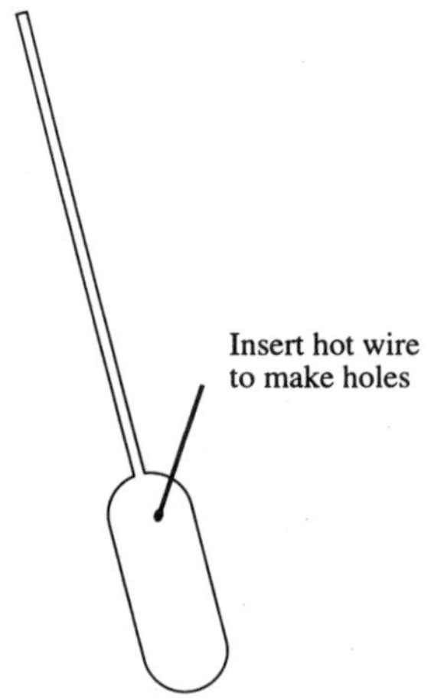


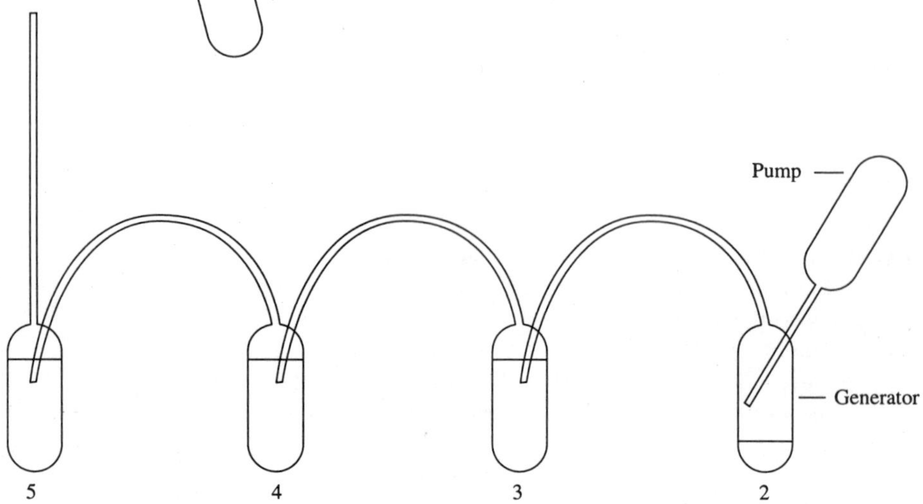

# SUPPLEMENTAL READINGS


"Atmosphere of Planet Earth." *CHEM MATTERS,* February 1983.

"Acid Rain." *EPA Journal* Special Supplement, OPA-86-009, September 1986.

"Why Isn't My Rain as Acidic as Yours?" *Journal of Chemical Education,* 1985, 158-159. Discusses how chemistry is often improperly applied to the acid rain issue.

"Balloon." *CHEM MATTERS,* February 1983.

"Calculating Chemistry--Gases." *CHEM MATTERS,* February 1983.

"Gas Again." *CHEM MATTERS,* February 1983.

"Gas Laws and Scuba Diving." *CHEM MATTERS,* February 1983.

"Getting a Lift." *CHEM MATTERS,* February 1983.

"European Concern about Acid Rain Is Growing." *Chemical & Engineering News,* January 28, 1985, 12-18.

*Acid Rain.* October 1982. Free 14 page pamphlet of the American Chemical Society, GRASP Office, 1155 16th St., NW, Washington, DC 20036.

"Acid Rain--How Great a Menace?" *National Geographic,* November 1981, 652-681.

"Acid Rain." *Scientific American,* October 1979, 43-51.

"Chlorofluorocarbons and Stratospheric Ozone." *Journal of Chemical Education,* 1987, 387-391. Fairly technical article for teachers.

"Build a Hot Air Balloon." *CHEM MATTERS,* December 1983.

"Hot Air Adventure." *CHEM MATTERS,* December 1983.

"How the Right Professor Charles Went Up in the Wrong Kind of Balloon." *CHEM MATTERS,* December 1983.

"Our Fragile Atmosphere: The Greenhouse Effect and Ozone Depletion." *EPA Journal,* December 1986, 12.

"Biosphere II: Out of Oxygen." *CHEM MATTERS,* February 1995.

"The Mass of the Earth's Atmosphere." *CHEM MATTERS,* February 1983.

"The Changing Atmosphere: Tending the Global Commons." *Chemical & Engineering News,* November 25, 1986, 14-64. Special report (reprint available from ACS) focusing on the struggle to understand and check the global warming and depletion of stratospheric ozone trends.

"Bubble Control." *CHEM MATTERS,* February 1984.

"The Pumphouse Incident." *CHEM MATTERS,* February 1984.

"What's That Fizz?" *CHEM MATTERS,* February 1984.

"Are We Close to the Road's End?" *Discover,* January 1986, 28-50. A special report on carbon dioxide and other gases affecting our climate.

"Carbon Dioxide and World Climate." *Scientific American,* August 1982, 35-43.

"Special Report on the Greenhouse Effect." *Newsweek,* July 11, 1988, 16.

"Are We Poisoning Our Air?" *National Geographic,* 170, April 1987, 502.

"The Silent Summer: Ozone Loss and Global Warming: A Looming Crisis." *Newsweek,* June 23, 1986, 107, 64.

"Tropospheric Chemistry." *Chemical & Engineering News,* October 4, 1982, 39-52. Special report (reprint available from ACS).

"Global Climate Change." *Scientific American,* 260, 1989, 36.

"The Planets: Between Fire and Ice." *National Geographic,* January 1985, 5-51. Profile of the planets, emphasizing the uniqueness of Earth for supporting life.

"Your Personal Greenhouse." *CHEM MATTERS,* December 1990.

"The Great Climate Debate." *Scientific American,* 263, 1990, 36.

"Balloon Pressure." *CHEM MATTERS,* October 1985.

"Blimp Flight." *CHEM MATTERS,* October 1985.

"Blimp Ride." *CHEM MATTERS,* April 1986.

"Bringing Helium Down to Earth." *CHEM MATTERS,* October 1985.

"Hydrogen and Helium." *CHEM MATTERS,* October 1985.

"Alternative Automobile Engines." *Scientific American,* July 1978, 39-49.

"Air Quality in the Home." *EPRI Journal,* March 1982, 7-14.

"The Exploding Cabin." *CHEM MATTERS,* October 1994.

"Indoor Air Pollution." *Science, 80,* March-April, 30-33.
*Cleaning Our Environment: A Chemical Perspective.* American Chemical Society, Washington, DC, 1978. Includes chapters on chemical analysis and monitoring, toxicology, air, water, solid waste, pesticides, and radiation.
"The Nitrogen Cycle." *The Science Teacher,* April 1996, 34-37.

# TEACHING SCHEDULE

| | TIME BLOCK 1 | TIME BLOCK 2 | TIME BLOCK 3 | TIME BLOCK 4 | TIME BLOCK 5 |
|---|---|---|---|---|---|
| Class Work | Read Intro +A<br>YD A | YT A.2<br>Read A.4<br>YT A.4 | SQ Part A<br>Read B.1-B.3<br>YT B.2 | Prelab B.4<br>Getting Ready for B.4<br>LA B.4 | Read B.5-B.6<br>YT B.6<br>CQ B.6 |
| Homework | Read A.1-A.2 | Read A.5 | | | Getting Ready for LA B.7 |
| | **TIME BLOCK 6** | **TIME BLOCK 7** | **TIME BLOCK 8** | **TIME BLOCK 9** | **TIME BLOCK 10** |
| Class Work | Prelab B.7<br>LA B.7 | LA B.7 cont.<br>SQ Part B | Review Parts A&B<br>Quiz Parts A&B | Discuss C.1<br>YT C.2 | Prelab C.4<br>LA C.4 |
| Homework | | | Read C.1 | Read C.3 | |
| | **TIME BLOCK 11** | **TIME BLOCK 12** | **TIME BLOCK 13** | **TIME BLOCK 14** | **TIME BLOCK 15** |
| Class Work | Read C.5<br>YT C.5<br>SQ Part C | YT D. 1<br>CQ D.1<br>Read D.2-D.3 | Prelab D.4<br>LA D.4 | Read D.5-D.6<br>CQ D.5 | Prelab D.7<br>LA D.7 pt. 1 |
| Homework | Read D.1 | Getting Ready for D.4 | | Getting Ready for D.7 | |
| | **TIME BLOCK 16** | **TIME BLOCK 17** | **TIME BLOCK 18** | **TIME BLOCK 19** | **TIME BLOCK 20** |
| Class Work | LA D.7 pt.2 | Read D.8<br>SQ Part D | YD E.2 | YD E.3 | Read E.5-E.6<br>SQ Part E |
| Homework | | Read E.1-E.2 | Read E.3 | Read E.4 | |
| | **TIME BLOCK 21** | **TIME BLOCK 22** | **TIME BLOCK 23** | **TIME BLOCK 24** | **TIME BLOCK 25** |
| Class Work | Review C,D,E<br>Quiz C,D, E | PIAT F.1-F.2<br>YTs F.2 (2) | PIAT F.3<br>YD F.3<br>Read F.4 | Review Health unit | Exam Health unit |
| Homework | Read F.1-F.2 | | | | |

**LA**=Laboratory Activity; **CQ**=ChemQuandary; **YT**= Your Turn; **YD**=You Decide; **PIAT**=Putting it All Together; **SQ**=Summary Questions; One **Time Block**=Approximately 50 minutes

# SUPPLY LIST

## Expendable Items and Unusual Materials

| Quantity | Unit | Item | Sections used |
|---|---|---|---|
| 24 | sheets | acetate | D.4 |
| 2.25 | g | amylase | B.7 |
| 6 | small pieces | apple, boiled | B.4 |
| 6 | small pieces | apple, fresh | B.4 |
| 300 | mL | Benedict's solution | B.7 |
| | | boiling chips | B.7 |
| 60 | mL | buffer solution, pH 10 | B.7 |
| 60 | mL | buffer solution, pH 2 | B.7 |
| 60 | mL | buffer solution, pH 4 | B.7 |
| 60 | mL | buffer solution, pH 7 | B.7 |
| 150 | mL | buffer solution, pH 8 | B.7, D.7 |
| 24 | | hair bundles | D.7 |
| 1 | | hair dryer | D.7 |
| 1 | mL | hydrochloric acid, conc | C.4 |
| 120 | mL | hydrogen peroxide, 3% | B.4 |
| 6 | small pieces | liver, boiled | B.4 |
| 6 | small pieces | liver, fresh | B.4 |
| 10 | mL | mineral oil | D.4 |
| 90 | mL | permanent wave neutralizer | D.7 |
| 90 | mL | permanent wave solution | D.7 |
| 1 | pkg | pH paper | D.7 |
| 6 | small pieces | potato, boiled | B.4 |
| 6 | small pieces | potato, fresh | B.4 |
| 24 | small | rubber bands | D.7 |
| 0.84 | g | sodium hydrogen carbonate | C.4 |
| 0.4 | g | sodium hydroxide | C.4 |
| 24 | | splints, wood | D.7 |
| 0.15 | g | starch, soluble | B.7 |
| 12 | sheets | sun-sensitive paper | D.4 |
| 4 | containers | sunscreen, different SPF | D.4 |
| 25 | mL | universal indicator | C.4 |
| 6 | small pieces | yeast cake, boiled | B.4 |
| 6 | small pieces | yeast cake, fresh | B.4 |
| 5 | g | zinc oxide | D.4 |

Our society is extremely health conscious. Stores are full of exercise equipment and videos on working out, and health spas and exercise clubs are commonplace across the country. Labels warning of specific health risks are prevalent on drugs, alcohol, food, personal care items, and many other products commonly used in and around the home. Improved technology and advances in pharmacology have wiped out some diseases and tremendously improved the ways others are treated.

This Personal Chemistry and Choices unit presents some chemical reasons for good health and tells what individuals can do to improve their personal well-being. A knowledge of body chemistry will help students make informed choices to promote good health. A related topic, the chemistry of personal grooming, is also explored in this unit.

# LEARNING GOALS

Upon completion of this unit, the student will be able to

1. Develop a grid of risks and benefits for an activity under personal control. [A.1]
2. Using information found in reference materials, identify toxic and carcinogenic compounds. [A.2]
3. Explain how nicotine skin patches work to help individuals quit smoking. [A.3]
4. Provide examples of correlation, and determine the causal relationship between a given pair of events. [A.4]
5. Define homeostasis, and give examples of how it is related to maintaining good health. [B.1]
6. Identify the most abundant elements found in the human body, and describe their function in maintaining health. [B.2]
7. Explain how enzymes work, and list several factors that may alter their effectiveness. [B.3–B.7]
8. Describe cellular energy production and storage, including the role of ATP. [B.6]
9. Define and give examples of acids and bases, and use net ionic equations to describe the dissociation of acids and bases in water. [C.1–C.5]
10. Describe the components of a buffer, and explain how it maintains pH balance. [C.3–C.5]
11. Apply the concepts polarity, charge, and molecular shape, and the rule like-dissolves-like to skin cleansing and the function of soap. [D.1]
12. List the parts of human skin, and describe their functions. [D.2]
13. Describe the effect of sunlight on skin and how sunscreens protect the skin. [D.3, D.4]
14. Explain the relationship between lack of exposure to sunlight and the childhood disease rickets. [D.5]
15. Describe hair structure, the types of bonding in hair protein, and the effects of various hair treatment chemicals. [D.6–D.8]
16. Use the concept of receptors to account for drug specificity and for the action of narcotic analgesics. [E.1]
17. Contrast the benefits and burdens associated with aspirin use. [E.2]
18. Describe three risks associated with overconsumption of alcohol. [E.2, E.3]
19. Explain the role of the liver in the body's defense against toxic substances. [E.5]
20. Use the concept of synergism to explain the hazards of combining drugs and medicines. [E.6]
21. Use a systematic method to evaluate the perceived risk from participating in an activity or using a technology. [F.1–F.3]

# A. PERSONAL CHOICES AND RISK ASSESSMENT

Students are introduced to this unit with a collage of warning labels from products they might use. Students are introduced to a method of risk assessment for smoking. The chemical components of cigarette smoke are discussed. Students learn the difference between ***correlation*** and ***cause and effect***.

## YOU DECIDE: RISK ASSESSMENT AND SMOKING *(page 455)*

In this activity, students work in small groups to analyze the risks and benefits of smoking and not smoking. After about 10 minutes, groups should be prepared to share their opinions with the class. Do not expect closure or total agreement on the responses to the grid. Some students will know the risks or burdens and choose to smoke anyway. As a homework assignment, have students create a warning label or redesign an existing label for a product of their choice.

## A.1 CIGARETTE SMOKING *(page 456)*

This section explores the extent of smoking in the United States.

## A.2 CHEMICAL COMPONENTS OF CIGARETTE SMOKE *(pages 456–458)*

Some of the chemical components of mainstream and sidestream smoke are listed in Table 1. A Blackline Master of the formulas for the compounds is available at the end of this guide. Students should not be required to memorize these structures.

## YOUR TURN *(pages 457–458)*

The term *toxic* is defined, and students are asked to evaluate the toxicity of the compounds in the previous table. The Merck Index and MSDS sheets are useful sources of information on toxicity. *The Physicians' Desk Reference* also contains information about toxicity; if a copy is available, you may want to have it in the classroom. Some information about the toxicity of the compounds is summarized below.

| Compound | Toxicity | Carcinogen |
|---|---|---|
| Acetone | Toxic by ingestion or inhaling, $LD_{50}$ 10.7 mL/kg | no |
| Ammonia | Toxic by ingestion or inhaling, causes edema of respiratory tract | no |
| Aniline | Both an acute and a chronic poison, $LD_{50}$ 250 mg/kg | suspected |
| Benzene | Toxic by ingestion, inhaling, and skin absorption $LD_{50}$ 3.8 mL/kg | alleged |
| Benzo(α)pyrene | Powerful cancer causing agent | alleged |
| Formaldehyde | Gas is intensely irritating to mucous membranes | alleged |
| Hydrogen cyanide | Exposure to 150 ppm for 30 minutes can cause death. Average fatal dose is 50–60 mg | no |
| Methanol | Toxic by inhaling, ingesting, and skin absorption, can cause blindness, fatal dose 100–250 mL | no |
| Naphthalene | Toxic by inhaling, ingesting, and skin absorption, $LD_{50}$ 1,250 mg/kg | no |
| Nitric oxide | Gas masks mandatory when handled in a laboratory | no |
| Pyridine | Exposure causes depression of central nervous system, gastrointestinal disturbances, kidney and liver damage, $LD_{50}$ 890 mg/kg | no |

## A.3 SOME OTHER COMPOUNDS IN CIGARETTE SMOKE *(pages 458–459)*

Some compounds found in cigarette smoke cause diseases other than cancer. Here we have included the effect of carbon monoxide on the developing fetus, the role of tars in emphysema, and the addictive nature of nicotine. Skin patches, which were available only by prescription when the student text went to press, could be purchased over the counter by mid-1996.

## A.4 MAKING JUDGMENTS ABOUT RISK *(page 460)*

Students are asked to check the correlation in several pairs of everyday events to help clarify the difference between correlation and causation. The former is a necessary condition for the latter, but it is not proof of a definite cause-effect relationship. Also, correlation is generally easier to establish than is causation.

It is helpful to check for understanding by working through a few pairs of events with the entire class.

## YOUR TURN *Events: Related or Not?* *(pages 460–461)*

### Answers to Questions

1. Correlates with
2. Correlates with
3. Is not related to
4. Correlates with
5. Correlates with—causes (sometimes)
6. Correlates with—causes
7. Correlates with—causes (with other factors)
8. Correlates with—causes
9. Correlates with
10. Correlates with—causes
11. Correlates with
12. Correlates with—causes
13. Correlates with
14. May correlate with

## A.5 STUDYING HUMAN DISEASE *(page 461)*

Information about the methods used in medical research is given to illustrate the difficulty sometimes experienced in distinguishing between correlation and causation.

## CHEMISTRY AT WORK/Preventing Illness, Easing Suffering, Giving Hope *(page 462)*

A family physician routinely used concepts from biochemistry and pharmacology when treating patients with HIV and AIDS. He emphasizes that his work also has a large human and social dimension. An activity gives students an opportunity to search for and use current information in newspapers and magazines.

## PART A: SUMMARY QUESTIONS *(page 463)*

1. $CO(g) + 2\ O_2(g) \rightarrow 2\ CO_2(g)$
2. a. $2\ C_{10}H_{14}N_2(s) + 19\ O_2(g) \rightarrow 20\ CO(g) + 14\ H_2O(g) + 4\ NO(g)$

   b. $$\left(2 \text{ mol C} \times \frac{12.01 \text{ g C}}{1 \text{ mol C}}\right) + \left(14 \text{ mol H} \times \frac{1.008 \text{ g H}}{1 \text{ mol H}}\right) + \left(2 \text{ mol N} \times \frac{14.01 \text{ g N}}{1 \text{ mol N}}\right) = 162.23 \text{ g/mol}$$

c. $9 \text{ mg } C_{10}H_{14}N_2 \times \frac{1 \text{ g}}{1{,}000 \text{ mg}} \times \frac{1 \text{ mol } C_{10}H_{14}N_2}{162 \text{ g } C_{10}H_{14}N_2} = 5.6 \times 10^{-5} \text{ mol } C_{10}H_{14}N_2$

d. $5.6 \times 10^{-5} \text{ mol } C_{10}H_{14}N_2 \times \frac{6.02 \times 10^{23} \text{ molecules } C_{10}H_{14}N_2}{1 \text{ mol } C_{10}H_{14}N_2}$

$= 3.3 \times 10^{19} \text{ molecules } C_{10}H_{14}N_2$

3. On the first day the nicotine patch contains the amount of nicotine the smoker would get from a day's smoking. The patches have less and less nicotine in them as time passes until the smoker no longer craves the nicotine.
4. a. Any of the components listed in Table 1
   b. Refer to toxicities from the work done in A.2 *Your Turn.*

# B. YOUR BODY'S INTERNAL CHEMISTRY

Part B analyzes the function of cells, those remarkable chemical factories in our bodies, starting with the elements contained in the body, and then analyzing the mechanism of enzyme action in the cells. Students learn the role of ATP in energy storage, and study how an enzyme changes the rate of a reaction.

## B.1 BALANCE AND ORDER: KEYS TO LIFE *(pages 464–465)*

The importance of balance or homeostasis in body systems and the external environment is discussed. The risks and benefits of modern medicines are pointed out.

## B.2 ELEMENTS IN THE HUMAN BODY *(pages 465–467)*

If the body is viewed as a chemical factory, a key question concerns what elements are found in this factory, and what roles they play in promoting the body's proper functioning. Student understanding is tested in the *Your Turn* activity.

## YOUR TURN *Elements in the Body* *(pages 466–467)*

This activity allows students to review the composition, location, and function of elements in the body. The concept of limiting reactant arises in dealing with the effects of overconsumption and underconsumption of trace minerals and macrominerals.

### Answers to Questions

1.

| Element | Location in Body | Health Role |
|---|---|---|
| Oxygen | In hemoglobin, in molecules of carbohydrate, protein, fat and water | Oxidizing material for fuel, component of body substances |
| Hydrogen | In carbohydrate, protein, fat, and water, in stomach acid and blood, buffer systems | In body substances, aids digestion, transferred between ions and molecules in pH regulation |
| Carbon | In carbohydrate, protein, and fat molecules | For body structure and storage molecules |
| Nitrogen | In protein | For body structure and enzymes |
| Calcium | Bones, teeth, nerve and muscle cells | For bone, teeth structure, nerve transmission, muscle contraction |
| Sulfur | Some proteins | Necessary for synthesis of proteins |
| Phosphorus | Bones, energy transfer molecules (ATP, ADP) | Necessary for synthesis of energy transfer and other molecules |
| Sodium | Body fluids, cells | Needed for memory, appetite, fluid balance |
| Potassium | Body fluids, cells | Aids nerve function, regular heartbeat |
| Chlorine | Stomach acid, body fluids | Plays a role in neurotransmission |
| Magnesium | Cells | Maintains regular heartbeat |
| Iron | In hemoglobin and myoglobin | Transports oxygen molecules |

2. No. A trace element lacking in the diet can act as a limiting reactant to stop important body reactions or functions despite the presence of adequate supplies of macrominerals. While the human body needs larger amounts of calcium and phosphorus than iron and iodine, this does not mean that iron and iodine are less important.

3. a. Eating a normal range of foods containing adequate total calories (food energy) usually ensures that sufficient amounts of these are also consumed, since they are found in a wide variety of plant and animal sources. By contrast, sources of trace minerals are limited, suggesting that a restricted diet will more ikely lead to a deficiency of trace minerals.
   b. Any changes in the types and quantities of foods consumed could produce a large shift (large percentage change) in the intake of certain trace minerals. Therefore, overdoses of trace minerals are more common than of macrominerals. This is especially likely to occur when individuals take supplementary vitamins and minerals.

## B.3 CELLULAR CHEMISTRY *(page 467)*

The speed and precision of body functions and processes can be attributed largely to the functioning of enzymes. The examples given in this section about enzymes give some sense of the extent of enzyme catalysis, which is partially apparent in the following laboratory activity.

## B.4 LABORATORY ACTIVITY: ENZYMES *(page 468)*

This activity illustrates how enzymes function to speed up certain reactions. In the absence of a catalyst and sunlight, the decomposition of hydrogen peroxide proceeds slowly at room temperature. In this laboratory activity the rates of the catalyzed and uncatalyzed reactions are compared. Raw liver, apple, potato, and yeast are used as sources of enzyme catalysts. The decomposition reaction is summarized by the equation $2\ H_2O_2(aq) \rightarrow 2\ H_2O(l) + O_2(g)$.

### Time

30 minutes

### Materials

120 mL of 3% hydrogen peroxide ($H_2O_2$) in brown bottle
24 test tubes (16 × 125 mm)
6 small pieces of fresh raw liver
6 small pieces of boiled liver
6 small pieces of fresh apple
6 small pieces of boiled apple
6 small pieces of fresh potato
6 small pieces of boiled potato
6 small pieces of fresh (baker's) yeast cake
6 small pieces of boiled yeast cake

* cut pieces 1" or less
* less than 1 apple, 1 potatoe
* less than 1/4 lb liver
* 2 test tubes/group of 2
* 2 pieces of each substance per group.

### Safety

Students should wear safety goggles and aprons throughout the laboratory activity.

### Advance Preparation

Cut the required number of pieces of the fresh samples of potato, apple, liver, and yeast. The pieces must be small enough to fit into a 16 × 125 mm test tube. Boil a potato, an apple, a piece of liver, and a cake of yeast. Cool and cut into pieces.

### Pre-Lab Discussion

Ask students why hydrogen peroxide solutions purchased in drug stores are in brown bottles. As the bottle labels indicate, hydrogen peroxide decomposes with age and exposure to light, heat, or dust.

### Lab Tips

Provide containers for student disposal of raw and cooked samples, which may then be put into the garbage disposal in the home economics department or cafeteria.

## Answers to Questions

1. More bubbles of gas are given off, which means that the catalyst speeds up the reaction.
2. The boiled materials do not catalyze the reaction.
3. a. Answers should be the same.
   b. All of the materials catalyze the decomposition of hydrogen peroxide. Some may be better catalysts than others.
4. If commercial hydrogen peroxide did not contain preservatives, it would all decompose before it could be transported and sold.
5. Enzymes in the blood catalyze the decomposition of the hydrogen peroxide.

## Post-Lab Discussion

A recent report describes an incident where a person who drank 35% hydrogen peroxide, believing it to be water, died within minutes. Ask students why 35% hydrogen peroxide is so deadly. (It is more than 10 times more concentrated than the peroxide sold in drug stores. The stomach contents and stomach lining very efficiently catalyze the decomposition of the hydrogen peroxide, forming a deadly foam that interferes with breathing and sometimes causes stroke if the bubbles are absorbed into the blood stream. It is illegal to sell such concentrated hydrogen peroxide for medicinal purposes in most areas of the nation.) This example also points out why it is important to label all containers clearly and to store them in special places where they will not be mistaken for food or water.

## B.5 HOW ENZYMES WORK *(pages 469–470)*

This section presents information on the mechanism of the interaction between enzymes and substrates. It explains how enzymes are selective, catalyzing only certain reactions. The function of coenzymes is briefly described. For example, alcohol dehydrogenase speeds up the conversion of alcohol to acetaldehyde, which is responsible for many of the adverse effects of overconsumption of alcohol.

## B.6 HOW ENERGY IS RELEASED AND STORED *(pages 470–472)*

More than 20 enzymes participate in the more than 20 steps necessary to oxidize glucose to carbon dioxide and water in the body. The role of ATP as an energy-storage molecule is explained. It is more important for students to understand the basic concept than it is for them to be able to recite the steps in the process.

## YOUR TURN *Enzymes and Energy in Action* *(pages 471–472)*

This activity provides practice in solving problems about the production and use of body energy.

## Answers to Questions

1. a. $100.0 \text{ mol ATP} \times \dfrac{1 \text{ mol glucose}}{38 \text{ mol ATP}} = 2.63 \text{ mol glucose}$

   b. $2.63 \text{ mol glucose} \times \dfrac{180 \text{ g glucose}}{1 \text{ mol glucose}} = 473 \text{ g glucose}$

2. $454 \text{ g muscle} \times \dfrac{0.0010 \text{ mol ATP}}{1 \text{ g glucose}} \times \dfrac{1 \text{mol glucose}}{38 \text{ mol ATP}}$

   $= 0.0119 \text{ mol glucose}$

## CHEMQUANDARY *Chew on This Problem!* *(page 472)*

1., 2. Students should be able to detect a sweeter taste after prolonged chewing.

3. Amylase, the enzyme in saliva, has converted starch into glucose, which tastes sweeter than starch.

## B.7 LABORATORY ACTIVITY: AMYLASE TESTS *(pages 472–474)*

Students investigate the effects of temperature and pH on the decomposition of starch by amylase. The investigation can be expanded to include pepsin, if desired. Additional instructions are available in the Enrichment Materials section at the end of this unit.

### Time

Day 1—30 minutes, Day 2—45 minutes

### Materials (for a class of 24 working in pairs)

60 test tubes (16 × 125 mm)
60 stoppers for test tubes
12 10-mL graduated cylinders
12 laboratory burners
12 250-mL beakers
boiling chips
12 ring stands
12 rings
12 wire gauze
12 test tube holders
150 mL 0.1% starch suspension (0.15 g in 150 mL water; refrigerate)
300 mL Benedict's solution (see instructions below)
150 mL 0.5% amylase solution (2.3 g amylase in 150 mL solution)
60 mL buffer solution, pH 2
60 mL buffer solution, pH 4
60 mL buffer solution, pH 7
60 mL buffer solution, pH 8
60 mL buffer solution, pH 10
refrigerator

### Safety

Students should wear safety goggles and aprons throughout the laboratory activity. The buffer solutions of pH 2 and 4 and of 8 and 10, should be treated as acidic and alkaline solutions, respectively. Caution students to be careful with boiling water and handling hot rings and clamps. Boiling chips should never be added to a boiling or very hot liquid.

### Advance Preparation

**Buffer solutions**

Capsules or envelopes of dry powders can be purchased to make buffer solutions by simply adding the required volume of distilled water. If this is impossible, all the buffer solutions can be made from the following two solutions.

**Solution A:** Dissolve 12.4 g anhydrous boric acid ($H_3BO_3$) and 10.5 g citric acid hydrate ($C_6H_8O_7 \cdot H_2O$) in enough water to make 1.00 L of solution.

**Solution B:** Dissolve 38.0 g trisodium phosphate dodecahydrate ($Na_3PO_4 \cdot 12H_2O$) in enough water to make 1.00 L of solution.

To make up 200 mL of each buffer, measure the volumes of each solution as specified in the chart below.

**Buffer Solutions**

| pH | mL solution A | mL solution B |
|---|---|---|
| 2 | 195 | 5 |
| 4 | 155 | 45 |
| 7 | 99 | 101 |
| 8 | 85 | 115 |
| 10 | 54 | 146 |

**Benedict's solution**

Prepared Benedict's solution for qualitative determinations can be purchased. If this is not possible, the following method of preparation can be used.

Dissolve 100 g sodium carbonate monohydrate ($Na_2CO_3 \cdot H_2O$) and 173 g of sodium citrate dihydrate ($Na_3C_6H_5O_7 \cdot 2H_2O$) in 800 mL distilled water. Warm the mixture slightly to speed up the dissolving process.

Dissolve 17.3 g of copper(II) sulfate pentahydrate ($CuSO_4 \cdot 5H_2O$) in 100 mL distilled water. Slowly pour this solution into the first solution. Stir constantly, cool, and add enough water to make 1.00 L.

## Lab Tips

If storage of so many test tubes is a problem, consider dividing the class into groups of four rather than the usual pairs. Remind students to label their test tubes clearly on Day 1, so that they can be identified on Day 2. The starch solution must be freshly prepared and stored in the refrigerator to retard spoilage.

## Expected Results

In the presence of simple sugars, Benedict's solution produces a yellow, orange, or reddish precipitate of copper(I) oxide when the reagent is heated. The intensity of the color indicates the amount of sugar in the solution.

**Color of Benedict's Solution**

| Enzyme | pH | | | | |
|---|---|---|---|---|---|
| | 2 | 4 | 7 | 8 | 10 |
| amylase (room temp) | blue | blue | orange | orange | blue |
| amylase (cold) | blue | blue | lighter orange | lighter orange | blue |

## Answers to Questions

1. The amylase is most effective at pH 7 and 8.
2. Enzymes work best at body temperature. If the temperature is too low, very little happens; if too high, the enzymes are denatured, as was shown in the peroxide decomposition demonstration.

## Post-Lab Discussion

Pool the results from different groups. The amylase is normally found in saliva in the relatively neutral pH of the mouth. It is interesting that because pH changes starch, digestion stops when swallowed material reaches the stomach, and does not continue until enzymes in the intestine begin breaking the starch down further. The old adage about chewing one's food thoroughly has an element of truth. Chewing helps mix amylase with starch to begin digestion.

## PART B: SUMMARY QUESTIONS *(page 474)*

1. Homeostasis is the maintenance of a state of balance. In the body, this state of balance is necessary for normal functioning. When homeostasis is upset, sickness results and, in extreme cases, death occurs.
2. Eating a balanced diet, exercising regularly, keeping the body clean, keeping the environment as healthful as possible.
3. Hydrogen, oxygen, carbon, nitrogen, calcium.
4. a. Chlorine is found as chloride ions inside and outside of cells, particularly nerve cells.
   b. Sodium ions are also important in transmitting nerve impulses.
   c. Iron is contained in hemoglobin and many enzymes.
   d. Magnesium ions are important for proper heart function.
   e. Zinc is required for the activity of many enzymes.
5. Carbohydrates and fats
6. a. Cells carry out chemical reactions rapidly, selectively, and efficiently.
   b. Reactions must be fast so that organisms can respond rapidly to safety threats.

Selectivity is important so that the appropriate reaction can be selected from a variety of possible reactions. Efficiency enables the organism to survive with a minimum energy input.

7. a. Enzymes

   b. These compounds act to catalyze reactions, allowing them to occur faster and at lower temperatures than the uncatalyzed reactions.

8. The oxidation process occurs in a sequence of reactions that allow the cells to release the energy in a controlled, step-wise fashion. For instance, the oxidation of 1 mole of glucose produces 38 moles of ATP.

9. $$\frac{69{,}000 \text{ g albumin}}{1 \text{ mol albumin}} \times \frac{17 \text{ kJ}}{1\text{g albumin}} \times \frac{1 \text{ mol ATP}}{31 \text{ kJ}}$$

$$\times \frac{6.02 \times 10^{23} \text{ molecules ATP}}{1 \text{ mol ATP}} = \frac{2.28 \times 10^{28} \text{ molecules ATP}}{1 \text{ mol albumin}}$$

# C. ACIDS, BASES, AND BODY CHEMISTRY

Acids and bases are defined in terms of the ions that are generated when they dissolve in water: $H^+$ for acids, and $OH^-$ for bases. Equations are given for the dissociation of acids and bases in water. Neutralization reactions are presented as net ionic equations. A distinction is made between strong and weak acids and bases. The final sections describe the composition and function of buffers, which are tested in a laboratory activity.

## C.1 STRUCTURE DETERMINES FUNCTION *(pages 475–476)*

The classic Arrhenius definitions of acids and bases are given, along with equations showing the formation of the ions associated with each. Remind students of the structures of acids and bases presented in the Water unit.

## C.2 STRENGTHS OF ACIDS AND BASES *(pages 476–478)*

Strong and weak acids are defined in terms of the extent of dissociation. Equations with double headed arrows are introduced, showing that the reactants are not completely consumed.

### YOUR TURN *Acids and Bases* *(pages 477–478)*

Students have an opportunity to practice writing dissociation equations.

### Answers to Questions

1. a. $HCl(aq) \rightarrow H^+(aq) + Cl^-(aq)$
   b. hydrochloric acid — hydrogen ion — chloride ion
2. a. $H_2SO_4(aq) \rightarrow H^+(aq) + HSO_4^-(aq)$
   sulfuric acid — hydrogen ion — hydrogen sulfate ion
   b. $H_2CO_3(aq) \rightarrow H^+(aq) + HCO_3^-$
   carbonic acid — hydrogen ion — hydrogen carbonate ion
3. a. $Mg(OH)_2(s) \rightarrow Mg^{2+}(aq) + 2\ OH^-(aq)$
   magnesium hydroxide — magnesium ion — hydroxide ion
   b. $Al(OH)_3(s) \rightarrow Al^{3+}(aq) + 3\ OH^-(aq)$
   aluminum hydroxide — aluminum ion — hydroxide ion
4. a. $2\ H^+(aq) + 2Cl^-(aq) + Mg^{2+}(aq) + 2\ OH^-(aq) \rightarrow 2\ H_2O(l) + Mg^{2+}(aq) + 2Cl^-(aq)$
   b. $H^+(aq) + OH^-(aq) \rightarrow H_2O$

## C.3 ACIDS, BASES, AND BUFFERS IN THE BODY *(pages 478–479)*

This section defines buffers and describes how some of the buffer systems in the body help maintain pH balance. Equations are given for the reactions of the $H_2CO_3/HCO_3^-$ buffer with both acids and bases.

## C.4 LABORATORY ACTIVITY: BUFFERS *(pages 479–480)*

Students investigate the buffering ability of one of the body's two key buffer systems, the carbonic acid–hydrogen carbonate system. After making their own buffer solution, students test its effectiveness in neutralizing added acid and base. The macroscale version of this activity can be found in the Enrichment Materials section at the end of this unit.

### Time

30 minutes

## Materials (for a class of 24 working in pairs)

12 24-well wellplates
12 Beral pipets containing distilled water
12 Beral pipets containing 0.1 M $NaHCO_3$ (sodium hydrogen carbonate), 0.8 g $NaHCO_3$ in 100 mL solution
12 Beral pipets containing 0.1 M NaOH (sodium hydroxide), 0.4 g NaOH in 100 mL solution
12 Beral pipets containing 0.1 M HCl (hydrochloric acid) 0.8 mL concentrated (12 M) hydrochloric acid diluted to 100 mL
12 Beral pipets containing universal indicator solution
12 coffee stirrers (or Beral pipet tips)
universal indicator color charts
toothpicks

## Safety

Students should wear safety goggles and aprons throughout the laboratory activity. Add acid to water when diluting hydrochloric acid.

## Advance Preparation

Label and fill the pipets and place them upside down in the wellplate for distribution.

## Pre-Lab Discussion

Review the function of a buffer as described in section C.3.

## Lab Tips

All references to water in this activity are to distilled (or deionized) water.

Remind students not to share the straws used to blow air into the flask to prepare the carbonate buffer system. Used straws should be discarded.

Provide universal indicator color charts or tables so students can estimate the pH values of their solutions. Students should not try to match the intensity of the color, only the color itself.

## Expected Results

| Step | Starting material | Added | Number of drops | Color | Approximate pH |
|---|---|---|---|---|---|
| 1 | distilled water + universal indicator | | | pale green | 7 |
| 3 | buffer solution + universal indicator | | | | |
| 4 | distilled water + universal indicator | 0.1 M NaOH | | dark blue | 10–12 |
| 5 | buffer solution + universal indicator | 0.1 M NaOH | | | |
| 7 | distilled water + universal indicator | 0.1 M HCl | | red | 2, 3 |
| 8 | buffer solution + universal indicator | 0.1 M HCl | | | |

The number of drops used will vary because of the different amounts of time students blow into the flask. The buffer solutions should require significantly more drops to change the color of the indicator than the unbuffered solutions.

## Answers to Questions

1. a. Yes
   b. The buffer solution required much larger amounts of acid or base to change the pH.
2. a., b. In each case, more acid or base is needed with the carbonate buffer.
3. $H_2CO_3(aq) + OH^-(aq) \rightarrow HCO_3^-(aq) + H_2O(l)$

## Post-Lab Discussion

Pool the class data so all students gain the benefit of the observations of all groups. Each group should be able to observe that the buffer system, compared with water, strongly resists pH change.

## C.5 BALANCING pH *(pages 480–481)*

The serious consequences of even small changes in pH on the human body are described in this section. The *Your Turn* asks students to apply their knowledge of acids and bases to everyday situations.

## YOUR TURN **Conditions That Affect pH Balance** *(pages 481–482)*

This activity acquaints students with a number of conditions that can overload the body's buffering capacity. This activity also includes a review of the pH balance requirements of the body.

### Answers to Questions

1. a. Hyperventilation causes an increase in blood oxygen levels and a subsequent decrease in carbon dioxide levels.
   b. This lowers acidity and causes an increase in pH.
   c. The person rebreathes the exhaled carbon dioxide, thus increasing the carbon dioxide levels in blood.
2. a. Holding one's breath causes a rise in carbon dioxide levels.
   b. This causes an increase in acidity and a decrease in pH.
3. The accumulation of lactic acid lowers the blood pH to more acidic levels.
4. a. Cardiac arrest prevents the body from eliminating carbon dioxide; thus the blood pH decreases as the concentration of carbon dioxide (and thus carbonic acid) increases.
   b. The injection of $HCO_3^-(aq)$ from sodium hydrogen carbonate would counteract this effect and return blood pH to more normal levels.
5. Aspirin is acetylsalicylic acid. A large excess could cause acidosis.

## PART C: SUMMARY QUESTIONS *(page 482)*

1. a. An acid is able to produce hydrogen ions ($H^+$) in water; a base produces hydroxide ions ($OH^-$) in water. Many acids are molecular, covalently bonding compounds; most bases are ionic compounds.
   b. A strong acid is almost completely ionized in water, which means that it separates almost 100% into ions. A weak acid is only slightly ionized.
   c. A strong base is almost completely ionized in water, meaning it is almost 100% separated into ions. A weak base is only slightly ionized.
2. No, the species must be able to form a hydrogen *ion* in solution.
3. A buffer system contains both an acid and a base and can neutralize both an acid and a base. The acid in the buffer system provides $H^+$ to neutralize the added base, and the base in the buffer system neutralizes added $H^+$ ions.
4. Hyperventilation causes decreased carbon dioxide levels and raises blood pH. Holding one's breath causes increased carbon dioxide levels and lowers blood pH. Strenuous muscle activity raises the lactic acid concentration and lowers blood pH. Cardiac arrest raises carbon dioxide levels and lowers blood pH.
5. a. $H^+(aq) + Lac^-(aq) \rightarrow HLac(aq)$
   b. $HLac(aq) + OH^-(aq) \rightarrow Lac^-(aq) + H_2O$

## EXTENDING YOUR KNOWLEDGE (OPTIONAL) *(page 482)*

a. (1) HCl, a strong acid; $Cl^-$, a weak base
   (2) $H_2O$, a weak base or a weak acid; $OH^-$, a strong base
   (3) $HNO_3$, a strong acid; $NO_3^-$, a weak base
   (4) $HC_2H_3O_2$, a weak acid; $C_2H_3O_2^-$, a strong base

b. The best buffer system would be $HC_2H_3O_2/NaC_2H_3O_2$. A buffer must consist of a weak acid or base and one of its salts. All of the other choices contain either a strong base or a strong acid.

# D. CHEMISTRY AT THE BODY'S SURFACE

Part D covers the more cosmetic aspects of personal care. Continuing the theme of the relationship of structure and function, we have focused on skin and hair and the agents used to keep them clean. The current topic of the effects of sun on skin is explored. Finally students look at what happens to the structure of hair when it is given a permanent wave.

## D.1 KEEPING CLEAN WITH CHEMISTRY *(pages 483–488)*

The structure of soap, and the process by which it cleanses the skin, is an application of the "like-dissolves-like" concept developed in the Water unit. The discussion includes the polarity and hydrogen bonding capacity of water molecules. The molecular explanation of solubility of polar and nonpolar substances in water is explored. The text explains how the soap molecule's polar and nonpolar ends enable it to dissolve both in water and nonpolar substances (grease and oils).

### YOUR TURN *Polarity and Solubility* *(pages 484–485)*

Students are asked to apply what they know about bond polarity, charge, and molecular shape to predict solubility.

#### Answers to Questions

1. a. NaOH: ionic, water-soluble
   b. $CH_3CH_2$–O–$CH_2CH_3$: nonpolar, soluble in nonpolar solvents
   c. $CH_2OHCH_2OH$: polar, water-soluble
   d. $KNO_3$: ionic, water-soluble
2. a. Cholesterol is practically insoluble in water because the one –OH group is only a small portion of a very large, nonpolar molecule.
   b. One would expect it to be more soluble in nonpolar solvents because most of the molecule is nonpolar.
3. a. Glucose should dissolve in water. Each glucose molecule contains several –OH groups, which can form hydrogen bonds with water.
   b. The fatty oil should dissolve in gasoline. The molecule has no polar –OH groups and would probably dissolve in nonpolar solvents like gasoline.
   c. Glucose has a number of polar –OH groups.
4. It is soluble in both water and oil, given that one end is polar (–$OSO_3^-$, $Na^+$) and the other end is a long nonpolar hydrocarbon chain.

### CHEMQUANDARY *Benzoyl Peroxide and Acne* *(pages 487–488)*

1. Benzoyl peroxide has the ability to dissolve in the oils and then react with the bacteria, just as hydrogen peroxide does.
2. The antibiotics kill the bacteria, preventing the infections that lead to pimples.

## D.2 SKIN: IT'S GOT YOU COVERED *(pages 488–487)*

This section briefly describes the structure of the skin and how it is related to the functions of the skin. This background is useful in the subsequent discussion of skin protection products.

## D.3 PROTECTING YOUR SKIN FROM THE SUN *(pages 489–492)*

This section discusses the body's defense system against the ultraviolet component of the sun's radiation and how the use of sunscreens can prevent some of the damage. The terms *minimal erythemal dose* (MED) and *sunscreen protection factor* (SPF) are explained.

## D.4 LABORATORY ACTIVITY: SUNSCREENS *(pages 492–494)*

This activity is designed to illustrate the relative effectiveness of sunscreens with different SPFs and to use the information obtained in determining the SPF of an unknown sample of sunscreen.

### Time

20 minutes

### Materials (for a class of 24 working in pairs)

containers of sunscreen of four different SPFs
unlabeled samples of sunscreen for unknowns
24 acetate sheets (such as overhead transparencies)
12 sheets of sun-sensitive paper (see Lab Tips)
mineral oil
zinc oxide
12 droppers for mineral oil

### Safety

Students should wear safety goggles and aprons throughout the laboratory activity.

### Advance Preparation

You can use an ultraviolet light if it is necessary to do the activity indoors. If the ultraviolet light is used, remind students not to look directly at it.

### Pre-Lab Discussion

Discuss the importance of protecting the skin from too much sun. Getting an unnatural dark tan is likely to lead to skin cancer. Emphasize that wearing sunscreen does not mean that one can safely stay in the sun for unlimited time periods. In fact, some evidence suggests that sunscreens are somewhat ineffective in preventing skin cancer because people merely use them to justify staying out in the sun for longer periods of time. Encourage students to consider their own sunbathing habits in light of the known connection between excess exposure to sun and skin cancer.

### Lab Tips

Sun-sensitive paper is available in stores that sell educational games and activities for children. The acetate sheets and the sun-sensitive paper can be cut into smaller pieces to save resources. Posting the sheets at the front of the room is a good way to compare results of class groups.

### Expected Results

The degree of darkening of the sunprint paper decreases as the lotion sunscreen ratings increase.

#### Answers to Questions

1. Results will vary.
2. Possible trade-offs: the convenience of driving a car versus the risk of an accident, and the pleasure gained from cigarette smoking versus the increased probability of lung cancer.
3. Dark-skinned persons have a higher concentration of melanin in their skin, which screens out more of the more harmful ultraviolet rays.
4. All of the components of the sunscreen are "chemicals", whether they were obtained from natural sources or manufactured. A glance at the label of any sunscreen reveals a list of active or inactive ingredients.

### Post-Lab Discussion

Focus the discussion on student results, answers to questions, and applying the findings to students' lives.

## D.5 GETTING A "D" IN PHOTOCHEMISTRY *(page 495)*

Following the previous information about the potentially harmful effect of sunshine, it is important that students understand that—in moderation—sunshine is not only healthful, but essential in promoting the formation of vitamin D necessary for bone strength.

## CHEMQUANDARY *Vitamin D Solubility* *(page 495)*

Vitamin D is soluble in fat. Since the vitamin is practically insoluble in water, it is more difficult for the body to excrete it in urine, an aqueous solution.

## D.6 OUR CROWNING GLORY *(page 496)*

This section describes both the macro- and micro-structure of hair as a prelude to discussions on the role of chemistry in grooming.

Ask students if they agree with the first sentence of this section concerning the amount of time, money, and concern people have with hair. An interesting discussion usually results.

## D.7 LABORATORY ACTIVITY: CHEMISTRY OF HAIR *(pages 497–499)*

Students investigate the general characteristics of hair and test the effects of various hair cleansing and treating solutions on hair properties. Students work in groups of four, each person testing one solution: a pH 4 solution, a pH 8 solution, a permanent wave solution, or a permanent wave + neutralizer solution. In Part 2 all students test their own hair.

### Time

45 minutes

### Materials (for a class of 24 working in pairs)

24 hair bundles (3-cm diameter, 15-cm length) from barber shop or hair salon
24 wood splints
90 mL pH 4 hydrochloric acid (approximately 10 drops of 0.1 M HCl/500 mL solution)
90 mL pH 8 sodium hydroxide (dilute 0.1 mL 0.1 M NaOH to 1.00 L)
180 mL permanent wave solution
90 mL permanent wave neutralizer
1 package Hydrion pH paper
distilled water
24 rubber bands, orthodontist's type, or cut pieces from rubber tubing
12 25-mL graduated cylinders
24 small test tubes (18 × 150 mm)
6 test tube racks
6 scissors
light source or hair dryer (optional)

### Safety

Students should wear goggles and aprons throughout the laboratory activity. The hair samples should be discarded in the trash, not in the sink.

### Advance Preparation

Get some hair from a barber shop or beauty salon. Arrange ahead of time with a local hair stylist to save some hair that is at least 15 cm (6 inches) long for this activity. Ask students to bring hair dryers from home to speed up the drying process.

### Pre-Lab Discussion

Have students bring in advertising claims and ingredient lists for various hair care products. Discuss the relative merits of the claims. Assign students to groups of four and let them decide which member of the group will investigate Treatment A, B, C, and D. Students should read the procedure carefully before beginning the activity so they can allocate their time efficiently. If possible, Part 1 activities should be completed in one class period. Allowing the hair to dry out overnight may change some of its characteristics.

## Lab Tips

All tests by a given team are performed on identical hair samples.

If the hair is left out overnight, students may find textural differences.

## Expected Results

In general, hair tensile strength decreases with increased pH; the samples at pH 8 and the permanent wave solution (without neutralizer) will be observed to break most easily.

## Answers to Questions

1. Coarse hair contains more strands of protein than does fine hair.
2. The cuticle overlaps, like roof shingles. In healthy hair, rubbing one's fingers down the hair catches intact cuticle and hair bunches up. If the cuticle is damaged (broken off), the hair will be smooth and will not bunch up.
3. (Answers will vary.)
4. Permanent wave solution with neutralizer
5. Permanent wave solution with neutralizer
6. The pH 8 and permanent wave (without neutralizer) solutions make the hair most brittle.
7. The pH 8 solution with heat caused the most damage.
8. Weaker bonds are broken and reformed with water, causing the curl to be temporary.
9. The basic solution is better at breaking the S–S bonds.
10. Sodium hydroxide solutions are more caustic and damaging to the hair.
11. After swimming and before sunbathing, one should rinse the hair to remove the chlorine and other water-borne substances. Swimming pools are generally kept slightly basic. The heat from the sunbathing increases the amount of damage.

## Post-Lab Discussion

After discussing the questions, encourage students to apply what they have learned to the advertising claims discussed in the pre-lab session.

## D.8 HAIR STYLING AND CHEMICAL BONDING *(pages 499–502)*

This section explains the previous laboratory activity using bonding principles and attractions between chemical groups. Be sure to connect the discussion to the observations made in Laboratory Activity D.7.

## PART D: SUMMARY QUESTIONS *(page 502)*

1. a. During skin cleansing, perspiration, skin-moistening oils, dead cells, and external dirt and oil are removed.
   b. To prevent them from being oxidized to odoriferous compounds by skin bacteria
   c. Excessive cleansing or the use of strongly basic agents can dry the skin and cause premature aging.
2. a. Ionic—water-soluble
   b. Slightly polar (due to presence of –OH)—water-soluble
   c. Nonpolar—need detergent
   d. Polar (presence of –OH)—water-soluble
3. Soaps or detergents act to connect polar water molecules and nonpolar grease or oil molecules. They consist of long molecules with one end that is polar (and therefore soluble in water) and the other end nonpolar (and therefore soluble in grease).
4. The skin acts as both a barrier and an exchange conduit between the organism and its environment. Vital body fluids are kept in and bacteria are kept out. Additionally, it is

the chief organ for the sense of touch, can communicate emotions, and plays a major role in heat regulation (maintaining a constant body temperature) by sweating.

5. Tanning is the body's defense mechanism to minimize the influx of damaging UV radiation. Melanin is formed.
6. Suntanning is usually associated with UV radiation, which has sufficient energy to cause disruption of some molecular bonds within skin cells. If DNA molecules are disrupted and the mutations reproduced, skin cancer may result.
7. Sunscreens contain molecules that can absorb UV radiation.
8. a. Hair is a polymer (long-chain molecule). Rubber is also a polymer.
   b. Water molecules can break some of the hydrogen bonds between hair strands.
9. Permanent wave solutions cause disulfide bonds between protein chains to break, rearrange, and then reform in a new arrangement. Water causes a temporary rearrangement of hydrogen and ionic bonding in protein chains.

## EXTENDING YOUR KNOWLEDGE (OPTIONAL) *(page 502)*

Dark skin (containing extra melanin) is better able to prevent high-energy, UV radiation from reaching the sensitive underlying skin cells. Therefore, dark-skinned persons are less likely to suffer from sunburn and skin cancer than are light-skinned persons.

# E. CHEMICAL CONTROL: DRUGS AND TOXINS IN THE HUMAN BODY

In this section, students learn how drugs can be used to help the body regain its chemical balance when something goes wrong, and how drugs (used improperly) and other toxins can upset the body's chemical balance. Once again, students learn that they can make certain choices to help keep their bodies in balance.

## E.1 A GLIMPSE OF DRUG FUNCTION *(pages 503–504)*

The difference between a drug and a toxin is explained. Drug function is explained in terms of receptor sites. Shapes of molecules are important. Narcotic analgesics provide an example of how drugs can pass through the body's systems and mimic body chemistry to produce certain desired effects.

## E.2 YOU DECIDE: PROS AND CONS OF ASPIRIN *(pages 505–507)*

This section is devoted to the drug aspirin—its history, modification, uses, and side effects. It considers aspirin as a specific case in which chemists started with a compound provided by nature and improved upon it. The last part of the section provides information about the drug alcohol, its uses and abuses.

The history of the development of aspirin is an interesting sidelight. Willow bark, which contains salicylic acid—a precursor of aspirin— was used as early as 400 B.C. by Hippocrates to ease the pain of childbirth. Native American cultures in what is now the Western United States chewed bark from the aspen tree (a botanical relative of the willow) to reduce pain and fever.

Salicylic acid was found to be the active ingredient responsible for pain and a fever reduction. In the later 19th century, synthesis of this compound made it more widely available. However, the acidic nature of salicylic acid also makes it highly irritating to the stomach. A German chemist, Felix Hoffmann, began to make slight changes in the molecule and tested the resulting products on his father, who suffered from arthritis. By 1899 the German Bayer company began to market one of Hoffmann's products—acetylsalicylic acid, or aspirin.

Since its introduction, aspirin has grown to be the best-selling drug in the world. Over 20 billion aspirin tablets (under a variety of trade names) are consumed annually in the United States.

### Answers to Questions

1. Student answers will differ. Evaluate the reasons for their choices.
2. Options that might be available to counteract the long-term side effects are taking another drug with the aspirin to counteract the effect or taking coated (enteric) aspirin, which changes the site where the medication dissolves. Taking aspirin with food also minimizes some of the side effects.
3. a. Effectiveness of the medication, dosage necessary, side effects, results of long-term exposure, effect on other drugs, effect on food absorption in the body, possible effects on fetuses

   b. Manufacturers' information, personal physician, pharmacist. Pharmacists are now required to provide patient counseling and verbal or written information about the drug.
4. Medications have the potential to upset the chemical balance of the body, and many have side effects that cause or contribute to other problems. It is also possible for resistant strains of bacteria to develop so that an antibiotic is no longer effective.

## E.3 YOU DECIDE: EFFECTS OF ALCOHOL *(pages 507–509)*

This activity is designed to emphasize the fact that alcohol is a drug with specific physiological effects. Divide the students into groups to answer the questions before beginning the class discussion. A guest speaker, such as a law enforcement official, a rehab counselor, or an emergency room worker might begin with a discussion of the laws of the state regarding alcohol use. A "preachy" tone, seldom effective with teenagers, should be avoided. In *Chem Matters*, February 1985, p. 8, there is an article on alcohol metabolism and dietetic and behavioral effects.

/9

### Answers to Questions

2 pts 1. a. 0.05%
   b. 0.05% – 0.015% – 0.015% = 0.02%

2 pts 2. Reaction time slowed by 15–25%, visual sensitivity reduced up to 32%, headlight recovery 7–32 seconds longer, lowered alertness, reduced coordination

2 pts 3. No, the person's blood level would be 0.095% one hour later, just under the legal limit in most states.

2 pt 4. This person approaches the 0.35% blood level, thus approaching the level where death can occur.

1 pt 5. The alcohol level does not exceed 0.05%, and drops to 0.02% in two hours. Given that the probability of causing an accident doubles when the driver's blood alcohol was just 0.06%, the longer the wait the better.

## E.4 NOTES ON SOME ILLICIT DRUGS *(pages 509–510)*

This section provides objective information about the effects of illicit street drugs including marijuana, cocaine, crack, ice, LSD, and anabolic steroids. Value judgments are avoided.

## E.5 FOREIGN SUBSTANCES IN THE BODY *(pages 510–513)*

This section discusses how the body defends itself against harmful substances through detoxification, particularly by the liver. The development of the concept of enzymatic catalysis is continued, using alcohol, benzene, and sulfite ions as examples. However, these detoxification processes are limited, and continued abuse causes liver damage. Additionally, the body is not able to protect itself from some disease-causing agents, such as the common cold and flu viruses.

## CHEMISTRY AT WORK/Using Chemistry to Bring Criminals to Justice *(pages 511)*

A forensic scientist solves problems for law enforcement. She lists the important qualifications for her work as familiarity with the laboratory, using an analytical mind, and public speaking ability. An activity suggests that students write a mystery involving a chemical clue.

## E.6 DRUGS IN COMBINATION *(page 513)*

This section points out the risks of combining drugs, which may cancel the therapeutic benefits of each and can even result in unexpected, hazardous effects. It is especially hazardous to combine drugs, such as alcohol and tranquilizers, that have similar biological results.

/20

## PART E: SUMMARY QUESTIONS *(page 514)*

1pt 1. a. Narcotic analgesics (such as morphine) are believed capable of relieving pain through their ability to block key receptor sites or nerve cells.

1pt b. The brain's own painkillers, endorphins and enkephalins, have several parts of their molecules that are similar to the narcotics, and would thus fit the same receptor sites.

2. a. Alcohol depresses the transmission of nerve signals by reducing the concentration of calcium ions at nerve endings, and thus prevents the release of the messenger molecules that are involved in neural transmissions.

   b. Cocaine functions as an anesthetic by reducing the sensitivity of the nerve membranes to return of messengers.

3. (1) Different medications may counteract each other; therefore the net effect of the combined treatment may be zero.

   (2) Different medications may interact in a manner that can harm or even kill the patient.

4. Yes, many medicines specify whether they should be taken on a full or an empty stomach to avoid gastrointestinal irritation and/or aid in absorption. Alcohol interacts synergistically with many other medicines, and, if instructions are not followed, life threatening reactions can result.

5. Chemical processes in the liver include: (1) separation of useful and harmful substances, (2) storage of glucose as glycogen, (3) conversion of amino acids to proteins, and (4) the detoxification of harmful materials.

6. Chemical toxins may be made harmless by reactions in the liver. Biochemical agents such as viruses, bacteria, fungi, and parasites may be attacked and destroyed by antibodies.

7. a. $$\frac{0.050 \text{ g alcohol}}{1 \text{ L blood}} \times \frac{1 \text{ mL alcohol}}{0.79 \text{ g alcohol}} = \frac{0.63 \text{ mL alcohol}}{1 \text{ L blood}}$$

   b. $$\frac{4.0 \text{ g alcohol}}{1 \text{ L blood}} \times \frac{1 \text{ mL alcohol}}{0.79 \text{ g alcohol}} = \frac{5.1 \text{ mL alcohol}}{1 \text{ L blood}}$$

   c. $$\frac{0.12 \text{ g alcohol}}{100 \text{ mL blood}} \times \frac{1 \text{ mL alcohol}}{0.79 \text{ g alcohol}} = \frac{0.15 \text{ mL alcohol}}{100 \text{ mL blood}}$$

## EXTENDING YOUR KNOWLEDGE *(page 514)*

- Chemists have identified specific molecules that are known to have therapeutic properties in some materials used in folk remedies. For example, willow bark contains a precursor of aspirin.
- Students can search the Internet for information on food-drug interactions. Health magazines are also good sources of such information.

# F. PUTTING IT ALL TOGETHER: ASSESSING RISKS

In Part F, students again examine risks and benefits—this time evaluating the effect of certain actions on personal health and safety. They learn new techniques for evaluating risk, and apply them in two *Your Turn* activities and a *You Decide* activity. This series of activities shows that although different individuals perceive risk differently, they can often come to some consensus on the relative risks of certain activities.

## F.1 RISK ASSESSMENT *(pages 515–516)*

Students assess risk by considering the various factors that influence the perception of risk among different individuals.

## F.2 RISK ASSESSMENT AND PERCEPTIONS *(pages 516–518)*

### YOUR TURN *What's the Risk?*

Students, in small groups, identify risks associated with sixteen activities or technologies. The groups then compare the different lists and explain why they might be different. Individual students first have the experience of listening to others in their group and then to the rest of the class. At the conclusion of the exercise, it should be clear to all that different people perceive risk differently.

### YOUR TURN *How Risky?*

In this activity, groups of students classify the risk of the activities and technologies using the graph in Figure 25. Answers to the questions will vary with groups.

## F.3 YOU DECIDE: LIVING WITH RISKS *(pages 518–519)*

The sixteen items are considered in yet another way. This time students can add up to three new items as long as they eliminate a number of items equal to the number added. The sixteen items are ranked from most risky to least risky. The ranking by experts and nonexperts that is presented below was adapted from an article by Paul Slovik in *Science*, April 1987, 236: 280. The adult experts included informed citizens, such as members of the League of Women Voters. It is important to emphasize that both lists are compiled on the basis of perceived risk. No data or numbers were considered in either case. These data were collected approximately a decade ago. It is interesting to speculate how these lists might differ if the data were collected in 1997.

**Rank order of perceived risk of activities and technologies**

| Adult Experts | College Students |
|---|---|
| Motor vehicles | Nuclear power |
| Smoking | Handguns |
| Alcoholic beverages | Smoking |
| Handguns | Pesticides |
| Surgery | Motor vehicles |
| Motorcycles | Motorcycles |
| X rays | Alcoholic beverages |
| Pesticides | Police work |
| Swimming | Surgery |
| Food preservatives | Food preservatives |

| | |
|---|---|
| Bicycling | Commercial aviation |
| Commercial aviation | X rays |
| Police work | Hunting |
| Nuclear power | Bicycling |
| Hunting | High school football |
| High school football | Swimming |

## F.4 LOOKING BACK *(page 519)*

The widespread role that chemistry plays in our lives has been the focus of this unit. In keeping with a major theme of *ChemCom*, students should see chemistry in their everyday lives, in this case their own bodies. Once again, students are able to make more informed choices about risks and benefits of certain activities after investigating the pertinent chemical topics.

## B.7 LABORATORY ACTIVITY: ENZYMES

The enzyme pepsin can also be tested by adding the following.

### Materials

240 mL pepsin solution (1.2 g of pepsin powder in 240 mL water)
Hard-boiled egg white

### Procedure

**Day 1**

1. Label four test tubes near the top with the temperature of your assigned test and a pH value of 2, 4, 7, or 8.
2. Using solutions provided by your teacher, place 5 mL of pH 2, 4, 7, and 8 solutions in the appropriately labeled tubes.
3. Place 1.5 g of egg white (the size of a large kidney bean) in each tube.
4. Add 5 mL of 0.5% pepsin solution to each tube. Continue with steps 5–8 in the procedure for testing amylase.

**Day 2**

1. Observe the test tubes for cloudiness (turbidity).
2. Record your observations.
3. Share your data with other members of your group.
4. Wash your hands thoroughly before leaving the laboratory.

### Expected Results

| | pH | | | |
|---|---|---|---|---|
| **Enzyme** | **2** | **4** | **7** | **8** |
| pepsin (room temp) | turbid | slightly turbid | clear | clear |
| pepsin (cold) | slightly turbid | very slightly turbid | clear | clear |

## MACROSCALE LABORATORY ACTIVITIES

Additional instructions for the teacher are found in the corresponding section of this guide. For teachers preferring this macroscale procedure, student instructions are printed on separate pages, which are ready for duplication. Larger quantities of substances and solutions are often needed for macroscale activities; in addition, some of the macroscale activities require significantly more time than the small-scale versions. Questions for students to answer are found in the student text following the small-scale version of the activity.

## C.4 LABORATORY ACTIVITY: BUFFERS

### Teachers' Directions

### Time

45 minutes

### Materials

250 mL 0.50 M hydrochloric acid (11 mL conc. HCl/250 mL solution)
250 mL 0.50 M sodium hydroxide (5 g NaOH/250 mL solution)
500 mL 0.10 M sodium hydrogen carbonate (4.2 g $NaHCO_3$/500 mL solution)
50 mL universal indicator solution (in dropper bottles)
distilled water
12 dropper bottles
24 125-mL Erlenmeyer flasks

12 50-mL burets
12 buret stands or ring stands with buret clamps
12 50-mL graduated cylinders
12 straws
12 50-mL beakers

## Expected Results

In general, adding acid or base to the buffered systems causes no observable pH change until the buffering capacity of the system is exceeded. Then the pH drops or rises, approaching the pH of the added acid or base. Actual pH values will be somewhat variable.

## C.4 LABORATORY ACTIVITY: BUFFERS

**Getting Ready**

In this activity, you will prepare and test a carbonate buffer. Adding acid and base to water will be compared with adding acid and base to the buffered solution.

Your teacher will assign you to an "acid" group or a "base" group. Read the procedures below, and prepare a data table. The table should contain five columns: original pH, pH after five drops of acid (base), initial volume of acid (base) in buret, final volume of acid(base) in buret, volume of acid (base) added.

### Procedure

**Acid Groups**

WATER PLUS ACID

1. Fill a clean buret above the zero mark with 0.50 M HCl. Briefly open the valve, letting solution run into a beaker until no air is left in the buret tip and the liquid level is at or below the 0-mL mark. Record the volume.
2. Place 40 mL of deionized (or distilled) water in a clean 125-mL Erlenmeyer flask. Add 10 drops of universal indicator. Compare the color of the flask contents to the indicator scale. Record the pH.
3. Place the flask under the buret. Slowly open the buret valve and add five drops of acid to the flask. Swirl the flask. Record the pH of the unbuffered acidic solution. Retain this flask as a color standard for comparison with the buffered solution.

CARBONATE BUFFER PLUS ACID

4. Add 40 mL of 0.10 M $NaHCO_3$ solution to a clean 125-mL Erlenmeyer flask. Use a clean straw to blow your breath (containing carbon dioxide) into the flask for 3 min. Be careful not to draw the solution into your mouth. You have now prepared the carbonate buffer. Discard the straw.
5. Add 5 drops of universal indicator. Swirl the flask. Record the solution color and pH.
6. Record the initial buret volume.
7. Add five drops of 0.50 M HCl from the buret to the flask and swirl. Record the color and pH.
8. Continue to add HCl solution until the color and pH are identical to those in the color standard flask from the water plus acid procedure above. Record the final buret volume.
9. Determine the volume of HCl solution added. Record this value.
10. Drain the HCl solution from the buret. Rinse the buret by pouring deionized or distilled water through it, allowing the water to drain into a beaker. Repeat this rinse step.
11. Wash your hands thoroughly before leaving the laboratory.

**Base Groups**

WATER PLUS BASE

1. Fill a clean buret above the zero mark with 0.50 M NaOH. Briefly open the valve, letting solution run into a beaker until no air is left in the buret tip and the liquid level is at or below the 0-mL mark. Record the volume.
2. Place 40 mL of deionized (or distilled) water in a clean 125-mL Erlenmeyer flask. Add 10 drops of universal indicator. Compare the color of the flask contents to the indicator scale. Record the pH.
3. Place the flask under the buret. Slowly open the buret valve and add five drops of base to the flask. Swirl the flask. Record the pH of this unbuffered basic solution. Retain this flask as a color standard for comparison with the buffered solution.

CARBONATE BUFFER PLUS BASE

4. Add 40 mL of 0.10 M $NaHCO_3$ solution to a clean 125-mL Erlenmeyer flask. Use a clean straw to blow exhaled breath (containing carbon dioxide) into the flask for 3 min. Be careful not to draw the solution into your mouth. You have now prepared the carbonate buffer. Discard the straw.

5. Add 5 drops of universal indicator. Swirl the flask. Record the solution color and pH.
6. Record the initial buret volume.
7. Add five drops of 0.50 M NaOH from the buret to the flask and swirl. Record the color and pH.
8. Continue to add NaOH solution until the color and pH are identical to those in the color standard flask from the water plus base procedure above. Record the final buret volume.
9. Determine the volume of NaOH solution added. Record this value.
10. Drain the NaOH solution from the buret. Rinse the buret by pouring deionized or distilled water through it, allowing the water to drain into a beaker. Repeat this rinse step.
11. Wash your hands thoroughly before leaving the laboratory.

## Questions

Answer the questions in the text.

# SUPPLEMENTAL READINGS

"Clean, Clean, Clean!" *CHEM MATTERS,* April 1991.
"Mystique of Musk." *CHEM MATTERS,* April 1991.
"Aspirin." *CHEM MATTERS,* February 1993.
"Pills That Compete with Aspirin." *Consumer Reports,* August 1982, 395-399.
"Aspirin: Is the Warning Necessary?" *Discover,* August 1982, 18-23.
"Is Bayer Better?" *Consumer Reports,* July 1982, 347-349.
"Medicinal Aspects of Aspirin and Related Drugs." *Journal of Chemical Education,* 1979, 331-333.
"Antacids." *CHEM MATTERS,* April 1983, 6. This issue focuses on acids and bases in everyday life.
"Lead Poisoning." *CHEM MATTERS,* December 1983.
*Folk Medicine: The Art and the Science.* American Chemical Society, 1986, 215 pages. This paperbound book explores the medical practices of nonwestern cultures to establish a scientific basis for the successes of folk remedies. It explains why western medical researchers are increasingly turning their attention to folk medicine for new drugs.
"Drugs to Be Sought from Chinese Herbs." *Chemical & Engineering News,* July 4, 1983, 16.
"Herbs for All Seasons." *National Geographic,* March 1983, 386-409.
"Interesting Reviving in the Use of Medicinal Plants." *Chemical & Engineering News,* November 16, 1981, 88.
"Nature's Gift to Medicine." *National Geographic,* September 1974, 420-440.
"How Soap Works." *CHEM MATTERS,* February 1985.
"Soap." *CHEM MATTERS,* February 1985.
"Soapuzzle." *CHEM MATTERS,* February 1985.
"Household Soaps and Detergents." *Journal of Chemical Education,* 1978, 596-597. Describes the components and molecular structure of soaps and detergents and the way they work.
*Nutrition and Aerobic Exercise.* American Chemical Society, 1986, 160 pages. This clothbound book examines the growing popularity of aerobic exercise as the focal point of daily health maintenance programs. It addresses the principal questions concerning the interaction of nutrition and aerobic exercise on muscle energy requirements, changes in nutritional needs, and long-term effects on body weight and composition. It features a discussion of the optimal plan to reduce body fat.
"The Interface Series: Energy and Exercise." *Journal of Chemical Education,* 1978: I. "How Much Work Can a Person Do?" July, 456-458; II. "Caloric Costs of Mass Transport." August, 526-528; III. "Heart Work." September, 586-587; IV. "Energy Storage Problems." October, 659-660; V. "Living Reagent." November, 726-727; and VI. "Reactions for All Occasions." December, 796-797.
"Blood Markers." *CHEM MATTERS,* April 1992.
"Matching Gifts." *CHEM MATTERS,* February 1992.
"The Chemistry of Fitness: Active Activities." *The Science Teacher,* March 1991, 28-32.
"Exercise." *Discover,* August 1982, 8-11.
"Pumping Oxygen." *CHEM MATTERS,* February 1984, 6-9. Exercise and oxygen metabolism.
"Alcohol." *CHEM MATTERS,* February 1985, 8-11.
"Metabolism of Alcohol." *Scientific American,* March 1976, 25-33.
"Dog Gone." *CHEM MATTERS,* February 1986.
"Smoking." *CHEM MATTERS,* February 1986.
"Nicotine Patches." *CHEM MATTERS,* October 1992.
"Test for Catalase." *CHEM MATTERS,* February 1986.
"The Taste Effect of Sodium Laurel Sulfate." *CHEM MATTERS,* April 1995.
"Toothpaste." *CHEM MATTERS,* February 1986.
"Yogurt." *CHEM MATTERS,* October 1989.
"Effects of Ethanol on Nutrition." *Journal of Chemical Education,* 1979, 532-533. Discusses alcohol-related malnutrition including the replacement of nutritious foods, the inhibition of vitamins, the increase in excretion of valuable minerals, and toxicity.
"Cocaine." *Scientific American,* March 1982, 128-141.
"Some Biochemistry of Sedatives." *Journal of Chemical Education,* 1979, 402-404.
"Testing Employees for Illegal Drugs." *Chemical & Engineering News,* June 2, 1986, 7-14.

"Denatured Alcohol." *CHEM MATTERS,* December 1990.
*The Journal of the American Medical Association (JAMA),* 253 (20), May 24/31, 1985. Focuses on the health effects of smoking and includes a reprint of the classic 1950 paper that first established a link between cancer and tobacco smoking.
"Unwitting Guinea Pigs." *CHEM MATTERS,* October 1985.
"The Cigarette Century." *Science 80,* September/October 1980, 36-43.
"Penicillin." *CHEM MATTERS,* April 1987.
"An Anti-Cancer Diet." *Discover,* June 1984, 23+.
"Skin Deep." *CHEM MATTERS,* December 1987.
"The Treatment of Diseases and the War against Cancer." *Scientific American,* November 1985, 51-59.
"Abnormal Insulin." *CHEM MATTERS,* February 1988.
"Artificial Sweeteners." *CHEM MATTERS,* February 1988.
"Dioxin (Part 1) from Plant Hormones to Agent Orange." *CHEM MATTERS,* February 1988.
"Dioxin (Part 2) Past War, Future Risk?" *CHEM MATTERS,* April 1988.
"Special Report: Cancer." *Discover,* March 1986.
"Fossil Molecules." *CHEM MATTERS,* April 1988.
"Regulation of Cancer-Causing Substances." *Chemical & Engineering News,* September 6, 1982, 25-32.
"Horses and Heroin." *CHEM MATTERS,* October 1988.
"Risk--Playing the Odds: A Worrier's Guide to the 20th Century." *Science 80,* October 1985, 29-47. Contains a wealth of useful charts, data tables, and statistics on relative risks in everyday life.
"Saving Arnold." *CHEM MATTERS,* December 1988.
"The Science of Decision Making: How We Make Up Our Minds, and Why We're So Often Wrong." *Discover,* June 1985, 22-31.
"Distance Running." *CHEM MATTERS,* February 1989.
"The Sun Worshippers." *CHEM MATTERS,* April 1984, 4-7.
"The Dark Side of the Sun." *Newsweek,* June 9, 1086, 60-64. Discusses the relationship between tanning and rising rates of skin cancer.
"Poison Ivy." *CHEM MATTERS,* October 1990.
"The Immunological Functions of Skin." *Scientific American,* June 1985, 46-53.
"The New and Improved Chemistry of Cosmetics." *Science 80,* April 1982, 54-61.
"Chemistry and Cleaning." *Journal of Chemical Education,* 1979, 610-611.
"Chemistry of Cosmetics." *Journal of Chemical Education,* 1978, 802-803. Discusses the chemistry of the skin, the structure of the hair, and the chemical basis for cosmetics.
"Household Soaps and Detergents." *Journal of Chemical Education,* 1978, 596-597.
"pH and Hair Shampoo." *CHEM MATTERS,* April 1983, 8-9. Focuses on acids and bases in everyday life.
"The pH of Hair Shampoos." *Journal of Chemical Education,* 1977, 553-554.
"The Bald Truth about Growing Hair." *Discover,* June 1986, 72-81.
"The Next Step: 25 Discoveries that Could Change Our Lives." *Science 85,* November 1985.
"The New Biology." *National Geographic,* September 1976 (special issue): "I. The Awesome Worlds within a Cell" (355-395), "II. The Cancer Puzzle" (396-399), and "III. Seven Giants Who Led the Way" (401-407).
"Delivery of Drugs: New Systems Prolong Their Effectiveness." *Chemical & Engineering News,* April 1, 1985, 30-48. [Reprints available from ACS.]
"Biomaterials in Artificial Organs." *Chemical & Engineering News,* April 14, 1986, 31-48. [Reprints available from ACS.]
"Biochemical Studies of Aging: Fitting More Pieces into a Far-From-Finished Mosaic." *Chemical & Engineering News,* August 11, 1986. Reprints available from ACS.
"Our Immune System: The Wars Within." *National Geographic,* June 1986, 702-736.
"Pepper Power." *CHEM MATTERS,* April 1995.
"Seeds of Doubt." *CHEM MATTERS,* April 1995.
"The Toxic Patient." *CHEM MATTERS,* October 1995.
"Beyond Supermouse: Changing Life's Blueprint." *National Geographic,* December 1984, 818-847. Details the promise of biotechnology in agriculture, animal husbandry, medicine, and energy.
"Genetic Engineering: A Special C&EN Report." *Chemical & Engineering News,* August 13, 1984,

1-65. Discusses the science, economics, and ethics of the research to date.
"Biotechnology Moves into the Marketplace." *Chemical & Engineering News,* April 16, 1984, 11-19.
"In-Vitro Methods May Offer Alternatives to Animal Testing." *Chemical & Engineering News,* November 12, 1984, 25-28.
"Lost in Space: Apollo 13's Fight for Survival." *CHEM MATTERS,* February 1994.
"Risky Business." *CHEM MATTERS,* December 1993.
"Chemicals in Everyday Life." *Journal of Chemical Education,* 64, 1987, 63-68.

# TEACHING SCHEDULE

| | TIME BLOCK 1 | TIME BLOCK 2 | TIME BLOCK 3 | TIME BLOCK 4 | TIME BLOCK 5 |
|---|---|---|---|---|---|
| Class Work | Read Introduction +A<br>Discuss Introduction<br>YT A.1 | YT A.2<br>SQ Part A<br>Read B | Prelab B.1<br>LA B.1<br>Read B.2 | LA B.1 cont. | Discuss B.2<br>YT B.2<br>Prelab B.3<br>LA B.3 |
| Homework | Read A.2 | Getting Ready for LA B.1 | | Read B.2 | Read B.4-B.5 |
| | **TIME BLOCK 6** | **TIME BLOCK 7** | **TIME BLOCK 8** | **TIME BLOCK 9** | **TIME BLOCK 10** |
| Class Work | Discuss B.4-B.5<br>YT B.4<br>CQ B.5 | Discuss B.6<br>YT B.6<br>YT B.7 | Review A&B<br>SQ Part B | Quiz A&B | Prelab C.1<br>LA C.1 |
| Homework | Read B.6 | | | Getting Ready for LA C.1 | |
| | **TIME BLOCK 11** | **TIME BLOCK 12** | **TIME BLOCK 13** | **TIME BLOCK 14** | **TIME BLOCK 15** |
| Class Work | Read C.2<br>CQ C.2<br>YT C.2 | Prelab C.3<br>LA C.3 | Read C.4<br>SQ Part C | Review Part C<br>Quiz Part C | YD D<br>YT D.3 |
| Homework | | | | Read D-D.3 | SQ Part D |
| | **TIME BLOCK 16** | **TIME BLOCK 17** | **TIME BLOCK 18** | **TIME BLOCK 19** | **TIME BLOCK 20** |
| Class Work | YD E.1 | PIAT<br>"Town Council Meeting"<br>Planning | PIAT<br>"Town Council Meeting"<br>Presentation | Review Industry Unit | Industry Unit Exam |
| Homework | Read E.2 | | Read E.3 | | |

**LA**=Laboratory Activity; **CQ**=ChemQuandary; **YT**= Your Turn; **YD**=You Decide; **PIAT**=Putting it All Together; **SQ**= Summary Questions; One **Time Block**=Approximately 50 minutes

# SUPPLY LIST

## Expendable Items and Unusual Materials

| Quantity | Unit | Item | Sections used |
|---|---|---|---|
| 24 | small | alligator clips with fine wire leads | C.1 |
| 325 | mL | ammonium molybdate reagent | B.3 |
| 85 | g | ammonium nitrate | B.1, B.3 |
| 20 | g | ascorbic acid | B.3 |
| 1.2 | g | barium chloride dihydrate | B.1 |
| 12 | 9-volt | batteries | C.3 |
| 12 | | battery clips | C.3 |
| 12 | squares | cobalt-blue glass | B.1 |
| 12 | wide | copper strips, 1 cm x 3 cm x 0.1 mm | C.1 |
| 12 | narrow | copper strips, 0.25 cm x 3 cm x 0.1 mm | C.1 |
| 12 | medium | copper strips, 0.5 cm x 3 cm x 0.1 mm | C.1 |
| 12 | 10-cm pieces | copper wire | C.3 |
| 1.2 | g | copper(II) nitrate trihydrate | C.1 |
| 6.2 | g | copper(II) sulfate pentahydrate | C.3 |
| 10 | L | distilled water | B.3 |
| 24 | | graphite electrodes | C.3 |
| 25 | mL | hydrochloric acid, conc | B.1 |
| 12 | | iron strips, 0.5 cm x 3 cm | C.1 |
| 1.4 | g | iron(II) sulfate heptahydrate | B.1 |
| 4 | g | iron(III) nitrate nononitrate | B.1, C.1 |
| 1 | pkg | litmus paper, red | B.1 |
| 1.3 | g | magnesium nitrate hexahydrate | C.1 |
| 12 | 3-cm lengths | magnesium ribbon | C.1 |
| 12 | large | paper clips | C.3 |
| 12 | small | paper clips | C.3 |
| 100 | mL | potassium chloride solution, saturated | C.1 |
| 0.5 | g | potassium nitrate | B.1 |
| 0.5 | g | potassium thiocyanate | B.1 |
| 60 | g | sodium hydroxide | B.1 |
| 4.3 | g | sodium nitrate | B.1 |
| 380 | g | sodium phosphate dodecahydrate | B.3 |
| 1.9 | g | sodium phosphate dodecahydrate | B.1 |
| 1.6 | g | sodium sulfate decahydrate | B.1 |
| 12 | pieces | steel wool | C.3 |
| 50 | mL | sulfuric acid, conc | B.1 |
| 60 | | toothpicks | B.1 |
| 12 | | voltmeters | C.1 |
| 12 | 10-cm | wire loops, platinum or nichrome | B.1 |
| 24 | | wood splints | B.1 |
| 1.5 | g | zinc nitrate hexahydrate | C.1 |
| 12 | narrow | zinc strips, 0.25 cm x 3 cm x 0.1 mm | C.1 |
| 12 | medium | zinc strips, 0.5 cm x 3 cm x 0.1 mm | C.1 |
| 12 | wide | zinc strips, 1 cm x 3 cm x 0.1 mm | C.1 |

*ChemCom* ends as it began—with a newspaper story about a major event in Riverwood. This time, two chemical industries, EKS and WYE, are vying for the privilege of locating a plant in Riverwood. EKS produces ammonia and related products, while WYE manufactures aluminum. In examining the two companies and their potential effects on the community, students learn about the general characteristics of the chemical industry in the United States and, about more specific characteristics of the two industries. The unit also examines how citizens and corporations jointly deal with the problems of pollution and protecting the environment.

Many chemical principles encountered in previous units are revisited. In addition, the qualitative characteristics of dynamic equilibria and electrochemistry are explored. A town meeting, where students have an opportunity to apply many of these principles, is a fitting conclusion to the *ChemCom* course.

# LEARNING GOALS

Upon completion of this unit the student will be able to

1. Identify issues to be considered in making a decision whether to locate a chemical plant in a community. [A.1, A.2]
2. Analyze a fertilizer sample for its major components, and describe the components' importance in agriculture. [B.1, B.2]
3. Use colorimetry to quantify phosphate content in fertilizer samples. [B.3]
4. Apply oxidation-reduction concepts to nitrogen fixation in the Haber process. [B.4]
5. Use electronegativity values to determine oxidation states. [B.4]
6. Describe effects of changing temperature and pressure to the equilibrium system representing the synthesis of ammonia. [B.5]
7. Trace the history and development of explosives, including the contributions of Alfred Nobel. [B.6]
8. Explain how voltaic cells can be used as an energy source. [C.1,C.2]
9. Use half-reactions to describe commercial electrochemical cells, including their charging and discharging reactions. [C.2]
10. Demonstrate the technique of electroplating. [C.3]
11. Describe the industrial applications of electrolysis represented by brine decomposition and aluminum production. [C.4]
12. Outline the types of products produced by the chemical industry. [D.1]
13. Describe factors that must be considered in scaling up a laboratory-scale reaction to an industrial level. [D.1]
14. Contrast responsibilities of the public and of industry in preserving the quality of life in a community. [D.2]
15. Describe the different types of jobs that are available in the chemical industry. [D.3]
16. Evaluate the positive and negative impact of a chemical industry on a community. [E.1, E.2]

# A. THE CHEMICAL INDUSTRY

Part A presents the risks and benefits that must be considered before either company EKS or company WYE is invited to locate in Riverwood. On one side, the chemical industry and its range of products are described. On the other, the community of Riverwood has criteria to be met before establishing a relationship with an industrial partner. The remainder of the unit explores the possible match between Riverwood and the two companies.

## A.1 CHEMICAL PRODUCTS *(pages 524–526)*

Different types of products, and the industrial methods for making them, are described.

## YOUR TURN *Chemical Processing in Your Life* *(page 526)*

Students answer questions designed to lead to an understanding of the impact of the chemical industry on their lives. Student answers to the questions will vary. Items on their list of products used in their natural state may include fruits, vegetables, wood products, wool and cotton fabrics, and metal products. Closer examination, however, will reveal that even most so-called natural materials have been treated or modified by the chemical industry. Even those few that are unchanged may be packaged in manufactured materials. Sometimes manufactured or synthetic substitutes are superior in performance to the natural substances they replace. For example, treated cotton fabrics are more wrinkle-resistant than are totally natural fibers.

## A.2 WHAT DOES RIVERWOOD WANT? *(pages 526–528)*

This section examines Riverwood's criteria that are intended to be met by any new industry. These criteria were partly based on the Toxics Release Inventory (TRI) and the guiding principles of the Chemical Manufacturers Association (CMA) Responsible Care Program. The TRI was established to provide citizens with information regarding chemical hazards in their community and to encourage planning on how to respond to chemical accidents. Addresses for obtaining more information about TRI are on page 228 of this Guide. More information about the Responsible Care Program can be obtained from the Chemical Manufacturers Association, 1300 Wilson Boulevard, Arlington, VA 22209. CMA has a home page on the Web (http://www.cmahq.com/) and also publishes a magazine suitable for student use, *ChemEcology*, which can be obtained free by high school teachers.

## YOUR TURN *Living with a Chemical Plant* *(page 528)*

This *Your Turn* provides a preliminary perspective for the town council meeting to be held at the end of this unit. This activity is best accomplished by assigning students to work in small groups. Student responses will vary. Remind them to hold further questions until the end of the unit.

## PART A: SUMMARY QUESTIONS *(page 528)*

1. How will the location of this plant affect Riverwood's unemployment situation? Will this plant adversely affect the quality of the environment? Will this plant produce waste that must be dealt with? (Many other responses are possible.)
2. $NH_3(g) + HNO_3(aq) \rightarrow NH_4NO_3(aq)$
3. $4\ Al(s) + 3\ O_2(g) \rightarrow 2\ Al_2O_3(s)$
4. "Environmentally safe" may mean that the plant will be safe for workers and will not produce unsafe conditions (such as water or air pollution) for the local community. "Environmentally responsible manufacturing" may mean that the company will treat potential pollution problems before they arise, rather than trying to find a solution after the occurrence.

5. Two other major uses of manufactured chemical substances are in the fertilizer industry and in polymer and plastics manufacturing. (Many other responses are possible.)

## CHEMISTRY AT WORK/Searching for Solutions in Research Chemistry *(page 529)*

A research chemist, who happens to be confined to a wheelchair, synthesizes new compounds to treat muscle and joint inflammation. The process of gaining approval for a new drug is briefly outlined.

# B. THE CHEMISTRY OF SOME NITROGEN-BASED PRODUCTS

In Part B students examine the chemistry and technological difficulties of fixing nitrogen, learn how the concept of limiting reactants applies to the use of fertilizer by plants, and read about nitrogen's role in the explosives industry.

Once again, students consider the question of who is responsible for how scientific inventions are used—this time, in the context of earlier use of explosives in warfare. Alfred Nobel, the inventor of dynamite, felt anguish when his invention became a tool for war; he was thus motivated to establish the Nobel prizes. Students should learn that as scientists or citizens, they share responsibility for helping to determine how new inventions are used.

## B.1 LABORATORY ACTIVITY: FERTILIZER *(pages 531–535)*

Students play the role of technician as they test an unknown fertilizer sample for the presence of six ions.

### Time

50 minutes

### Materials (for a class of 24 working in pairs)

12 100-mL beakers
24 red litmus paper strips
12 10-cm nichrome or platinum (Pt) wire loops, each attached to a cork stopper
12 24-well wellplates
12 Bunsen burners
12 18 × 150 mm test tubes (for cleaning nichrome wires)
12 squares of cobalt-blue glass
12 ring stands
12 rings
12 wire gauze squares
60 toothpicks
24 wood splints soaked in 0.5 M KCl (3.73 g KCl/100 mL solution)
120 Beral pipets
50 mL 6 M HCl for cleaning nichrome wires (25 mL conc. HCl/50 mL solution)

### Test Solutions

50 mL 6.0 M HCl (25 mL conc. HCl/50 mL solution)
50 mL 3.0 M NaOH (60 g NaOH/500 mL solution)
50 mL conc. $H_2SO_4$ in dropper bottles
50 mL 0.1 M $FeSO_4$ (1.4 g $FeSO_4 \cdot 7H_2O$/50 mL solution)
50 mL 0.1 M ($BaCl_2$) (1.2 g $BaCl_2 \cdot 2\,H_2O$/50 mL solution)
50 mL 0.2 M KSCN (0.5 g KSCN/50 mL solution)

### Known Solutions

50 mL 0.1 M $Na_3PO_4$ (1.9 g $Na_3PO_4 \cdot 12H_2O$/50 mL solution)
50 mL 0.1 M $NaNO_3$ (4.3 g $NaNO_3$/50 mL solution)
50 mL 0.1 M $Na_2SO_4$ (1.6 g $Na_2SO_4 \cdot 10H_2O$/50 mL solution)
50 mL 0.1 M $KNO_3$ (0.50 g $KNO_3$/50 mL solution)
50 mL 0.1 M $NH_4NO_3$ (0.40 g $NH_4NO_3$/50 mL solution)
50 mL 0.1 M $Fe(NO_3)_3$ (2.0 g $Fe(NO_3)_3 \cdot 9H_2O$/50 mL solution)

### Unknown "Fertilizer" Solutions

Solutions in Beral pipets, containing one anion and one cation

## Safety

Students should wear safety goggles and aprons throughout the laboratory activity. **Concentrated sulfuric acid is extremely hazardous: students should use it only under strict supervision.** We recommend that teachers add the concentrated sulfuric acid in Step 3b. Barium compounds are toxic. Direct student attention to the Caution statement in the student text.

## Advance Preparation

Label Beral pipets and fill them with the test solutions (except for concentrated sulfuric acid) and the six known ions. Place the pipets in the wellplate with the points up to avoid spillage. If student teams are to use sulfuric acid, it should be supplied in dropper bottles made specifically for concentrated sulfuric acid. Prepare the unknown solutions and place them in Beral pipets labeled with a code. Instruct students to record the code in their laboratory report. You can use any available solutions that contain the appropriate ions. Samples of commercial fertilizers are not recommended for the unknowns because some contain insoluble matter, time-released substances, or ions that complicate and confound the simple identifying tests suggested in this procedure. Concentrations should be at least 0.1 M. Examples include (1) a solution that contains both 0.1 M $NH_4NO_3$ and 0.1 M $Na_3PO_4$ (testing positive for $NH_4^+$, $NO_3^-$, and $PO_4^{3-}$); (2) a solution of 0.1 M $Fe_2(SO_4)_3$ (testing positive for $Fe^{3+}$ and $SO_4^{2-}$). It is not necessary to prepare a different unknown for each pair of students.

## Lab Tips

Wellplates must be rinsed with distilled water before the same depression is used a second time during the activity. Remind students to clean the wellplates thoroughly when they finish the laboratory activity.

The burner setting should be adjusted to produce a hot, blue flame. Remind students that a hot platinum or nichrome wire will itself display a color. It is important to record the color that appears *before* the wire is red hot.

Remind students that red litmus paper turns blue (or slightly violet) in a base.

## Pre-Lab Discussion

Remind students of the earlier work (Food unit, page 261) involving connections among limiting reactants, fertilizers, and plant growth.

Stress the importance of keeping a careful record of each particular solution's location in the wellplate.

## Expected Results

**Data Table—Ion Tests**

| Ion | $NH_4^+$ | $Fe^{3+}$ | $K^+$ | $NO_3^-$ | $PO_4^{3-}$ | $SO_4^{2-}$ |
|---|---|---|---|---|---|---|
| Color | NC | Brown | NC | NC | NC | NC |
| NaOH | NC | Precipitate | NC | NC | NC | NC |
| $BaCl_2$ | NC | NC | NC | NC | Precipitate | Precipitate |
| $BaCl_2$ and HCl | NC | NC | NC | NC | Precipitate dissolved | Precipitate remained |
| Brown ring test | X | X | X | Brown ring | X | X |
| NaOH and litmus | Blue | NC | NC | X | X | X |
| Flame test | NC | Sparks | Violet | X | X | X |
| Flame test and cobalt glass | NC | Whitish | Pink | X | X | X |
| KSCN test | NC | Red | NC | NC | NC | NC |

NC: No change
X: The test is not performed.

## Answers to Questions

1. Student answers will vary, but must be consistent with the data.
2. A white precipitate in the barium chloride test indicates the presence of either phosphate or sulfate. If the precipitate then dissolves in 6 M hydrochloric acid, the unknown is a phosphate.
3. The ammonium ion could be present.
4. No, because a candle flame is not hot enough and the color of the flame could mask other colors. (The pale blue flame of the Bunsen burner is hot enough to excite valence electrons, resulting in the emission of color, but does not pose any serious masking problem.)
5. No, these tests provide only qualitative information. Quantitative information is necessary for deciding whether a fertilizer is suitable for a particular use.

## Post-Lab Discussion

Discuss the use of the data chart to determine the ions present in the unknown sample. Emphasize that in this activity, students played the role of an analytical technician, one of the positions that will be available at the new plant.

## B.2 FERTILIZER'S CHEMICAL ROLES *(pages 535–538)*

This section is about the roles of the primary plant nutrients—nitrogen, phosphorus, and potassium. In addition to focusing attention on the concept of limiting reactants, remind students that fertilizers are a good example of the chemical industry's role in producing goods that society needs. The use of manufactured fertilizers is important in enabling the United States to be a major food producer.

Avoid getting bogged down in details of the nitrogen cycle (Figure 2 on page 537); the main point is that a natural recycling system supplies a usable form of nitrogen to plants. Chemical fertilizers supplement the natural system, increasing growth rates and yields of crops.

## YOUR TURN *Plant Nutrients* *(page 538)*

These questions highlight the importance of using chemical technologies to work with nature.

## Answers to Questions

1. Legumes contain nitrogen-fixing bacteria in their roots. If such crops are rotated with plants that extract nitrogen compounds from the soil, the soil's store of nitrogen is replenished.
2. Non-harvested parts of plants contain nitrogen, phosphorus, potassium, and other nutrients, originally extracted from the soil. By plowing the plants back into the soil, farmers make these nutrients available for the next season's crops; less fertilizer is needed than would otherwise be the case.
3. If scientists developed plants capable of fixing nitrogen, energy costs would be reduced because less fertilizer would be needed, and less fuel would have to be burned to apply the nitrogen to the soil.
4. a. Nitrate ions are essential for the plant to synthesize amino acids and proteins. If there is an insufficient supply, plant growth will be stunted and the general health of the plant reduced.
   b. Phosphate ions are incorporated into ADP, ATP, RNA, and DNA. A deficiency would retard cellular growth and reproduction, as well as energy transfer.
   c. Potassium ions activate some enzymes, help regulate the water balance of the plant, and are essential for the conversion of carbohydrates from one form to another. A deficiency would show up as a generalized reduction in cellular health.

## B.3 LABORATORY ACTIVITY: PHOSPHATES *(pages 538–541)*

In this activity, students estimate the phosphate concentration in a fertilizer solution using colorimetric techniques.

## Time

50 minutes

## Materials (for a class of 24 working in pairs)

20 g ascorbic acid
20 g commercial fertilizer (20-20-20) or, as an alternative, a mixture of 80 g $NH_4NO_3$ and 380 g $Na_3PO_4 \cdot 12\ H_2O$
1.0 L 10.0 ppm standard phosphate buffer solution
325 mL ammonium molybdate–sulfuric acid reagent
10 L distilled water
60 test tubes (18 × 150 mm)
12 test tube holders
12 test tube racks
12 ring stands
12 rings
12 wire screens
36 400-mL beakers
12 10-mL graduated cylinders
12 50-mL or 100-mL graduated cylinders
12 stirring rods
12 Bunsen burners
balance

## Safety

Students should wear safety goggles and aprons throughout the laboratory activity.

## Advance Preparation

The standard phosphate solution is prepared by dissolving 1.00 g $K_2HPO_4$ in a small volume of water and then diluting to 1.00 L in a volumetric flask. After this solution is thoroughly mixed, dilute 20.0 mL to 1.00 L, which produces 1 L of standard (10.0 ppm) phosphate solution.

Prepare the ammonium molybdate-sulfuric acid reagent by dissolving 7.5 g of ammonium molybdate in 75 mL of distilled water. Set the solution aside to cool in an ice bath.

Slowly, with mixing, add 125 mL of ice-cold 18 M $H_2SO_4$ to 125 mL of ice-cold distilled water. **Caution: Diluting concentrated sulfuric acid generates considerable heat, and may produce splattering if not added slowly with continual stirring. Perform this dilution with considerable care. Add the acid to the water, not the water to the acid.**

Slowly add the cold sulfuric acid solution (with stirring) to the ice-cold ammonium molybdate solution. Keep the ammonium molybdate solution in an ice bath during this step.

For optimum results, the solutions should be made up a week ahead of time. The solution keeps indefinitely.

## Pre-Lab Discussion

Point out the connection of this activity to earlier qualitative work with fertilizer (page 531).

## Lab Tips

Intensity of the color is affected by the length of time the solution is allowed to boil; some consistent method of heating must be agreed upon to ensure comparable results by all teams. The point at which the first "bump" of boiling occurs is an acceptable reference point.

If a spectrophotometer is available, data can be collected and plotted to obtain an absorption curve for the phosphate standards. This curve can then be used to compare sample values with the standards. A wavelength of 610 nm gives maximum absorption.

## Expected Results

The actual value of phosphate as phosphorus(V) oxide ($P_2O_5$), using 20-20-20 fertilizer, is 20%. Student results will range between 17% and 25% phosphate.

## Answers to Questions

1. Coffee, tea, orange juice, and powdered drinks are possible answers.
2. The colorimeter is able to provide repeatable, quantitative data.
3. Because colorimetric analysis depends on comparing an unknown to a known series of concentrations, if the standards are not carefully prepared, the data will be erroneous.
4. Such a reaction involves finding the mass of the precipitate. Using a balanced chemical equation, the mass of the original ion can be calculated.
5. a. Farmers have soils analyzed to determine the deficiencies, which then can be corrected by applying the correct fertilizer formulation.
   b. Fertilizer is expensive; adding too much increases the costs of producing the crop. Too much fertilizer also damages some crops and causes environmental damage due to runoff into waterways and aquifers.

## Post-Lab Discussion

Center the discussion on the questions that emphasize the importance of colorimetric techniques and the importance of quantitative data to farmers.

## B.4 FIXING NITROGEN *(pages 541–544)*

In this section, the Haber-Bosch process for fixing nitrogen is used to elaborate on the concepts of oxidation and reduction first introduced in the Resources unit. The original definitions of oxidation and reduction are extended to include covalent compounds.

*ChemCom* ties the idea of positive and negative oxidation states to the concept of electronegativity. Students learn to use the concept of electronegativity to assign positive and negative oxidation states to atoms.

## YOUR TURN *Electronegativity and Oxidation State* *(pages 543–544)*

Using the electronegativity chart on page 542 simplifies these problems. Encourage students to look for patterns in the values and the arrangement of elements in the periodic table. For example, all of the metals have relatively low electronegativities.

## Answers to Questions

1.

| Compound | Positive oxidation state | Negative oxidation state |
|---|---|---|
| a. $SO_3$ | S | O |
| b. $N_2H_4$ | H | N |
| c. $H_2O$ | H | O |
| d. HCl | H | Cl |
| e. NaCl | Na | Cl |
| f. CO | C | O |
| g. $IF_3$ | I | F |
| h. $MnO_2$ | Mn | O |

2. a.

| Compound | Positive oxidation state | Negative oxidation state |
|---|---|---|
| (1) $PbF_2$ | Pb | F |
| (2) NaI | Na | I |
| (3) $K_2O$ | K | O |
| (4) NiO | Ni | O |
| (5) $FeCl_3$ | Fe | Cl |
| (6) PbS | Pb | S |

b. When a metal and a nonmetal form a binary compound, the metal has a positive oxidation state and the nonmetal has a negative oxidation state.

3. Nickel is oxidized because its oxidation state becomes more positive, while sulfur is reduced because its oxidation state becomes less positive.
4. This is oxidation because the oxidation state becomes more positive.

## B.5 INDUSTRIAL NITROGEN FIXATION AT RIVERWOOD *(pages 545–547)*

The synthesis of ammonia illustrates a reversible reaction and dynamic equilibrium. Most students find this information new, but readily understandable. The effects of temperature and catalysts are discussed without using a quantitative treatment.

## CHEMQUANDARY *What Price Survival?* *(page 547)*

The products of the chemical industry that make it possible to feed and house a rapidly growing world population can also create pollution and damage the environment. For example, burning fuels, which is necessary to produce food in modern agricultural systems, also produces carbon dioxide and thus contributes to the greenhouse effect.

## B.6 NITROGEN'S OTHER FACE *(pages 547–550)*

This section discusses the use of nitrogen compounds as explosives, emphasizing the work of Alfred Nobel and his family. Explosions are highly exothermic, rapid reactions (the products are at a lower energy state than the reactants). Emphasize that in an explosive reaction, the volume of the products (gases) is much greater and at a higher temperature than the solid or liquid reactants. In nitrogen-containing explosive reactions, the nitrogen returns to the stable, triple bonded $N_2$ molecules.

## YOUR TURN *Chemistry of Explosives* *(page 551)*

This activity reinforces the information discussed in the previous section.

### Answers to Questions

1. The reaction produces 44 mol of gaseous products for each 4 mol of TNT; therefore, one mol of TNT produces 11 mol of gases.
2. a. 11,000 "units" of gas.

   b. The factor is 44/21, approximately twice the volume of oxygen consumed.
3. a. $8 \times 2 = 16$-fold increase

   b. In addition to the energy produced in the reaction, this tremendous increase in volume exerts enormous pressure in the vicinity of the reaction, pushing violently apart any objects close by.
4. a., c, and d could be explosive because they are exothermic reactions that produce gaseous products from a solid or liquid reactant.

## B.7 YOU DECIDE: FOOD OR ARMS *(pages 551–552)*

The story of the two faces of nitrogen provides an excellent example of using a particular technology for either constructive or destructive purposes. In the questions that follow, students are asked to make decisions about banning the technology. Then they are asked to compare this scenario to a more current issue: nuclear energy.

### Answers to Questions

There are no unequivocal answers to these questions. The reasons students use to support their positions should be given primary consideration when evaluating their responses.

1. Student answers will vary.
2. Over time, millions of persons would undoubtedly have starved to death because world agricultural production would have been outstripped by population growth in

nearly all countries. The economic strength of nations (such as the United States) that depend on agricultural exports to maintain a favorable trade balance would be considerably lessened. And, most likely, war would still be with us, though without nitrogenous explosives.

3. a. Student answers will vary.
   b. Nuclear power (like ammonia technology) can be used either for constructive purposes (electric power generation; agricultural, medical and industrial applications) or destructive purposes (nuclear weapons). If the technology had been banned, we would not be able to benefit from its constructive uses.

## PART B: SUMMARY QUESTIONS *(pages 552)*

1. Fertilizers typically contain a mixture of nitrogen, phosphorus, and potassium compounds, since these three elements are most commonly the limiting reactants for the growth of plants in agricultural systems. By adding these compounds to the soil, additional growth can be achieved.
2. The stability of the triple bond in the nitrogen molecule makes this compound especially unreactive, and therefore unavailable to plants. Some method for fixing nitrogen (changing it to ammonia, nitrates, or nitrites) must be available to make it usable by plants.
3. Elements other than oxygen can attract electrons, causing oxidation to occur. An example from a previous question is $Ni(s) + S(s) \rightarrow NiS(s)$.
4. a. Oxidation state expresses the apparent degree of oxidation of an atom in a substance. "Oxidation state" is sometimes referred to as "oxidation number."
   b. The atom that is oxidized gains a more positive oxidation state, while the atom that is reduced acquires a less positive oxidation state.
   c. In the equation given, each nitrogen atom is oxidized, since it gives up control of electrons, and each oxygen atom is reduced because it gains control of electrons from nitrogen.
5. a. Nitrogen comes from the atmosphere; hydrogen comes from natural gas.
   b. The forward reaction is exothermic.
   c. The forward reaction is favored at low temperatures.
   d. The reaction is too slow at low temperatures; too little product is formed over a reasonable time interval.
   e. A catalyst speeds up the formation of both products and reactants equally. The benefit of a catalyst is that lower temperatures are required for the reaction, and that the time for product formation is reduced.

## EXTENDING YOUR KNOWLEDGE (OPTIONAL) *(page 553)*

- Chemical substances found include phosphorus(V) oxide ($P_2O_5$), potash ($K_2O$), and ammonium nitrate ($NH_4NO_3$). Fertilizer containers typically display a series of three numbers, such as 6-12-12, which refer, respectively, to the percent (by mass) of nitrogen, available phosphorus, and potassium. Grasses and other leafy materials typically have higher amounts of nitrogen. Trees and shrubs, including rosebushes, have higher amounts of phosphorus and potassium. Sometimes 20-0-0 fertilizer is sold. This contains ammonium nitrate, which is recommended for lawns.
- This discussion question invites student opinion and research.

# C. THE CHEMISTRY OF ELECTROCHEMISTRY

Attention now turns to the WYE Metals Corporation and their product—aluminum. Because aluminum ores are refined by electrolysis, Part C introduces students to electrochemical cells, batteries, and the process of electroplating.

## C.1 LABORATORY ACTIVITY: VOLTAIC CELLS *(pages 554–557)*

In this laboratory activity, students construct and test various electrochemical cells. The activity applies oxidation-reduction reactions to the concrete, practical production of electrical energy.

### Time

50 minutes

### Materials (for a class of 24 working in pairs)

12 24-well wellplates
48 Beral pipets
50 mL 0.1 M copper(II) nitrate (1.2 g $Cu(NO_3)_2 \cdot 3H_2O$/50 mL)
50 mL 0.1 M zinc nitrate (1.5 g $Zn(NO_3)_2 \cdot 6H_2O$/50 mL)
50 mL 0.1 M magnesium nitrate (1.3 g $Mg(NO_3)_2 \cdot 6H_2O$/50 mL)
50 mL 0.1 M iron(III) nitrate (2.02 g $Fe(NO_3)_3 \cdot 9H_2O$/50 mL)
12 wide copper strips, 1 cm × 3 cm × 0.1 mm
12 medium copper strips, 0.5 cm × 3 cm × 0.1 mm
12 narrow copper strips, 0.25 cm × 3 cm × 0.1 mm
12 wide zinc strips, 1.0 cm × 3 cm × 0.25 mm
12 medium zinc strips, 0.5 cm × 3 cm × 0.25 mm
12 narrow zinc strips, 0.25 cm × 3 cm × 0.25 mm
12 3-cm lengths of magnesium ribbon
12 iron metal strips, 0.5 cm × 3 cm, or small iron nails
12 voltmeters
24 small alligator clips with fine wire leads
36 strips of filter paper, 8 mm × 3 cm
saturated KCl solution (about 35 g/100 mL water)

### Safety

Students should wear safety goggles and aprons throughout the laboratory activity.

### Advance Preparation

Prepare the solutions listed above. Label and fill Beral pipets with the appropriate solutions. Place them in a wellplate with the points up to avoid spillage. A physics teacher may have the voltmeters and be willing to lend them to you.

### Lab Tips

Remind students to avoid allowing the voltmeter needle to deflect in a negative direction for any length of time. This damages the meter. Also remind them to clean and rinse the wellplates immediately after using. The metals may be collected in an appropriately labeled container.

### Pre-Lab Discussion

Check to see that students have prepared the required data table. Ask groups also to record their work on the board or on an overhead transparency so that values may be compared. Assign students to complete the answers to the questions before the post-lab discussion.

## Expected Results

| Negative Post | Positive Post | Volts |
|---|---|---|
| medium Cu | Mg | 2.7 |
| medium Cu | medium Zn | 1.1 |
| medium Cu | Fe | 0.8 |
| narrow Cu | Mg | 2.7 |
| narrow Cu | narrow Zn | 1.1 |
| narrow Cu | Fe | 0.8 |
| wide Cu | Mg | 2.7 |
| wide Cu | wide Zn | 1.1 |
| wide Cu | Fe | 0.8 |

## Answers to Questions

1. a. Mg/Cu, Zn/Cu, and Fe/Cu
   b. Mg, Zn, and Fe are all above Cu in the activity series, which means that they lose electrons more readily than does Cu. Mg, Zn, and Fe have decreasing electronegativities.
2. a. Less. Zn and Cr are closer in the activity series than are Zn and Cu.
   b. More. Zn and Ag are closer together.
   c. Less. Sn and Cu are closer together.
3. Neither increasing nor decreasing the size of the electrode has any observable effect.
4. There is not much difference in the tendencies of silver and gold to lose electrons, and the battery would be very expensive.

## Post-Lab Discussion

Ask students to draw a conclusion from their data about the factors affecting the electrical potential. Although results obtained by various groups may differ slightly, the rank order should be the same for all. The activity series arranges elements in order of decreasing tendency to lose electrons, and so the reporting of results in the activity is consistent with their arrangement in the activity series. Magnesium loses electrons more readily than does zinc, which in turn, loses electrons more readily than does iron. Be sure students understand at the conclusion of the discussion that in an electrochemical cell electrons flow spontaneously from one metal to another. The greater the electrical potential (potential difference), the greater the difference in the ability of the two metals in question to lose electrons.

The next section discusses a variety of different cells and batteries; be sure that students understand these relatively simple cells before proceeding. The *ChemCom* course does not introduce the calculation of $E^0$ values, and little would be gained by adding that skill at this time.

## C.2 ELECTROCHEMISTRY *(pages 557–561)*

Students apply their knowledge of electrochemical cells gained in the previous activity to understanding batteries and cells in common, everyday use. The principle is the same, but some of the reactions are quite complex. Help students to see the similarities without becoming overwhelmed by the details. A display of "dissected" batteries is instructive. Be cautious about allowing students to handle caustic and corrosive materials from batteries.

## CHEMQUANDARY *Electric Cars—Truly "Zero Emissions"?* *(page 561)*

Some automobile makers will have electric vehicles in limited service by 1997, but reasonably priced models probably will not be available until the turn of the century. Some economists and environmentalists doubt that widespread use of electric vehicles will have the desired impact on the environment because new cars with gasoline engines are actually very clean. Although electric vehicles emit no pollutants, the manufacture and disposal of lead batteries remains a problem, as does the pollution from coal power plants. This story continues to develop, and undoubtedly, some students will be very knowledgeable in this area.

## YOUR TURN *Getting a Charge from Electrochemistry* *(pages 561–562)*

In this activity, students apply their understanding of redox reactions to electrochemical cells. The student text gives a simple mnemonic to help remember terms. Here are some others that may help: LEO the lion goes GER (Loss of Electrons is Oxidation, Gain of Electrons is Reduction). RC Cola (Reduction occurs at the Cathode), and OIL RIG (Oxidation is Loss, Reduction is Gain).

### Answers to Questions

1. a. Al $\rightarrow$ Sn
   b. Mg $\rightarrow$ Pb
   c. Fe $\rightarrow$ Cu
2. a. $Sn^{2+}(aq) + 2\ e^- \rightarrow Sn(s)$ (reduction)
   $Cd(s) \rightarrow 2\ e^- + Cd^{2+}(aq)$ (oxidation)
   b. Cd loses electrons more readily.
3. 

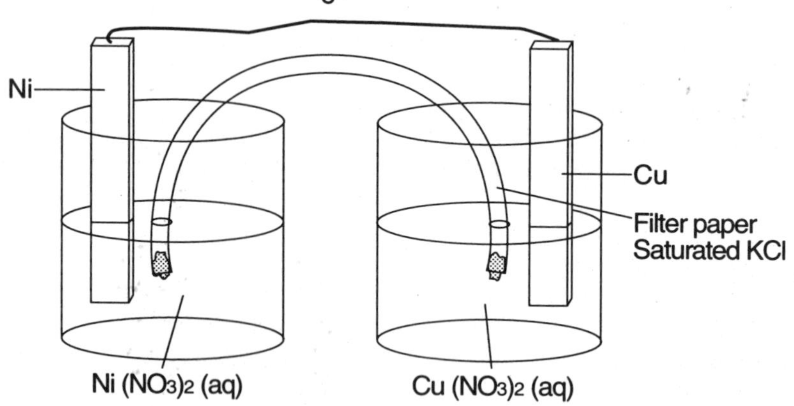


4. a. Anode = Zn, cathode = Cu
   b. Anode = Al, cathode = Zn
   c. Anode = Mg, cathode = Mn
   d. Anode = Ni, cathode = Au

## C.3 LABORATORY ACTIVITY: ELECTROPLATING *(pages 562–565)*

Students observe how electrical energy can be used to "drive" a nonspontaneous reaction.

### Time

50 minutes

### Materials (for a class of 24 working in pairs)

50 mL 0.5 M $CuSO_4$ (6.2 g $CuSO_4{\cdot}5H_2O$/50 mL)
12 9-volt batteries
12 battery clips
24 graphite electrodes
12 large paper clips
12 pieces steel wool
12 10-cm lengths of copper wire
12 small paper clips
12 24-well wellplates

### Safety

Students should wear safety goggles and aprons throughout the laboratory activity.

### Advance Preparation

Prepare the 0.5 M $CuSO_4$ solution described above. Prepare the graphite electrode–paper clip assemblies and the copper wire–paper clip assemblies, as shown in Figure 10 and Figure 11 in the student text.

## Lab Tips

Be sure that the paper clips are free of oils and rust before using.

An inexpensive carbon electrode can be crafted from a wooden pencil that has no eraser. Small pencils like those used to record golf or bridge scores are large enough. Sharpen the pencil at both ends and attach the alligator clip to one end. Remind students not to allow electrodes to touch each other.

## Pre-Lab Discussion

Ask students to think about the differences between this activity and the previous activity, C.1, where they constructed voltaic cells.

## Expected Results

The objects plate only at the negative electrode.

## Answers to Questions

1. a. The anode is the positive electrode (carbon electrode attached to the positive terminal of the battery).
   b. The cathode is the negative electrode (carbon electrode attached to the negative terminal of the battery).
2. The electrons flow from the anode (+) to the cathode (–).
3. A short circuit would be created. The object to be plated would not be in the circuit.
4. The copper wire is used to provide copper ions to replace those plated out on the paper clip.
5. a. $Cu(s) \rightarrow Cu^{2+}(aq) + 2\ e^-$
   b. $Cu^{2+}(aq) + 2\ e^- \rightarrow Cu(s)$
6. a. Attach the ring to the cathode (negative) electrode, the gold metal to the anode (positive) electrode, and immerse in a gold electrolyte solution.
   b. Attach the spoon to the cathode, silver metal to the anode, and immerse in a silver electrolyte solution.
   c. Use a sample of impure copper as the anode. Place a piece of pure copper (or graphite) at the cathode. Immerse both electrodes in a copper(II) ion-containing electrolyte. Pure copper metal will accumulate on the cathode; there is a net migration of copper from the impure anode to the cathode during this electrolysis.

## Post-Lab Discussion

Ask students to compare the equations that they wrote for the electroplating reaction with those written for the voltaic cell reactions. Be sure they understand the difference between the spontaneous reactions that occur in an electrochemical cell and nonspontaneous reactions that occur in electroplating. Remind students that the same terminology is used in both cases, since redox-type reactions are involved in both cases.

## CHEMISTRY AT WORK/A Career That's Good from Start to Finish *(page 564)*

Gene Kropp is a Technical Field Engineer and Salesman. Some of his work is routine, but there is enough unpredictability and problem solving to keep his job interesting and exciting. He has a degree from a two year college.

## C.4 INDUSTRIAL ELECTROCHEMISTRY *(pages 565–568)*

This section focuses on the electrolysis of brine (aqueous NaCl), and on the industrial production of aluminum using the Hall-Héroult process. Emphasize that although the scale of operations is vastly different, the same chemical reactions occur on the laboratory bench and in the industrial plant.

## PART C: SUMMARY QUESTIONS *(page 569)*

1. a. No
   b. Yes

2. They should be far apart on the activity series. The greater the difference in the ability to lose electrons, the greater the electrical potential.
3. a. Discharging
   b. The reverse reaction would consume energy (be endothermic) and would probably be nonspontaneous.
   c. Endothermic
   d. Charging
   e. Sulfuric acid is used up in discharging the battery. The greater the density of the sulfuric acid, the less the battery has discharged.
   f. When the battery is being discharged, sulfuric acid is consumed and water is produced (see equations in text). The battery liquid will be at maximum density when the battery is fully charged; battery liquid density will decrease as the battery discharges.
   g. The lead in the lead dioxide (spongy lead) is being reduced (gaining control of additional electrons); metallic lead is being oxidized (losing control of electrons).
   h. When it is being charged from an outside source, hydrogen gas may be produced as a by-product at the lead dioxide electrode, owing to the reduction of $H^+$ to $H_2(g)$.
   i. If a spark is present, an explosion can occur, represented by the familiar reaction: $2\ H_2(g) + O_2(g) \rightarrow 2H_2O + \text{energy}$.
4. The cell would produce a fairly low electrical potential because magnesium and aluminum are near each other in the activity series.
5. a. $Al \rightarrow Cr$
   b. $Mn \rightarrow Cu$
   c. $Fe \rightarrow Ni$
6.

| Voltaic cell | Electroplating cell |
|---|---|
| a. exothermic | endothermic |
| b. spontaneous | nonspontaneous |
| c. from more active metal to less active | from less active metal to more active |

7. a. Aluminum is more active than either iron or copper, suggesting that it donates electrons more readily than either of the other metals and would therefore be more likely to be found in compounds.
   b. Aluminum loses electrons more readily than does iron, which means it is more difficult for $Al^{3+}$ to gain electrons than it is for $Fe^{3+}$.
   c. In the Hall-Héroult process, aluminum oxide is dissolved in cryolite and electrolyzed in a cell with a carbon lining, which serves as the cathode. The anode is a carbon electrode. The cathode donates electrons to the aluminum ions, forming metallic aluminum as a liquid, which is periodically drawn off. The carbon anodes are oxidized and must be replaced.
8. a. $4\ Al^{3+}(\text{melt}) + 3\ C(s) + 6\ O^{2-}(\text{melt}) \rightarrow 4\ Al(l) + 3\ CO_2(g)$
   b. $378\ \text{g Al} \times \frac{1\ \text{mol Al}}{27.0\ \text{g Al}} \times \frac{3\ \text{mol electrons}}{1\ \text{mol Al}} = 42.0\ \text{mol electrons}$

# D. THE CHEMICAL INDUSTRY AND RIVERWOOD—A CLOSER LOOK

Part D takes a look at the chemical industry from its beginnings, when a scientist would work alone in a one-room laboratory, to modern industry, which provides a variety of jobs. Part D also tackles the problems of pollution and disposal of waste materials. Some of the ways that industry and government deal with these problems are described. This part forms an important transition to the town council meeting in which students will decide which, if any, industry the citizens of Riverwood should invite to locate in their town.

### YOUR TURN *The Top of the Chart* *(page 571)*

The questions in this *Your Turn* ask students to interpret information on the top 25 chemical substances found in Table 4. It may be necessary for students to look up formulas for some substances, using the *Merck Index* or other reference books.

#### Answers to Questions

1. The inorganic chemicals are sulfuric acid, nitrogen, oxygen, calcium oxide, ammonia, sodium hydroxide, chlorine, phosphoric acid, sodium carbonate, nitric acid, ammonium nitrate, and carbon dioxide. Most of the other substances have carbon atoms bonded to other carbon atoms. Methanol is considered to be organic because it is the first of a series (alcohols).
2. The organic substances are ethene, methyl *tert*-butyl ether, propene, 1,2-dichloroethane, urea, vinyl chloride, benzene, ethylbenzene, methanol, styrene, terphthalic acid, formaldehyde, and xylene. They contain carbon atoms bonded to other carbon atoms.
3. It appears that most of the nitric acid produced is used to make ammonium nitrate. However, only about half of the ammonia is used in this way.
4. The quantities of sodium hydroxide and chlorine are very nearly the same because they are manufactured in the same process.

### D.1 FROM RAW MATERIALS TO PRODUCTS *(pages 572–573)*

The many factors involved in scaling up laboratory reactions to industrial-size processes are described. The section also discusses some ways industry has dealt with the problem of waste materials.

### D.2 INDUSTRY AS A SOCIAL PARTNER *(pages 574–575)*

Here, the subject is the social partnership between industry and citizens. On one side, the chemical industry is obligated to produce useful products or services, while causing as little damage as possible to the environment and public health. On the other side, citizens need to protect their own health and safety, but should also understand the requirements of industry and the extent of society's need for its products and services. Such a partnership underscores the dual responsibilities of citizens and industry.

### D.3 JOBS IN CHEMICAL COMPANIES *(pages 575–577)*

In this section the various jobs promised by the EKS and WYE chemical companies are reviewed. Students will see that many positions can be successfully filled without a college major in chemistry.

### YOUR TURN *(page 578)*

Students are assigned to write a description of the position of chemical technician. Reference sources that may be consulted include the classified ads in local and state newspapers, job listings from the Internet, and databases that are probably available in the school counseling office. Many school and municipal libraries also have career sections and the capability to search electronically for information. Student answers to the questions will vary, but some typical answers are given below.

### Answers to Questions

1. The title of the position is chemical technician. (Some students may also list grades such as chemical technician II, etc.)
2. Generally such positions require the equivalent of an associate's degree in chemical technology from a community college or vocational school.
3. Requirements may include such competencies as the abilities to use instruments such as AA (atomic absorption spectroscopy), IR (infrared spectroscopy), UV (ultraviolet spectroscopy), and GC (gas chromatography), to use mathematics, to write accurate and understandable prose, and to use computer databases and spreadsheets.
4. The duties and responsibilities might include performing routine tests, calibrating pollution-reduction devices, entering data into a computer, analyzing graphical output from recordings of instrumental tests.
5. a. The educational requirements and general requirements listed in the response to Question 3 are probably similar for the two companies.
   b. WYE will probably require familiarity with electrochemical techniques and tests, while the successful applicant for the EKS position may need to know the chemistry of ammonia and be familiar with reactions that occur under pressure.

## PART D: SUMMARY QUESTIONS *(page 578)*

1. The chemical industry and society should be partners in reducing pollution and developing uses for waste materials.
2. Even excluding the petroleum and food industries, chemical industries earn roughly one of every ten manufacturing dollars, and employ more than one million U.S. citizens. Chemical industries are either directly or indirectly involved in nearly every product we purchase and nearly every kilojoule of energy we use.
3. a. b. (1) Sulfuric acid—battery acid, manufacturing fertilizer, petroleum refining;
   (2) nitrogen—manufacturing ammonia, liquid nitrogen for cooling;
   (3) oxygen—used in acetylene torches, used in oxygen tents for medical purposes;
   (4) ethene—fuel, raw material for polyethylene.
4. The four major factors are engineering, profitability, waste, and safe operation. Engineering must manage the heat generated by a reaction. The most efficient reaction path is sought because the industry must show a profit. Unwanted waste must be dealt with in an economical fashion. The workers must be safe in the plant and the plant must not endanger the surrounding population.
5. a. The analytical department checks on the purity of substances purchased, completes quality control tests on the product, and helps to locate problems in the production phase.
   b. The environmental department oversees the installation, operation, and maintenance of pollution control and waste management, and monitors the safety of the work environment.
   c. The public relations department communicates with the local and regional community, the general public, and government.
   d. Corporate management oversees the work of all departments and handles personnel, company policy, and finances.
   e. The research department invents new products and processes and finds ways to improve existing ones.
6. Unwanted materials can become the starting materials for new products.
7. "Green chemistry" focuses on preventing pollution problems before they occur, rather than cleaning them up afterward.

8. The ISO 14,000 is responsible for developing international standards for environmental management that can be used in place of often-conflicting national standards.

## EXTENDING YOUR KNOWLEDGE (OPTIONAL) *(page 578)*

- Students can interview the owner of a dry-cleaning establishment to find out about their operations.
- Information about the Toxics Release Inventory (TRI) is found at the end of Section D in the Teachers Guide for the Living in a Sea of Air: Chemistry and the Atmosphere unit (page 228).

# E. PUTTING IT ALL TOGETHER: A CHEMICAL PLANT FOR RIVERWOOD—CONFRONTING THE FINAL CHOICES

This part sets the stage for the concluding town council meeting and provides rules and suggestions for conducting the meeting. It begins with an article by the staff writer and an editorial urging citizens to attend and participate in the meeting.

## E.1 YOU DECIDE: ASSET OR LIABILITY? *(pages 579–581)*

The main purpose of this activity is not just to have students make a decision concerning the proposed plants, but to consider the issues affecting and affected by the decision. The text outlines a number of these; encourage students to think of others.

## E.2 THE TOWN COUNCIL MEETING *(pages 581–584)*

Detailed instructions for this activity are in *ChemCom's* student text. Because this is the second town meeting activity, students should be able to carry it out with little direction from the teacher. To maximize student participation, assign half the class to advocate the positive position and the other half to defend the negative position. Some class time should be allowed for groups to organize and assign roles within each group, but much preparation can be done as homework.

To what extent formal rules of debate are used depends on the composition of each class. Monitor the discussion so that one person or group does not dominate. Thirty to 45 minutes should be enough time to bring out the important points.

Be sure to allow time at the end for the town council to meet and discuss options. Voting or conducting a referendum provides closure for this activity and the unit, as well as for the *ChemCom* course.

## E.3 LOOKING BACK AND LOOKING AHEAD *(page 584)*

It is worthwhile to have students look back and compare the knowledge they now have with the background they brought to the course at its beginning. We hope that in addition to learning about chemistry, they now have a better idea of how important chemistry and technology are in their world. Students owe it to themselves as citizens to continue exploring and becoming informed about chemistry and other sciences in order to make responsible decisions requiring some familiarity with scientific topics.

# ENRICHMENT MATERIALS

Additional instructions for the teacher are found in the corresponding sections of this guide. For teachers preferring the macroscale activity, the student instructions are printed on separate pages, which are ready for duplication. Larger quantities of substances and solutions are often needed for the macroscale activity; in addition, some of the macroscale activities require significantly more time than the small-scale versions. Questions for students to answer are found in the student text following the small-scale version of the activity.

## C.1 LABORATORY ACTIVITY: VOLTAIC CELLS Teacher's Directions

### Materials (for 24 students working in pairs)

3.0 L 0.1 M $Cu(NO_3)_2$ (72 g $Cu(NO_3)_2 \cdot 3H_2O$ /3.0 L)
3.0 L 0.1 M $Zn(NO_3)_2$ (89 g $Zn(NO_3)_2 \cdot 6H_2O$ /3.0 L)
3.0 L 0.1 M $Mg(NO_3)_2$ (77 g $Mg(NO_3)_2 \cdot 6H_2O$ /3.0 L
3.0 L 0.1 M $Fe(NO_3)_3$ (120 g $Fe(NO_3)_3 \cdot 6H_2O$ /3.0 L)
12 zinc metal strips (8 cm × 15 cm × 0.25 mm)
12 copper foil strips (8 cm × 10 cm × 0.1 mm)
distilled water
24 alligator clips with fine wire leads
12 15-cm magnesium ribbon strips
12 iron metal strips (8 cm × 10 cm)
12 porous 75-mL cups (9 cm × 4 cm)
12 400-mL beakers
12 scissors
12 centimeter scales

### Safety

Students should wear safety goggles and aprons throughout the laboratory activity.

### Advance Preparation

Soaking porous cups in 0.1 M $NH_4NO_3$ or 0.1 M $NH_5Cl$ will lead to better results.

## C.3 LABORATORY ACTIVITY: ELECTROPLATING Teacher's Directions

### Materials

3.0 L 0.5 M $CuSO_4$ (400 g $CuSO_4 \cdot 5H_2O$ /3.0 L solution)
12 pieces copper foil (3 cm × 10 cm × 0.1 mm)
12 steel wool pads
12 metallic objects to plate
12 graphite sticks
12 6-volt (lantern) batteries or 6-volt DC power sources
12 400-mL beakers
24 alligator clips with fine wire leads (2/team)
12 tongs
paper towels

### Safety

Students should wear safety goggles and aprons throughout the laboratory activity.

# C.1 LABORATORY ACTIVITY: VOLTAIC CELLS

## Getting Ready

In the Resources unit (page 117) you learned that some metals release electrons (become oxidized) more readily than others—they are more active. The relative tendencies for metals in contact with water to release electrons can be summarized in an activity series of the metals (see Table 3 on page 555). A metal higher in the activity series will give up electrons more readily than one that is lower.

The differing tendencies of metals to lose electrons can be employed to obtain electrical energy from a chemical reaction. We do this by constructing a voltaic cell in which electrons flow spontaneously through a wire connecting the two metals. The flow of electrons is called current. In this laboratory activity you will study several voltaic cells.

When two metals of differing electron-releasing tendencies are connected in a voltaic cell, an electrical potential is created between the metals. The electrical potential, which is measured in volts (V), is somewhat like water pressure in a pipe. It represents the "push" that drives electrons through the wire connecting the two metals. The greater the difference in activity of the metals, the greater the "electron pressure," or electrical potential of the cell.

To provide both electrical conduction inside the cell and receptors for electrons leaving the wire or external circuit, a voltaic cell is prepared by immersing each metal in a solution of its ions. A porous cup allows the two solutions to maintain electrical contact. The cup must contain channels that are just large enough to permit the flow of dissolved ions without allowing gross mixing of the solutions.

## Procedure

Prepare an appropriate data table for the following procedures.

**Zinc–copper Electrochemical Cell**

1. Pour 75 mL of $Zn^{2+}$ solution into a clean 250-mL beaker.
2. Fill a porous cup about 3/4 full of $Cu^{2+}$ solution.
3. Using steel wool, shine strips of copper and zinc.
4. Avoiding spilling, carefully place the porous cup in the beaker. Be sure the top of the cup remains above the liquid level in the beaker.
5. Attach wire leads to the metal strips. Then attach one wire lead to the voltmeter. (If your voltmeter has selectable scales, select a scale with the approximate range of 0–2 V.) Very quickly and lightly touch the second wire lead to the other terminal of the voltmeter. If the needle moves in the positive direction, secure the wire to the terminal. If the needle moves in the negative direction, detach the first wire lead, reverse the positions of the leads, and attach them to the terminals.
6. Record the reading on the voltmeter dial.
7. Disconnect the wire leads, remove the metal strips, and dispose of the liquids as your teacher directs.

**Magnesium–copper Electrochemical Cell**

1. Refer to the activity series of metals and predict the change in your results for a magnesium–copper cell.
2. Repeat Steps 1–7 above, using a piece of magnesium ribbon and 75 mL of $Mg^{2+}$ in place of zinc and its solution.

**Iron-copper Electrochemical Cell**

1. Refer to the activity series of metals and predict the change in your results for an iron–copper cell.
2. Repeat Steps 1–7 above, using an iron strip and 75 mL of $Fe^{3+}$ solution in place of zinc and its solution.

**Studying Electrode Size**

1. Obtain zinc and copper strips half as wide as the original electrodes used earlier.
2. Repeat Steps 1–7 in the zinc–copper electrochemical cell procedure.
3. Disconnect the wire leads. Remove the zinc and copper electrodes from the system. Replace them with a new zinc and copper strip half again as wide.
4. Repeat Steps 1–7 in the zinc–copper procedure.
5. Wash your hands thoroughly before leaving the laboratory.

## C.3 LABORATORY ACTIVITY: ELECTROPLATING

### Getting Ready

All reactions you observed on pages 554–557 were spontaneous. Materials were arranged so electrons would spontaneously move through wires. Voltaic cells involving such reactions generate electrical energy to run motors and perform other useful work.

By contrast, electrical energy can be used to "force" nonspontaneous or difficult oxidation-reduction reactions to occur. Charging a lead storage battery in an automobile is one example. Electroplating is another.

In electroplating, electrical energy from a battery or other source provides electrons to convert metal ions to atoms. Electroplating produces a thin layer of metal deposited on another surface to protect the surface or make it more beautiful. For example, coating inexpensive jewelry with a very thin layer of gold makes it more attractive.

An electroplating cell used to accomplish these chemical changes consists of two electrodes (anode and cathode), a solution of ions, and a source of electricity. In this laboratory activity you will electroplate a metallic object of your choice. The object will become one of the electrodes; a piece of copper will serve as the other electrode.

### Procedure

**Part 1: Preliminary Electroplating Test**

1. Place 250 mL of 0.5 M $CuSO_4$ in a 400-mL beaker.
2. Attach the alligator clips of two wire leads to the terminals of a 6-volt battery.
3. Attach the other ends of the leads to two carbon (graphite) sticks, which will serve as electrodes.
4. Dip the carbon electrodes into the solution in the beaker. Do not allow the electrodes to touch each other. Record your observations.
5. Allow the current to flow until you observe a change in one of the electrodes. Record your observations.
6. Disconnect the leads from the battery. Remove the carbon electrodes and rinse them with running tap water.

**Part 2: Electroplating an Object**

1. Place 250 mL of 0.5 M $CuSO_4$ in a 400-mL beaker.
2. Clean the object to be plated with the steel wool pad. Decide which wire lead should be attached to the copper strip and which to the object, based on Part 1 observations. Connect the wires to the copper and to the object.
3. Immerse the object and the copper strip in the plating solution. Do not let the electrodes touch each other. Attach the wire lead alligator clips to appropriate battery terminals.
4. Allow current to flow until copper plating is observed.
5. Record your observations.
6. Disconnect the leads from the battery. Remove the object from the plating solution with tongs. Rinse the object under running tap water. Dry it with a paper towel.
7. Examine the object and record your observations.
8. Wash your hands thoroughly before leaving the laboratory.

# SUPPLEMENTAL READINGS


"Manufacturing Memories." *CHEM MATTERS,* February 1990.

"The Tell-Tale Bullet." *CHEM MATTERS,* February 1990.

"The World of Industrial Chemistry." *Journal of Chemical Education* series beginning in February 1979 issue (edited by W. Conrad Fernelius and Harold W. Wittcoff). Particularly useful articles include: "The Chemical Industry: A Historical Perspective." February 1979, 114-115; "The Chemical Industry: What Is It?" April 1979; "An Acid Can be Basic--Sulfuric Acid," August 1979, 529; "Chloro-Alkali Industry," September 1980, 640-641; "Chemical Processing: Patch or Continuous," Part I September 1982, 766-768; Part II October 1982, 860-862; "A Summary Chart of the Manufacture of Important Chemicals," May 1983, 411-413.

"When Push Comes to Shove: Disturbing the Equilibrium." *CHEM MATTERS,* February 1985.

"Industrial Chemistry Bibliography." *Journal of Chemical Education,* April 1985, 331-332.

"Facts and Figures for the Chemical Industry." *Chemical & Engineering News.* Annual issue, which includes, the top 50 chemicals and the associated production figures for the 1975-1985 period; the top 100 chemical producers; information on foreign chemical industries; and a wealth of other information and statistics.

"Facts and Figures for Chemical R & D." *Chemical & Engineering News.* Annual issue, which includes, an overview of where R & D funds come from and go to; and specific information for the federal government, industry, and university/colleges.

"Industrial Analytical Chemistry: The Eyes, Ears, and Handmaiden to Research and Development." *Journal of Chemical Education,* 1986, 237. Discusses the role of industrial analytical chemists, prerequisite training, and research challenges.

"The B.S. Chemist in Industry." *Journal of Chemical Education,* 1985, 734-735.

"Pharmacology: A Place for Chemists." *Journal of Chemical Education,* 1982, 231.

"Chemistry in Action: How to Plan a Visit to the Chemical Industry." *Journal of Chemical Education,* 1982, 582.

"Sunken Treasure." *CHEM MATTERS,* April 1987.

"Forensic Chemistry." *Journal of Chemical Education,* 1982, 41.

"Career Choice: Chemist or Chemical Engineer." *Journal of Chemical Education,* 1981, 494.

"The Industrial Research Chemist." *Journal of Chemical Education,* 1981, 269.

"Chemistry and Public Policy." *Chemical & Engineering News,* March 9, 1987, 20-29. Reprint available from ACS.

"Leaf Jewelry." *CHEM MATTERS,* December 1987.

*The Chemical Balance: Benefiting People, Minimizing Risks.* Chemical Manufacturers Association. Free booklet discusses how the chemical industry is addressing the issues of product safety, worker safety, chemical transportation safety, and managing chemical waste.

*ChemEcology.* Chemical Manufacturers Association. Free monthly newsletter describing the chemical industry's efforts in the environment, energy conservation, and occupational health.

"Ernie's Amazing Journey." *CHEM MATTERS,* February 1990.

"Delaware: Who Needs to Be Big?" *National Geographic,* April 1983, 171-197.

"Electrochemistry: State of the Art Symposium." Special issue of the *Journal of Chemical Education,* April 1983.

"Opportunities in Chemistry: Executive Summary." *Chemical & Engineering News,* October 14, 1985.

"Harnessing Light by a Thread: Miracles of Fiber Optics." *National Geographic,* October 1979, 516-535.

"The Chip: Electronic MiniMarvel." *National Geographic,* October 1982, 421-457.

"California's Silicon Valley." *National Geographic,* October 1982, 459-477.

"The Laser--A Splendid Light." *National Geographic,* March 1984, 335-363.

"The Wonders of Holography." *National Geographic,* March 1984, 364-377.

"Employment in the Chemical Industry: Chemical Employment Growth Continues." *Chemical & Engineering News,* June 19, 1989, 67-73.

# TEST ITEMS

This section of your Teacher's Guide contains Multiple-Choice Questions and Free Response Questions for each of the units. An effort was made to match the questions to the objectives, but as always, you should exercise your professional judgement concerning what is appropriate for your students. Some objectives, particularly those referring to activities, can be better assessed by methods other than paper-pencil test items. Consider laboratory practical activities and assessing classroom work as it is in progress. The answers to the multiple-choice items and points to consider for the free response items are given. Again, exercise your professional judgement and knowledge of your students when evaluating these items.

The unique nature of the *ChemCom* curriculum creates special challenges in evaluating student performance. To help with end of course evaluation, a standardized examination is available from the ACS DivCHED Examinations Institute. It incorporates novel test item formats, such as multiple-response options, linked event-decision style questions, and grid questions, which permit measurement of student achievement that is difficult to assess with single-answer, multiple-choice questions.

# SUPPLYING OUR WATER NEEDS

## Multiple-Choice Questions

Choose the letter corresponding to the best answer for each of the following.

1. A vitamin supplement tablet contains 0.06 g of vitamin C. How else can this mass be expressed?
   a. 6 mg  b. 60 mg  c. 600 mg  d. 0.006 kg

2. How many 250-mL servings can be poured from a 2.0-L bottle of soft drink?
   a. 2  b. 4  c. 8  d. 12

3. Most of the earth's supply of water is found in
   a. rivers
   b. oceans
   c. glaciers
   d. water vapor

4. If the instructions on a bottle of medicine say "shake before using," the material inside is probably a
   a. solution  b. colloid  c. suspension  d. pure liquid

5. Which of the following is a solution?
   a. distilled water  b. a cola drink  c. Italian dressing  d. cooking oil

6. The label on a solid tomato fertilizer says "ammonium phosphate, ammonium sulfate, muriate of potash, sulfate of potash, sulfate of potash-magnesia, manganese oxide, iron oxide." The fertilizer is a
   a. solution  b. compound  c. mixture  d. pure substance

7. Some water softening units use ion exchange resins. What occurs in such a unit?
   a. Calcium ions are exchanged for magnesium ions.
   b. Magnesium ions are exchanged for chloride ions.
   c. Calcium ions are exchanged for iron ions.
   d. Calcium ions are exchanged for sodium ions.

Ethylene glycol ($C_2H_6O_2$) and ethanol ($C_2H_6O$) have very similar structures as shown below.

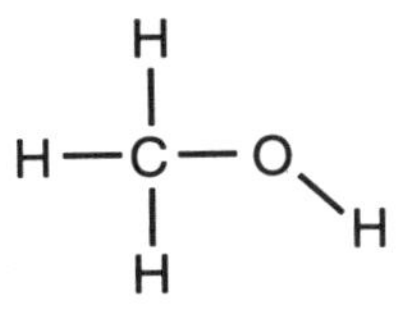

ethanol

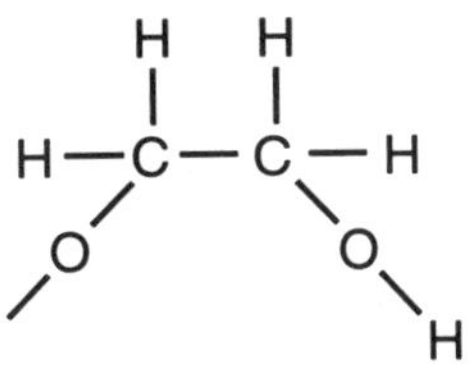

ethylene glycol

Ethylene glycol has a boiling point of 182 °C; ethanol has a boiling point of 78 °C. Ethanol (as you found in the solvents Laboratory Activity) is very soluble in water.

8. Which one of the following statements would you expect to be true?
   a. Ethylene glycol dissolves in water and ethanol.
   b. Ethylene glycol dissolves in ethanol but not in water.
   c. Both ethylene glycol and ethanol are nonpolar molecules.
   d. Ethylene glycol does not dissolve either in water or ethanol.

9. The most effective way to separate a mixture of ethylene glycol and ethanol would be
   a. osmosis.
   b. filtration.
   c. charcoal adsorption.
   d. distillation.

## Free Response

Answer each of the questions as thoroughly and completely as possible.

10. The average person in the United States uses 340 L of water daily. The total U.S. consumption of water per person per day is much higher than this value. Account for the difference.

11. In a laboratory activity, foul water was purified by oil-water separation, sand filtration, charcoal adsorption/filtration, and distillation. Which one of these steps would not be practical to purify a city's water supply? Give reasons for your answer.

12. Calculate the mass of water and the mass of sugar in a sugar solution that has a total mass of 55 g and is labeled 20.0% sugar.

13. Milk of Magnesia has a pH of 10.5, eggs are 7.8, and rainwater is around 5.8. Identify each as acidic, basic, or neutral.

14. Write the names of the following compounds.
   a.____________________ NaOH
   b.____________________ NaCl
   c.____________________ HCl(aq)

15. You are marooned on a sandy island surrounded by millions of liters of ocean water. The only water available on the island is from a stagnant pond of murky-looking water. In your survival kit you have the following items:

| | |
|---|---|
| one nylon jacket | one knife |
| one plastic cup | one liter-bottle of liquid bleach |
| two plastic bags | one 5-liter glass bottle |
| one length of rubber tubing | one bag of salted peanuts |

   Describe a plan to produce drinkable water using as much of the equipment as necessary.

16. Write the formula and name of a compound composed of each of the following ions.
   a. $NH_4^+$, $Cl^-$

   b. $Na^+$, $CO_3^{2-}$

   c. $Fe^{3+}$, $Cl^-$

17. List and briefly describe two unusual properties of water. Explain each property using the concept of polarity.

18. Carbon tetrachloride, $CCl_4$, is a nonpolar liquid that can be used in dry cleaning.
   a. Is it an element or a compound? How do you know?
   b. When 5.0 g of bromine, $Br_2$, a reddish-brown liquid, is added to 75 g of clear, colorless $CCl_4$, the mixture has a reddish-brown color. Over time, no settling out is observed. When the mixture is filtered, no residue is observed on the filter paper. When a light beam is directed at the mixture, its path through the mixture is not visible. Is the mixture a solution, colloid, or suspension? Explain your answer.
   c. What is the mass percent of $Br_2$ in the mixture?
   d. A $CCl_4$ sample was measured in the laboratory and found to have a mass of 19.64 g and a volume of 12.36 mL. Calculate the density of $CCl_4$. Show your work.

   e. Describe a piece of laboratory equipment that could be used to measure the 19.64 g and another piece of equipment that could be used to measure the 12.36 mL.

19. Explain each of the following statements using what you know about solubility.
   a. A paint brush used to apply oil-based paint cannot be cleaned up with water.
   b. A salad dressing composed of vinegar (a solution of acetic acid in water) and oil separates into two layers.

c. Pens sold as "laundry markers" can make marks on clothing that will not come off in the wash.

Farms may pollute nearby waterways if rain washes some of the fertilizer from the fields. The fertilizer may enter waterways as potassium nitrate, $KNO_3$. Below is a solubility graph for $KNO_3$.

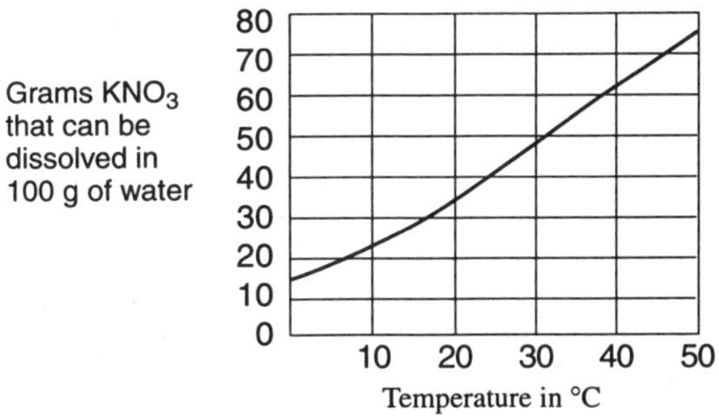


20. What is the maximum mass of $KNO_3$ that can be dissolved in each 100 g of water if the temperature of the water is 20 °C?

21. A solution is made by dissolving 50 g of $KNO_3$ in 100 g of water at 40 °C. Should this solution be called saturated, unsaturated, or supersaturated? What would happen if a small crystal of solid $KNO_3$ were added to this solution?

22. Prepare a graph of the following data and then answer the questions that follow it.
Change in Solubility of Oxygen with Change in Temperature

| Temp, °C | Dissolved Oxygen, mg/1,000 g water |
|---|---|
| 2 | 12.7 |
| 7 | 11.0 |
| 9 | 10.4 |
| 19 | 9.2 |
| 20 | 9.1 |
| 21 | 8.9 |

a. From your graph determine how much oxygen gas can be dissolved in water at 5 °C.
b. Make a statement that relates dissolved oxygen to the temperature of water.

## Answers to Questions

**Multiple Choice**

| | | | | |
|---|---|---|---|---|
| 1. b | 2. c | 3. b | 4. c | |
| 5. b | 6. c | 7. d | 8. a | 9. d |

**Free Response**

Student answers will vary. Some points to consider follow.

10. Much water is used indirectly—manufacturing, agriculture, etc.
11. Oil-water separation and distillation are not easily used for large quantities. In addition the energy requirements are too high.
12. $55 \text{ g} \times 20.0\ \% = 11 \text{ g}$
13. Milk of Magnesia—basic
eggs—basic
rainwater—acidic
14. a. sodium hydroxide
b. sodium chloride
c. hydrochloric acid (hydrogen chloride is also acceptable)
15. (Student answers will vary.)
16. a. ammonium chloride, $NH_4Cl$

b. sodium carbonate, $Na_2CO_3$
c. iron(III) chloride, $FeCl_3$

17. Water has an unusually high freezing point and boiling point and high surface tension (other answers are possible). Water molecules are polar and stick to each other.
18. a. compound—two elements
b. solution—properties described are those of solutions
c. 5.0 g + 75 g = 80 g % Br = 5.0 g/80 g × 100 = 6.2%
d. 19.64 g/12.36 mL = 1.589 g/mL
e. A balance is used to measure mass, and a graduated cylinder is used to measure volume. (Drawings are acceptable.)
19. a. Nonpolar oil-based paint is not soluble in polar water.
b. Oil is not soluble in water, while polar vinegar is soluble in water.
c. The ink is not water soluble.
20. 34 g
21. The solution is unsaturated; the crystal will dissolve.
22. a. about 11.5 mg $O_2$/1,000 g water
b. The amount of dissolved oxygen decreases as the temperature of water increases.

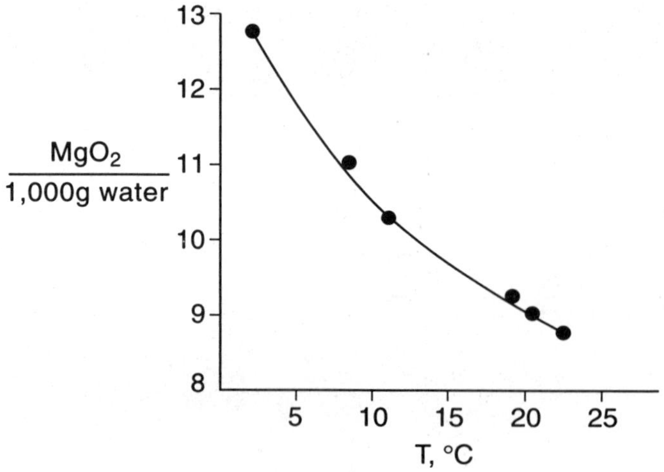

# CONSERVING CHEMICAL RESOURCES

*A periodic table is required for this test.*

## Multiple-Choice Questions

Chose the letter corresponding to the best answer for each of the following.

1. We are in danger of exhausting our supplies of copper because
   a. copper atoms are being destroyed in chemical processes.
   b. copper ores are being converted to copper metal.
   c. copper atoms are being widely distributed in the environment.
   d. environmentalists want strict air pollution regulations.

2. The likely formula for a compound formed from the elements aluminum and sulfur is
   a. $Al_2S_3$ b. $Al_3S_2$ c. AlS d. $Al_6S_3$

3. The elements within a particular family of elements are likely to
   a. have the same melting point and boiling point.
   b. progress from metal to metalloid to nonmetal as their atomic numbers increase.
   c. have the same atomic number.
   d. have similar chemical properties.

4. In a laboratory activity to test the reactivity of metals, a student used three metals and three solutions of compounds of the metals. Each metal was placed in certain solutions as shown in the data table below. The following observations were made (NR - no reaction, R-reaction, — experiment not done).

| Metal | Solution | | |
|---|---|---|---|
| | $X(NO_3)_2$ | $Y(NO_3)_2$ | $Z(NO_3)_2$ |
| X | — | NR | R |
| Y | R | — | R |
| Z | NR | NR | — |

   The arrangement of the metals in order from most reactive to least reactive is
   a. X, Y, Z b. Z, X, Y c. Y, Z, X d. Y, X, Z

5. What is conserved in a balanced chemical equation?
   a. mass b. atoms c. charge d. all of these

6. How many atoms of iron can be found in one formula unit of $Fe_2O_3$?
   a. 1 b. 2 c. 3 d. 6

7. How many atoms of carbon are needed to react completely with one formula unit of $Fe_2O_3$ to form Fe and CO?
   a. 1 b. 2 c. 3 d. 6

8. Chemists have devised a measurement unit called the mole. Which of these units is most similar in concept to the mole?
   a. dozen b. centimeter c. liter d. kilograms per liter

9. In the electrolysis of molten sodium chloride, sodium ions are attracted to the negative electrode, where metallic sodium is formed by the gain of electrons. This change in sodium can be classified as
   a. oxidation c. reduction
   b. leaching d. roasting

10. Corrosion of metals can be prevented by sealing the surface with a nonreactive substance. Which gas is most important to be kept from reacting with the metal surface by this method?
a. helium b. hydrogen c. oxygen d. nitrogen

## Free Response Questions

Answer each of the questions as thoroughly and completely as possible.

11. What is the difference between a *renewable resource* and a *nonrenewable* resource? Give an example of each.

12. Identify each chemical property of calcium that is described in the following paragraph. Calcium is a lustrous silver-white metal. It acquires a bluish-gray tarnish on exposure to moist air. It is softer than aluminum or magnesium, but harder than sodium. Its melting point is 850 °C and its boiling point is 1440 °C. Calcium's density is 1.54 g/cm$^3$. When the metal is added to water, alcohol, or dilute acids, hydrogen gas is produced. Calcium is insoluble in kerosene.

13. Compare and contrast the properties of metals and nonmetals. Consider such properties as appearance, heat and electrical conductivity, ductility, and malleability.

14. The "four R's" of resource conservation and management are *rethink, replace, reuse, recycle.* Explain what each term means and provide one example of a way you, as an individual, can contribute to each effort.

15. The following metallic activity series will help you answer this question: Mg, Zn, Cu, Ag. The charge on the most common ion of Mg, Cu, and Zn is 2+. The most common charge on an Ag ion is 1+.
a. Which one of the elements is most likely to be found free in nature?
b. Which one of the following reactions is more likely to occur?
$Cu^{2+}(aq) + 2\ Ag(s) \rightarrow Cu(s) + 2\ Ag^{+}(aq)$
$Cu(s) + 2\ Ag^{+}(aq) \rightarrow Cu^{2+}(aq) + 2\ Ag(s)$
c. A magnesium (Mg) strip is placed in an aqueous solution of copper(II) nitrate, $Cu(NO_3)_2$. What do you expect to observe?
d. Write a chemical equation to support your answer in part c.
e. In part c, is the magnesium metal reduced, oxidized, or neither? Explain your answer.

16. Balance the following equations:
a. ___$Fe_3O_4(s)$ + ___$H_2(g)$ → ___$Fe(s)$ + ___$H_2O(g)$

b. ___$Al_2(SO_4)_3(aq)$ + ___ $Ca(OH)_2(aq)$ → ___ $Al(OH)_3(s)$ + ___ $CaSO_4(aq)$

17. Three different processes are used to refine ores: electrolysis, heating in the presence of carbon monoxide or carbon, and roasting in air.
a. What do the three processes have in common?
b. What types of metallic ores are usually processed by each method?

18. Discuss three factors important in making a decision whether or not to mine a specific ore.

19. Calculate the mass percent of copper in chalcopyrite, $CuFeS_2$.

20. Here are formulas for some known compounds: KI, $CaCl_2$, MgO, $AlCl_3$, $CO_2$. Predict the formulas for compounds formed from
a. Na and Br
b. Ba and F
c. Be and O
d. Ga and Br
e. C and S

21. The melting point of fluorine ($F_2$) is -219 °C while that of bromine($Br_2$) is -7.3 °C. Predict the melting point of chlorine ($Cl_2$). Explain how you decided.

22. New materials can successfully substitute for metals in some products. Explain why each of these substitutions is successful.
    a. Ceramics replace steel in diesel or turbine engines.
    b. Plastics replace steel in automobile bumpers.
    c. Optical fibers replace copper in phone wires.

23. In what way are phrases such as "using up" and "throwing away" inaccurate from a chemical viewpoint? What is the actual meaning of these phrases?

24. If municipal solid waste generates 12,000,000 tons of glass containers and the recycling rate is 40.0%, how many tons of glass are recycled?

## Answers to Questions

### Multiple Choice

| | | | | |
|---|---|---|---|---|
| 1. c | 2. a | 3. d. | 4. d | 5. d |
| 6. b | 7. c | 8. a | 9. c | 10. c |

### Free Response

11. renewable—replenished by natural processes in a reasonable amount of time—wood
    nonrenewable—not replenished by natural processes—petroleum

12. acquires a bluish-gray tarnish, produces hydrogen with water, alcohol, and dilute acids

13.

| | Appearance | Heat conductivity | Electrical conductivity | Ductility, malleability |
|---|---|---|---|---|
| Metal | shiny | good | good | yes |
| Nonmetal | dull | poor | poor | no |

14. rethink—reconsider habits and practices
    replace—use another material for the same purpose
    reuse—use again in the same form
    recycle—reprocess and use to manufacture new items

15. a. Ag
    b. $Cu(s) + 2\ Ag^{+}(aq) \rightarrow Cu^{2+}(aq) + 2\ Ag(s)$
    c. reddish-brown solid forms, blue color fades
    d. $Mg(s) + Cu^{2+}(aq) \rightarrow Cu(s) + Mg^{2+}(aq)$
    e. oxidized

16. a. $Fe_3O_4(s) + 4\ H_2(g) \rightarrow 3\ Fe(s) + 4\ H_2O(l)$
    b. $Al_2(SO_4)_3(aq) + 3\ Ca(OH)_2(aq) \rightarrow 2\ Al(OH)_3(s) + 3\ CaSO_4(aq)$

17. a. The metal ions are reduced.
    b. electrolysis—very active metals, Li, K, Na, Ca
       heating with CO—Mn, Zn, Cr, Fe
       roasting—inactive metals, Cu, Ag, Au

18. percent of metal in ore, how difficult it is to separate the ore, energy requirements

19. $CuFeS_2$: molar mass = 63.55 g/mol + 55.85 g/mol + 2 (32.07) g/mol = 183.54 g/mol
    % Cu = 63.55 g/mol/ 183.54 g/mol × 100 = 34.62%

20. a. NaBr     b. $BaF_2$     c. BeO     d. $GaBr_3$     e. $CS_2$

21. The boiling point of chlorine is estimated to be -113 °C. This value is between the boiling points of fluorine and bromine.

22. a. Ceramics have high melting points and are strong at high temperatures.
b. Plastic are flexible and absorb impact from collision.
c. Optical fibers can transmit voice or electronic messages.

23. Atoms must be conserved and therefore cannot really be thrown away or used up. Such phrases refer to having resources spread out or out of place.

24. 12,000,000 tons × 0.400 = 4,800,000 tons

# PETROLEUM: TO BURN? TO BUILD?

## Multiple-Choice Questions

Choose the letter corresponding to the best answer for each of the following questions.

1. Ibuprofen, like aspirin, is an anti-inflammatory drug. Below is its structural formula. The functional group in this compound indicates that it is

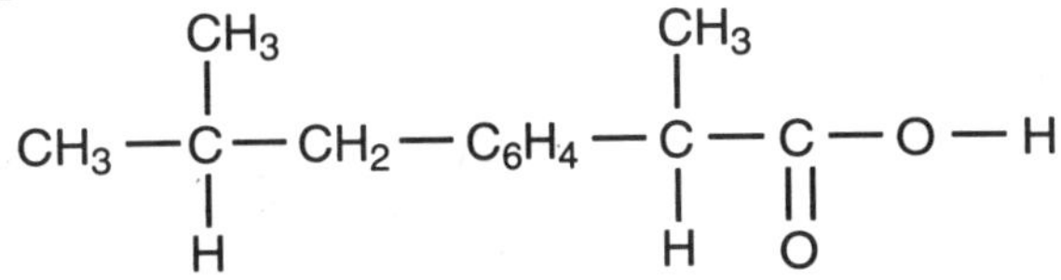

a. an acid    b. an ester    c. a base    d. an alcohol

2. Which statement about bond breaking and bond making is true?
   a. Bond breaking is endothermic and bond making is exothermic.
   b. Bond breaking is exothermic and bond making is endothermic.
   c. Bond breaking and bond making can be either endothermic or exothermic depending on the bonds.
   d. Bond breaking is either endothermic or exothermic, but bond making does not involve a change in energy.

3. Saturated hydrocarbons can be converted into unsaturated hydrocarbons by a process that
   a. removes some hydrogen atoms from the hydrocarbon molecules.
   b. removes sulfur and sulfur compounds from the petroleum.
   c. bonds additional carbon atoms to the saturated hydrocarbon molecules.
   d. breaks bonds and removes carbon atoms from the hydrocarbon molecules.

4. Which is the correct structural formula for butane?

a. 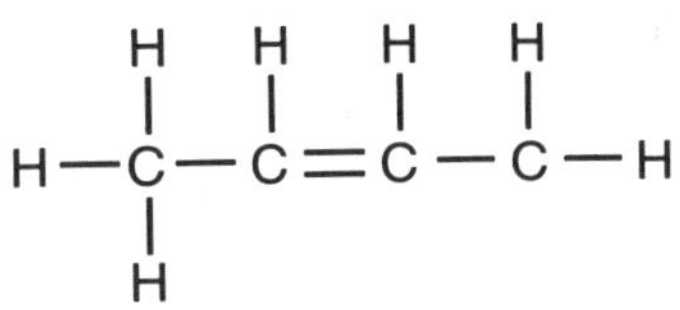

c. 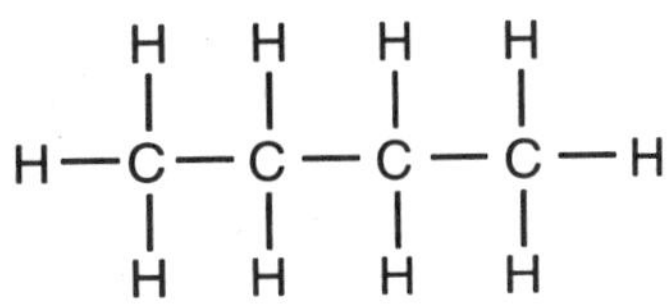

b. 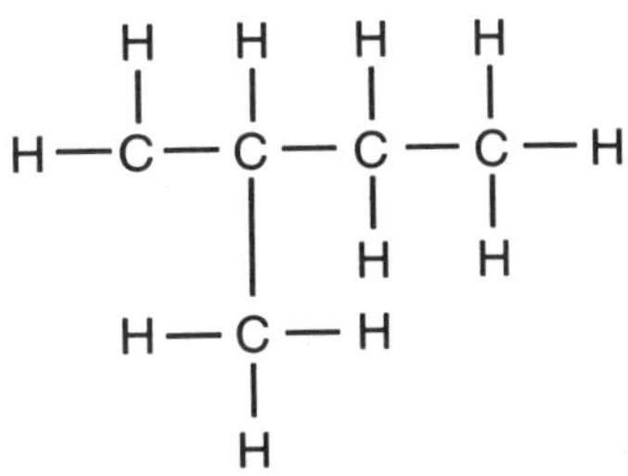

d. 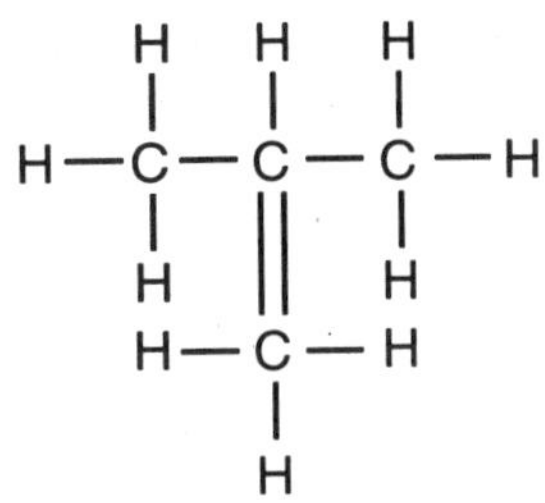

5. Which hydrocarbon has the lowest boiling point?
   a. methane, $CH_4$    c. hexane, $C_6H_{14}$
   b. pentane, $C_5H_{12}$    d. octane, $C_8H_{18}$

6. When ethene is polymerized into polyethylene,
   a. smaller molecules combine to form larger ones.
   b. larger molecules break apart to form smaller ones.
   c. the atoms in molecules are rearranged to produce isomers.
   d. the different molecules in a mixture become separated into fractions.

7. Compounds can be represented by chemical formulas but mixtures cannot. Which material, obtained from petroleum, is an example of a mixture rather than a pure compound?
   a. heptane    b. cyclohexane    c. gasoline    d. octane

8. In the fractionation of crude oil, hydrocarbons are separated by their respective boiling points. If the crude oil enters the bottom of the tower where it is heated, which of the following compounds would be collected *lowest* in a fractionating tower?
a. $C_{15}H_{32}$ b. $C_3H_8$ c. $C_8H_{18}$ d. $C_{24}H_{50}$

9. Which pair of elements would you predict would bond together covalently?
a. H & O b. C & H c. H & H d. All of them bond covalently.

## Free Response

Answer each of the following questions thoroughly and completely.

10. The supply of petroleum for both building and burning is limited. Several alternatives for each are discussed in the text. Choose one substitute that can be used for burning and one substitute that can be used for building. Discuss the advantages and problems in using the substitutes you have chosen.

11. In a laboratory activity, you determined the heat of combustion of candle wax. The following data were recorded:

| | |
|---|---|
| volume of water in can | 100.0 mL |
| initial temperature of water in can | 15 °C |
| final temperature of water in can | 35 °C |
| mass of candle wax lost | 0.28 g |

a. Calculate the heat of combustion per gram of candle wax. Be sure to show your work and use correct units on all numbers.
b. If the molar mass of candle wax is 352 g/mol, calculate the molar heat of combustion.
c. In the laboratory activity, the values obtained by different teams were not all the same. Give some typical reasons for the differences. (Calculation errors don't count!)

12. $C_2H_6O$ has two isomers, one with a boiling point of 78 °C and the other with a boiling point of −24 °C.
a. Sketch the structural formula of each isomer.
b. Which isomer has the higher boiling point? Explain your choice.

13. Write a balanced chemical equation, including the energy term, for the combustion of propane, $C_3H_8$. The molar heat of combustion is 2,200 kJ/mol.

14. The figure below shows total energy consumption per person per year for the years 1850-1990.

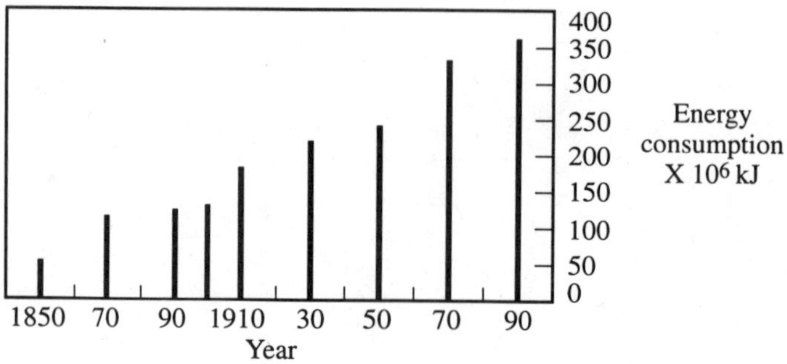


a. How many times higher was petroleum consumption per person in 1990 than it was in 1850?
b. How much energy was consumed in the United States in 1990 if the total population is assumed to be 250,000,000?
c. If 1 barrel is equivalent to $5.8 \times 10^7$ kJ, what is the energy consumption per person equivalent in barrels for 1990?
d. 37% of the energy consumed in 1990 was from oil. How many barrels of oil were consumed per person in that year?

15. Compare North America and the Middle East in terms of population, petroleum reserves, and petroleum consumption.

16. A solution was placed in a distillation apparatus and heated until one half of the solution was distilled. The temperature was recorded every 30 seconds and the data taken were graphed as shown below.

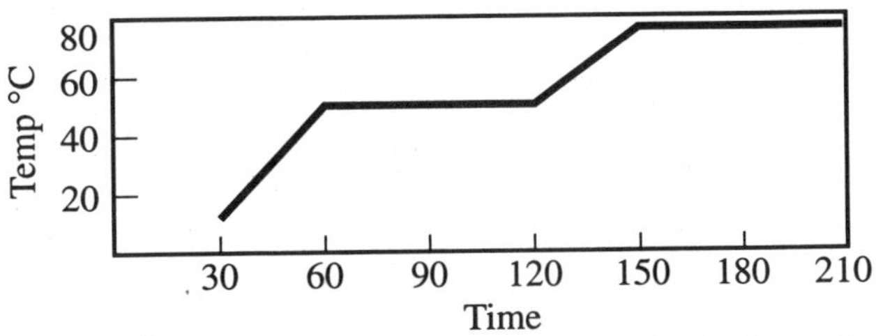


a. What is the smallest number of substances in the original solution?
b. At what temperature(s) did each substance or substances distill?
c. Can you conclude whether the liquid left in the flask is a solution or a pure substance? Why?

17. Trace the energy changes from the coal used by a city's steam generator to the fan on the air conditioner used in a house.
18. Describe two current ways that the octane number of gasoline is increased in the United States.
19. Identify each of the following structural formulas as an acid, alcohol, aromatic compound, ester, or *trans* isomer.

a. 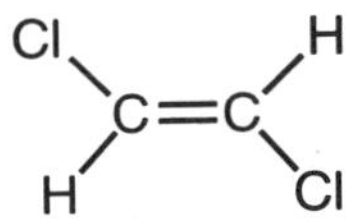

d. 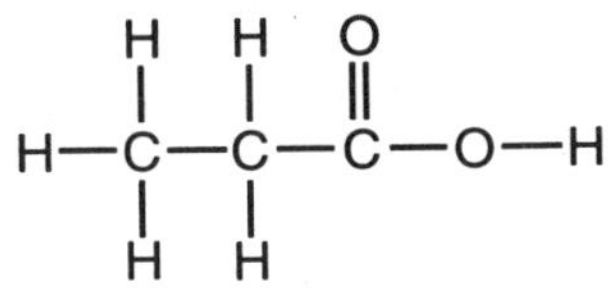

b. 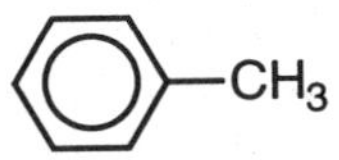

c. 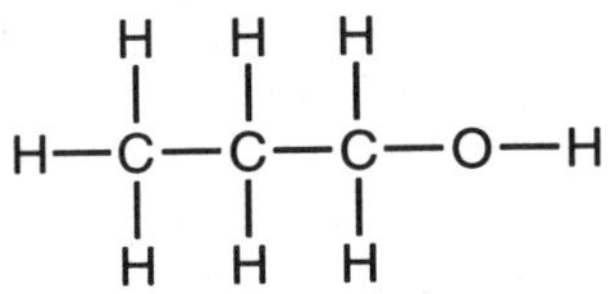

e. 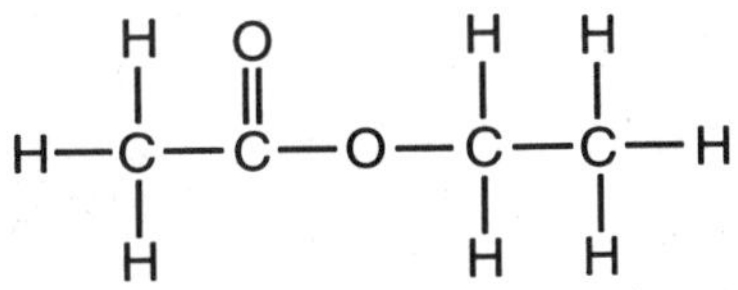

20. Petroleum and metals (studied in the Resources unit) are both natural resources. How are they alike and how are they different?

## Answers to Questions

**Multiple Choice**

| | | | | |
|---|---|---|---|---|
| 1. a | 2. a | 3. a | 4. c | 5. a |
| 6. a | 7. c | 8. d | 9. d | |

10. (Student answers will vary. See sections E.1 and E.2 in the student text.)

11. a. 100 mL = 100 g    35 °C – 15 °C = 20 °C
    100 g × 20 °C = 2,000 cal
    2,000 cal/0.28 g = 7,140 cal/g
  b. 7,140 cal/g × 352 g/mol = 2.5 × 106 cal/mol
  c. The heat that is lost to the air and can is not measured.

12. a.

#1 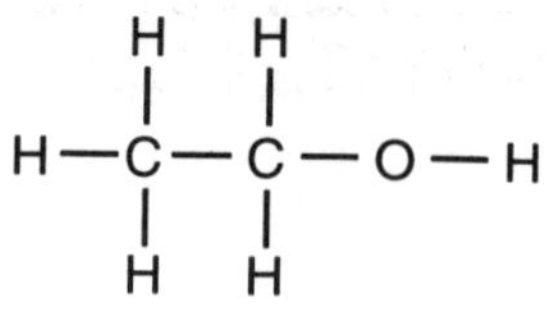

#2 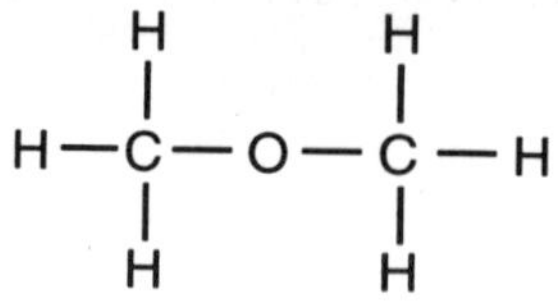

b. Isomer 1. It can hydrogen bond to other molecules.

13. $C_3H_8(g) + 5\ O_2(g) \rightarrow 3\ CO_2(g) + 4\ H_2O(g) + 2{,}200$ kJ

14. a. six times
   b. $3.60 \times 10^8$ kJ/person $\times\ 2.5 \times 10^8$ persons = 9.0 persons = $9.0 \times 10^{16}$ kJ
   c. $3.60 \times 10^8$ kJ $\times$ 1 barrel/$5.8 \times 10^7$ kJ = 6.2 barrels
   d. $9.0 \times 10^{16}$ kJ $\times$ 1 barrel/$5.8 \times 10^7$ kJ $\times\ 0.37 = 5.7 \times 10^8$ barrels

15.

| | Population | Petroleum Reserves | Petroleum Consumption |
|---|---|---|---|
| North America | large | small | high |
| Middle East | small | large | low |

16. a. 2
   b. 55 °C, 82 °C
   c. You can make no conclusion about the number of substances present in the remaining solution. Another distillation (or other type of test) is required to determine if more than one substance is present.

17. coal → steam → electricity → fan

| coal | → | steam | → | electricity | → | fan |
|---|---|---|---|---|---|---|
| chemical | | thermal | | electrical | | mechanical, sound |

18. Cracking to make more branched chain hydrocarbons; using additives such as methanol, ethanol, or MTBE

19. a. *trans* isomer b. aromatic compound c. alcohol d. acid e. ester

20. Petroleum and metal ores are both nonrenewable resources that must be treated to be used to make many different products. Crude petroleum is a liquid, while metals are found in solid ores. Petroleum is separated by distillation, while metal ores require chemical change (reduction) to free the metal. Petroleum is a mixture of many different chemical compounds, while metal ores usually contain only a few compounds. Petroleum can be used to make a wider variety of products.

# UNDERSTANDING FOOD

## Multiple-Choice Questions

Choose the letter corresponding to the best answer for each of the following.

1. A "high energy" food is one that is high in
   a. vitamins   b. calories   c. proteins   d. minerals

2. Which class of compounds provides the basic building blocks for our bodies?
   a. minerals   b. fats   c. proteins   d. carbohydrates

3. One (1) Calorie is equal to
   a. 1,000 joules   c. 1,000 calories
   b. 1,000 kilojoules   d. 1,000 kilocalories

4. Which set of elements is included in the major biomolecules to support human nutrition?
   a. Ca, Mg, S   c. C, H, O, N
   b. Na, P, K   d. Cl, Cu, Fe, Zn

5. Here are the ingredients needed to make a simple cake.

   | | | |
   |---|---|---|
   | 2 cups flour | 1 cup sugar | 1/3 cup oil |
   | 2 eggs | 1.5 tbsps baking powder | 1 cup water |

   Suppose you have in storage 14 cups of flour, 26 eggs, 15 cups sugar, 15 tbsps of baking powder, 40 cups of oil, and access to unlimited water. What is the greatest number of these simple cakes you can make?
   a. 7   b. 10   c. 13   d. more than 13

6. Which category of nutrients is the body's most efficient source of energy?
   a. fats   c. proteins
   b. carbohydrates   d. vitamins

7. Which procedure can be used to determine the quantity of thermal energy released from a food sample?
   a. distillation   c. titration
   b. boiling   d. calorimetry

8. A person can lose body weight if total energy intake
   a. is greater than total energy expended or used.
   b. is less than total energy expended or used.
   c. is equal to total energy expended or used.
   d. is spread out over time, regardless of how much energy is expended or used.

9. The starting substance that becomes "used up" first during a chemical reaction is known as a
   a. biochemical   c. limiting reactant
   b. metabolite   d. catalyst

10. Which statement about malnourishment is true?
    a. A malnourished person cannot build all of the required new body structures.
    b. Undernourishment always accompanies malnourishment.
    c. Inefficient burning of fuel is the major problem with malnourishment.
    d. No decomposition of $H_2O$ molecules can take place, so dehydration occurs.

11. One cup of milk contains about 300 mg of calcium. If a 16-year-old female wants to meet the Recommended Daily Dietary Allowance for calcium, which is 1,200 mg, from drinking milk, what is the fewest number of cups of milk she should drink?
    a. 1 cup   b. 3 cups   c. 4 cups   d. 8 cups

## Free Response

Answer each of the questions as thoroughly and completely as possible.

12. Why is it wise to consider both risks and benefits when discussing the banning of a particular food additive? Use either sodium nitrite or aspartame to discuss your reasoning.

13. a. In reference to an individual's diet, explain the wisdom of the adage "variety is the spice of life".
    b. Is simply "counting Calories" an adequate way to approach dieting? Why?

14. A student ate a quick noon lunch of 1 cup reduced-fat chicken soup, 8 crackers, 2 slices cheese, 1 cup skim milk, and 1/2 cup canned peaches. The labels from the containers gave the following information. The Average Daily Values are given for a 125 lb person eating a 2000 Cal diet.

| Nutrient | Amount per serving | | | | | Average Daily Value |
|---|---|---|---|---|---|---|
| | Soup 1/2 cup | Crackers 4 | Cheese 1 slice | Skim Milk 1 cup | Peaches 1/2 cup | |
| Total Fat | 4 g | 3 g | 4 g | 0 g | 0 g | 65 g |
| Total Carbohydrates | 9 g | 9 g | 2 g | 12 g | 24 g | 300 g |
| Protein | 3 g | 1 g | 3 g | 8 g | 0 g | 48 g |
| Sodium | 950 mg | 160 mg | 340 mg | 125 mg | 10 mg | 2,400 mg |

a. Calculate the total fat, carbohydrate, protein, and sodium in this meal.
b. Which nutrient exceeds the Daily Value in this single meal?

15. A certain brand of orange juice contains 24 g sugar (6 teaspoons) per serving. When burned, this amount of sugar can heat 1,760 g water from 20 °C to 87 °C.
    a. How many Calories are contained in the sugar in one serving of orange juice? (1 Calorie = 1,000 calories)
    b. How much energy is contained in one teaspoon of sugar?
    c. A 125 lb person bicycling requires 6.1 Cal/min. How many minutes must the person bicycle to burn up the sugar in the serving of orange juice?

16. Identify the major uses by the body for each of the following categories of nutrients.
    a. fats
    b. carbohydrates
    c. proteins
    d. vitamins
    e. minerals

17. Explain each of the following statements from the text.
    a. A person can be malnourished without being undernourished.
    b. Proteins are polymers of amino acids.
    c. Trace minerals in the diet are just as important as macrominerals.

18. In a laboratory activity, dyes were separated by paper chromatography. Three dyes produced chromatograms as shown on next page.
    a. What is the smallest number of different dyes that could be present in sample 2?
    b. Do any of the samples have the same dye(s) in them? How do you know?
    c. Calculate the $R_f$ of the dye spots in sample 3.

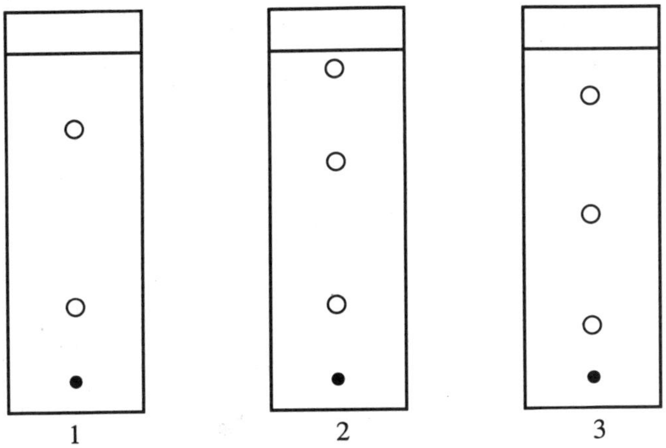


## Answers to Questions

| | | | | | |
|---|---|---|---|---|---|
| 1. b | 2. c | 3. c | 4. c | 5. a | |
| 6. b | 7. d | 8. b | 9. c | 10. a | 11. c |

12. (Student answers will vary. See section D.6 in the student text.)

13. a. Eating a variety of foods increases the likelihood of an adequate balance of nutrients.
    b. This is not a wise approach because the diet may not provide the variety of nutrients needed by the body.

14. a. Total fat: 18 g
    Total carbohydrate: 76 g
    Total protein: 18 g
    Total sodium: 3035 mg
    b. Sodium

15. a. 1,760 g × 67 °C = 120,000 calories or 120 Calories
    b. 120 Cal/6 tsp = 20 Cal/tsp
    c. 120 Cal × 1 min/6.1 Cal = 20 min

16. a. energy
    b. energy
    c. building
    d. necessary in small amounts for growth, reproduction, and general health
    e. help enzymes do their job

17. a. Malnourished means lacking in essential nutrients, while undernourished means taking in insufficient Calories. A person could take in enough Calories for fuel without getting all the essential nutrients (see answer 13b).
    b. Many amino acids can combine to make large protein molecules.
    c. Trace minerals can act as limiting reactants.

18. a. three
    b. Samples 1 and 2; the first spot has the same $R_f$ value in samples 1 and 2
    c. $\frac{0.5}{3.6} = .14$ $\quad \frac{1.8}{3.6} = 0.50$ $\quad \frac{3.2}{3.6} = 0.89$

# NUCLEAR CHEMISTRY IN OUR WORLD

*A periodic table is required for this test.*

## Multiple-Choice Questions

Choose the letter corresponding to the best answer.

1. Which of the following types of electromagnetic radiation have the most energy?
   a. radiowaves  b. gamma rays  c. ultraviolet rays  d. microwaves

2. Which process theoretically yields the greatest energy per gram of fuel?
   a. burning natural gas  c. fission of uranium-235
   b. burning coal  d. fusion of hydrogen

3. Cathode rays are now known to be
   a. X rays  c. streams of electrons
   b. visible light  d. alpha particles

4. In his famous lead-block experiment, Ernest Rutherford proved that pitchblende emitted three types of radiation. What was the composition of radiation that was not deflected by the magnet in this experiment?
   a. relatively heavy, positively charged subatomic particles
   b. negative ions
   c. electromagnetic radiation
   d. electrons

5. Which statement about atomic structure is correct ?
   a. The nucleus occupies about one ten-thousandth the diameter of each atom.
   b. The molar mass of neutrons is about ten thousand grams.
   c. Isotopes of the same element have different atomic numbers.
   d. The mass number of an element equals the number of electrons outside of the nucleus of one of its atoms.

6. Which statement is correct regarding the symbol $^{14}C$?
   a. It represents an isotope of carbon with 8 neutrons.
   b. It represents an isotope of carbon with a mass number of 12.
   c. It represents an isotope of carbon that contains 14 protons.
   d. It represents the most common isotope of carbon.

7. An isotope has an atomic number of 10 and a mass number of 20. It must have
   a. a relative mass greater than 20.
   b. ten protons.
   c. ions that are neutral or have no charge.
   d. atoms whose nuclei each contain 10 electrons.

8. The half-life of radium-226 is 1,620 years. What fraction of an original sample of radium-226 will remain after 4,860 years have passed?
   a. one-half  b. one-third  c. one-fourth  d. one-eighth

9. Which is the most penetrating type of radiation given off by radioactive atoms?
   a. alpha particles  c. gamma rays
   b. beta particles  d. ultraviolet light

10. Which nuclear particle is emitted if an atom of uranium $^{238}U$ changes to an atom of thorium $^{234}Th$?
    a. a proton  c. an electron
    b. a neutron  d. an alpha particle

11. Which of the three isotopes in the table below is most likely to occur naturally on Earth?
a. A b. B c. C d. all equally likely

| Isotope | Half-Life | Radiation emitted |
|---|---|---|
| A | $6.3 \times 10^6$ years | $\alpha$ |
| B | 3.25 years | $\beta$ |
| C | 4.6 seconds | $\beta$ |

## Free Response

Answer each of the following questions as thoroughly and completely as possible.

12. Fill in the rest of the information in this table.

| Isotope | Nuclear notation | Number of protons | Number of neutrons |
|---|---|---|---|
| ? | ? | 36 | 52 |
| xenon-133 | $^{133}_{54}X$ | ? | ? |
| potassium-? | ? | ? | 21 |

13. Balance the following nuclear equations.

a. $^{2}_{1}H + ^{3}_{1}H \rightarrow ^{4}_{2}He +$ ______

b. ______ $+ ^{14}_{7}N \rightarrow ^{17}_{8}O + ^{1}_{1}H$

c. $^{87}_{37}Rb \rightarrow ^{87}_{38}Sr +$ ______

14. A chemistry student measures the half-life of a radioactive isotope. The following table summarizes the results.

| Elapsed Time(hours) | Percent Remaining |
|---|---|
| 0.0 | 100 |
| 1.1 | 80 |
| 4.0 | 45 |
| 8.1 | 20 |
| 13 | 8.0 |
| 20 | 2.0 |

a. Plot a graph showing the decay of this isotope.
b. Determine the half-life of the isotope from the graph.
c. Based on your graph, how many hours have elapsed when 5.0% of the isotope remains?

15. Discuss the risks and benefits of food irradiation.

16. What is ionizing radiation? Why can it cause serious damage to living organisms?

17. Discuss two factors that determine the extent of biological damage from radiation.

18. Describe the properties of radioisotopes that make them useful in medicine as tracers and for irradiation.

19. a. How is nuclear fission different from radioactive decay?
b. How is nuclear fission different from nuclear fusion?
c. Why isn't nuclear fusion used as a controlled source of energy?

20. Nuclear waste disposal is a major unsolved societal problem.
a. Distinguish between high-level and low-level wastes.
b. Describe one solution to the problem of nuclear waste disposal.
c. Discuss one risk and one benefit associated with your solution.

## Answers to Questions

**Multiple Choice**

1. b 2. d 3. c 4. c 5. a
6. a 7. b 8. d 9. c 10. d 11. a

**Free Response**

12.

| Isotope | Nuclear notation | Number of protons | Number of neutrons |
|---|---|---|---|
| krypton-88 | $^{88}_{36}Kr$ | 36 | 52 |
| xenon-133 | $^{133}_{54}Xe$ | 54 | 79 |
| potassium-40 | $^{40}_{19}K$ | 19 | 21 |

13. a. $^{1}_{0}n$

b. $^{4}_{2}He$

c. $^{0}_{-1}e$

14. a.

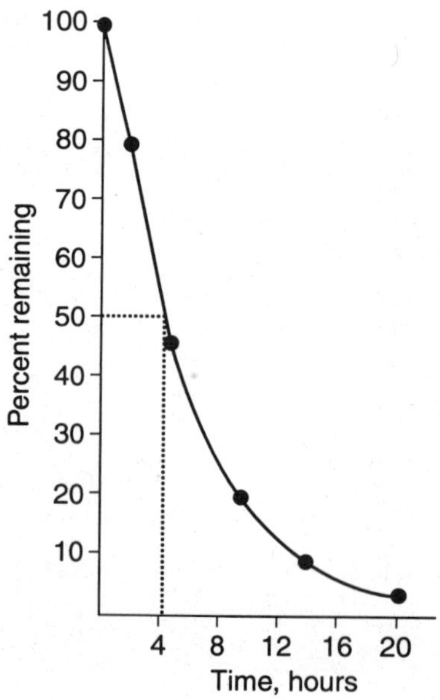


b. 4.1 hrs
c. 17 hrs

15. Benefits—Microorganisms that cause food to spoil are killed, which reduces medical costs and lost productivity.
Risks—Potentially harmful species may be formed during the radiation process. (Irradiation does not make foods radioactive.) The costs may be higher.

16. Ionizing radiation has high energy and can cause ions to form when electrons are ejected from molecules. These highly reactive ions can cause serious damage to living systems.

17. The two factors are radiation density and the dose. See Sections B.3–B.5.

18. Radioisotopes emit radiation (alpha, beta, gamma), which makes it possible for the doctor to find out how the tracer is distributed in the body. Isotopes with short half-lives are used. In cancer

treatment the tumor is exposed to ionizing radiation, which kills the abnormal cancerous cells. Great care must be taken in the administration of irradiation to minimize damage to normal cells.

19. a. In nuclear fission a nucleus is bombarded with neutrons, and it splits into relatively large fragments and a number of neutrons, which can begin a chain reaction. Radioactive decay emits small particles and does not start a chain reaction.
    b. In nuclear fission a large nucleus splits into two or more parts; in nuclear fusion two small nuclei are joined to make a larger nucleus.
    c. It is too difficult to maintain the incredibly high temperatures necessary to sustain a controlled reaction.

20. a. High-level wastes are products of nuclear fission or transuranic products formed from U-235. Low-level wastes include laboratory protective clothing, diagnostic radioisotopes, and other wastes with only low levels of radioactivity.
    b. (Student answers will vary. See Section E.2.)
    c. (See Section E.2.)

# LIVING IN A SEA OF AIR: CHEMISTRY AND THE ATMOSPHERE

## Multiple-Choice Questions

Choose the letter corresponding to the best answer.

1. Which is responsible for the largest amount of air pollutants in the United States?
   a. industrial processes
   b. fuel burning for space heating and electricity
   c. transportation
   d. solid waste disposal

2. As you go higher in altitude, the air pressure_________, and the volume of a gas sample at constant temperature will_________.
   a. increases, increase
   b. decreases, decrease
   c. decreases, increase
   d. increases, decrease

3. If the average human adult inhales 14 times each minute, and the average volume of gas inhaled each time is 500 mL, how many liters of gas are inhaled in one hour?
   a. 840 L  b. 420 L  c. 60 L  d. 0.5 L

4. Which statement supports the idea that equal volumes of gases at the same temperature and pressure contain equal numbers of molecules?
   a. Elements such as argon are found in small quantities in the air.
   b. All gases have the same density at standard conditions.
   c. The space between the molecules in a gas sample is great compared to the size of the molecules themselves.
   d. Two liters of hydrogen gas combine exactly with one liter of oxygen gas to form water.

5. Which statement is true for the reaction represented by this equation? Assume that all gases are at standard temperature and pressure.
   $2\ NO(g) + O_2(g) \rightarrow 2\ NO_2(g)$
   a. $O_2$ is always the limiting reactant.
   b. 60 g of NO ideally will react with 16 g of $O_2$.
   c. 2 liters of NO ideally will react with 2 liter of $O_2$.
   d. Two moles of $NO_2$ formed will occupy a volume of about 44.8 L.

6. The pressure of the air can be measured by a
   a. barometer
   b. calorimeter
   c. thermometer
   d. photometer

7. Based on your knowledge of the properties of gases, predict what will happen to the density of 100 g of helium gas if the sample is heated from room temperature to 60 °C at constant pressure.
   a. remain unchanged
   b. increase
   c. decrease
   d. depends on the air pressure

8. A 5.00 L sample of gas is at 35 °C and 1.00 atm pressure. What will the volume become if the pressure is increased to 4.00 atm and the temperature remains constant?
   a. 20.0 L  b. 8.00 L  c. 1.25 L  d. 15.0 L

9. You are at the beach on a sunny day. You observe that the sand gets very hot by midday, while the water temperature remains about the same. What is the best explanation for this observation?

a. The cool breeze makes the water feel colder than the sand.
b. Sitting on the hotter sand makes the water feel cooler.
c. Sand both absorbs and gives off heat more readily than does water.
d. Hotter sand causes a sea breeze to be produced, which cools the water.

10. Electrostatic precipitators reduce air pollution because
a. many pollutants exist as charged ionic particles, or can form charged particles.
b. precipitators utilize water as an ionizing catalyst.
c. precipitators eliminate the need for waste disposal.
d. some useful products can be obtained from the pollutants collected by precipitators.

11. Which matches the phrase "a hydrogen ion concentration of $10^{-7}$ mol/L"?
a. a pH of 7
b. an acidic water solution
c. a basic water solution
d. pure rain water

12. Which of the following would lower the pH of a neutral aqueous solution?
a. ammonia
b. milk
c. lemon juice
d. sugar

## Short-Answer

Answer the following questions as thoroughly and completely as possible.

13. Write a balanced chemical equation to represent each of the following statements. Include (s), (l), (g) and (aq) symbols for physical states.
a. Natural gas, $CH_4$, reacts with oxygen from air to produce carbon dioxide gas and water vapor.
b. Carbon monoxide gas reacts with oxygen gas to produce carbon dioxide gas.
c. Aqueous hydrogen peroxide, $H_2O_2$, in the presence of a catalyst, decomposes to liquid water and oxygen gas.

14. Three problems made worse by human activities are an increase in greenhouse gases, endangering the ozone layer, and photochemical smog. Choose one of the three problems and answer the questions below.
a. Describe the problem.
b. Discuss human activities that increase the magnitude of the problem.
c. Propose actions to reduce the impact of human activities.

15. Briefly explain how geographic and weather conditions in locations such as Los Angeles can increase the severity of air pollution problems.

16. A student in a welding class wants to purchase 100 L of oxygen and 40 L of acetylene. Is that sufficient information to help the student find the best buy? If not, what other information is needed? Why?

17. Indoor air pollution is a subject that is getting increasing attention.
a. Describe the three common indoor air pollutants.
b. Explain a method to reduce each of the air pollutants listed in part a.

18. a. Explain why catalysts are sometimes used for chemical reactions.
b. Below is an energy diagram for a chemical reaction. Show what the diagram would look like if a catalyst were used.

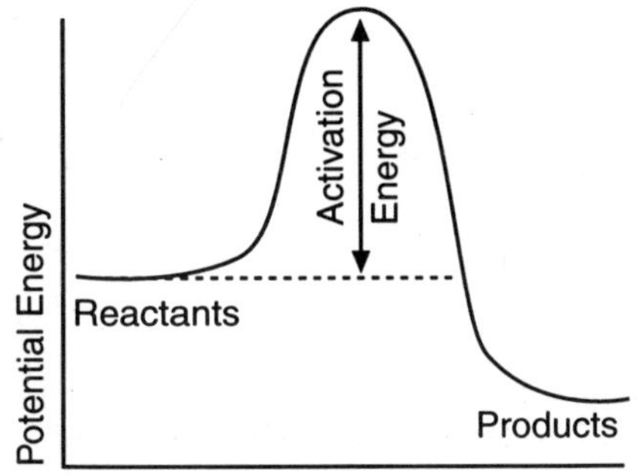

c. Explain why all new cars are equipped with catalytic converters.

19. Use the kinetic molecular theory to explain each of these observations about gases.
    a. If the volume of a gas sample is constant, increasing the gas temperature increases its pressure.
    b. If a scuba diver rises quickly to the water surface (pressure 1 atm) from 10 meters (pressure 2 atm) lung damage can result.
    c. The odor from the shop baking cinnamon rolls spreads quickly throughout the mall.
    d. Gases at 1 atm can be compressed so that they occupy a tiny fraction of their original volume.
    e. The atmosphere exerts pressure on our bodies, yet our bodies are not crushed.

## Answers to Questions

### Multiple Choice

| | | | | |
|---|---|---|---|---|
| 1. c | 2. c | 3. b | 4. d | 5. d |
| 6. a | 7. c | 8. c | 9. c | 10. a |
| 11. a | 12. c | | | |

### Short Answer

13. a. $2\ CH_4(g) + 3\ O_2(g) \rightarrow 2\ CO_2(g) + 2\ H_2O(g)$
    b. $2\ CO(g) + O_2(g) \rightarrow 2\ CO_2(g)$
    c. $2\ H_2O_2(aq) \rightarrow 2\ H_2O(l) + O_2(g)$

14. Greenhouse gases—See Section C.4
    Ozone layer—See Section C.8
    Photochemical smog—See Section D.3, D.6

15. A city such as Los Angeles is not only located in a valley, but also experiences temperature inversions that often trap pollutants near the surface.

16. The information is not sufficient; the student must also specify the temperature and pressure or buy the gases by mass.

17. a. The four indoor air pollutants discussed in the text are radon, formaldehyde, second-hand smoke and asbestos. See Section E.2.
    b. (Student answers will vary.)

18. a. Catalysts are used to speed up chemical reactions without changing the products.
    b.

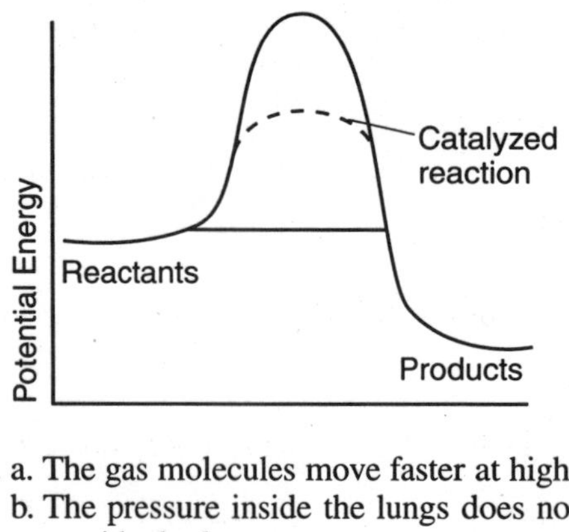


19. a. The gas molecules move faster at higher temperatures.
    b. The pressure inside the lungs does not have sufficient time to become equal to the pressure outside the lungs.
    c. Gas molecules are in constant motion.
    d. There is much space between molecules of a gas at 1 atm pressure.
    e. The pressure inside and outside is the same.

# PERSONAL CHEMISTRY AND CHOICES

## Multiple-Choice Questions

Choose the letter corresponding to the best answer.

1. A combination of alcohol and a barbiturate has an effect up to 200 times that of either drug taken alone. Such an effect is called
   a. activation
   b. chemical dependency
   c. synergism
   d. synthesis

2. Which term means the overall balance of body chemistry required for good health?
   a. homeostasis
   b. ecosystems
   c. equilibrium
   d. side effects

3. Which combination of elements makes up all but about 2% of body mass?
   a. C, Ca, O, P, S
   b. C, Ca, H, N, O
   c. C, H, I, Na, O
   d. C, Ca, Cl, O, Si

4. Which is the best explanation for why hair protein is different from skin protein or enzyme protein?
   a. Each protein has a distinctive sequence of amino-acid units, which creates a large number of possibilities for folding and interaction.
   b. Each protein has a unique shape.
   c. Ribosomes "direct" the order of amino acids in each protein.
   d. Globular proteins are water soluble, while fibrous proteins are tough and often rope-like in structure.

5. Which of these could you take to counteract a mild and temporary condition of excess stomach acidity?
   a. carbonated soft drink
   b. sodium bicarbonate
   c. aspirin
   d. large amounts of distilled water

6. Why don't oils excreted by the body wash away with water alone?
   a. Water molecules are ionic, while oil molecules are not.
   b. Enzymes from the body prevent "oily dirt" from mixing with water.
   c. Bacteria react with the secreted oils, causing them to break down into polar molecular substances.
   d. Water molecules are more strongly attracted to one another than to oil molecules.

7. Acne can be controlled through careful washing followed by moderate chemical treatment with benzoyl peroxide and antibiotics. This procedure causes
   a. the killing of harmful bacteria that live on natural oils.
   b. the production of glycerol when soap is used to wash away the oils.
   c. a reaction between soap and water.
   d. the elimination of dead skin.

8. Why doesn't "burning" of carbohydrates in living systems such as the human body occur at high temperatures as it does in a forest fire?
   a. ADP molecules serve as a mechanism for capturing large quantities of energy.
   b. ATP + $H_2O$ form as energy is released.
   c. Relatively small numbers of ATP molecules are involved in storing and releasing energy.
   d. Cells oxidize the fuel in a series of steps involving many different enzymes.

9. Which of these would work best as a soap?
   a. $CH_3–CH_2–CH_2–CH_2–CH_2–CH_2–SO_3^-Na^+$

b. $CH_3–CH_3$
c. glucose
d. gasoline

10. Why does the structure of hair make it possible to style it?
   a. The hair cuticle consists of overlapping layers of flat cells that are capable of sliding past each other.
   b. Oils and natural acids coat and protect the structure of hair microfibrils, but the coating is easily lost.
   c. The weak chemical interactions holding protein chains to one another can be shifted in location through the action of water and other substances.
   d. Excess heat dries hair oils, causing protein chains in hair to separate from one another.

11. The addition of an enzyme to the reaction that forms ATP would
   a. lower the activation energy.
   b. cause it to require more energy.
   c. decrease the rate of the reaction.
   d. cause all of the above.

12. Melanin, a skin pigment, protects the dividing dermal cells by absorbing what type of radiation?
   a. visible light   b. ultraviolet radiation   c. gamma rays   d. X rays

13. Alcohol affects the nervous system by
   a. slowing down the transmission of nerve signals.
   b. speeding up the transmission of nerve signals.
   c. completely blocking transmission of nerve signals.
   d. dissolving nerve cells.

## Short-Answer

Answer each of the following questions as completely and thoroughly as possible.

14. Explain the meaning of the term "homeostasis" in the context of blood pH.

15. Too much or too little exposure to the sun's rays can lead to negative health effects. What are they?

16. Explain the difference between the terms *toxic* and *carcinogenic.*

17. Some smokers select skin patches to help them stop smoking. Explain the principle by which these patches work.

18. Enzymes are a very important class of biological molecules.
   a. Explain their importance to the proper functioning of the human body.
   b. Explain, in general, how enzymes and substrates interact.
   c. Briefly describe the role of enzymes in energy storage and release.

19. A particular substance could be either a drug or a toxin. Explain why this is true.

20. A variety of over-the-counter pain relievers, commonly called NASIDs, are available to treat chronic pain such as that associated with arthritis. What kind of information would you want to have before you decided on a particular medication? Where or from whom would you obtain this information?

21. Overconsumption of alcohol has both short-and long-term effects. Describe three long-term health problems that may result from alcohol abuse.

22. Describe two ways that the liver eliminates harmful substances from the body.

## Answers to Questions

**Multiple Choice**

| | | | | | |
|---|---|---|---|---|---|
| 1. c. | 2. a | 3. b | 4. a | 5. b | 6. d |
| 7. a | 8. d | 9. a | 10. c | 11. a | 12. b |
| 13. a | | | | | |

**Short Answer**

14. (See Section C.5.)

15. Too much exposure to sun causes sunburn and, eventually, skin cancer. Too little exposure to sun causes rickets.

16. *Toxic* means harmful to the human body. *Carcinogenic* means causes cancer. All toxins do not cause cancer.

17. Nicotine skin patches deliver the nicotine to the smoker. Slowly, over a period of weeks, the amount of nicotine is reduced until the smoker is able to overcome the addiction.

18. a. Enzymes catalyze reactions necessary for the body to properly function.
    b. The substrate fits into the enzyme active site. Bonds are weakened, which makes the reaction more energetically favorable. The product departs from the active site, and the enzyme surface is available for another reaction.
    c. Enzymes catalyze over 20 reactions, which result in the oxidation of glucose which produces carbon dioxide,water, and energy.

19. The difference is in the dose. Nearly all drugs become toxins if too much is taken.

20. One would want to know the dosage, the side effects, and the cost. Both doctors and pharmacists would be helpful in providing this information.

21. Overconsumption of alcohol can cause high blood pressure; diseases of liver, pancreas, and intestines; blackout periods; vitamin deficiencies; heart problems; neuroses; and accidents.

22. The liver separates useful materials from some harmful materials and eliminates the harmful ones in the bile. Other harmful materials are detoxified, or chemically modified before they are excreted.

# THE CHEMICAL INDUSTRY: PROMISE AND CHALLENGE

## Multiple-Choice Questions

Choose the letter corresponding to the best answer.

1. The electrical potential (in volts) produced by an electrochemical cell depends on
   a. the type of metals used
   b. the size of the cells
   c. the type of metals and size of cells
   d. the presence of an anode only

2. A student dips a clean nichrome wire into a solution of $KNO_3$. The coated wire is placed into a candle flame. The candle flame remains yellow instead of violet, as expected. Which statement explains this result?
   a. Nitrates produce bright yellow flames.
   b. The candle's bright yellow flame masks the violet color.
   c. The candle flame is too hot to produce the violet color.
   d. Dipping the nichrome wire into $KNO_3$ will always produce a yellow flame with heat.

3. In the nitrogen molecule ($N_2$), each atom is in a zero oxidation state. An atom of nitrogen in a molecule of ammonia ($NH_3$) is
   a. in a positive oxidation state.
   b. in a negative oxidation state.
   c. in a more strongly bonded state.
   d. also in a zero oxidation state.

4. Which pair of substances is commonly used as car battery electrodes?
   a. Zn and $CuSO_4$
   b. Pb and Mg
   c. Pb and $PbO_2$
   d. Ag and Au

5. Which statement is correct about these two half reactions?

   $Zn(s) \rightarrow Zn^{2+}(aq) + 2e^-$
   $Cu^{2+}(aq) + 2e^- \rightarrow Cu(s)$

   a. Zinc atoms are reduced.
   b. Zinc atoms gain electrons.
   c. Copper ions lose electrons.
   d. Copper ions are reduced.

6. Electrical energy is required to release metallic aluminum from its minerals. Which statement best explains why?
   a. No common substances give up electrons easily enough to reduce $Al^{3+}$ ions.
   b. $Na_3AlF_6$ is a very unstable mineral.
   c. Molten $Na_3AlF_6$ is a solvent for $A1_2O_3$.
   d. Carbon electrodes are used in the electrolytic process, which produces the aluminum from $A1_2O_3$.

7. This is the equation for the formation of $NH_3$ from $N_2$ and $H_2$ in the Haber process.

   $N_2(g) + 3\ H_2(g) \Leftrightarrow 2\ NH_3(g)$

   What information does the double arrow symbol ($\Leftrightarrow$) convey?
   a. Ammonia is a more stable substance than is either $N_2$ or $H_2$.
   b. As $NH_3$ forms, some of it reverts back to $N_2$ and $H_2$.
   c. The reaction involves both oxidation and reduction.
   d. The yield of $NH_3$ in this process is 100%.

**Electronegativity**

| Element | Symbol | Value |
|---|---|---|
| Oxygen | O | 3.5 |
| Nitrogen | N | 3.0 |
| Chlorine | Cl | 3.0 |
| Sulfur | S | 2.5 |
| Carbon | C | 2.5 |
| Hydrogen | H | 2.1 |
| Manganese | Mn | 1.5 |

**Activity Series of Common Metals**

Na(s)
Mg(s)
Al(s)
Zn(s)
Sn(s)
Pb(s)
Cu(s)
Ag(s)
Au(s)

8. Using the activity series above, decide which statement is true for a voltaic cell consisting of silver metal (Ag) in a solution of silver ions and magnesium metal (Mg) in a solution of magnesium ions. The two half cells are connected by a salt bridge.
   a. Magnesium metal is produced at the cathode.
   b. Both magnesium metal and silver metal are produced at the cathode.
   c. Silver metal is produced at the cathode.
   d. Silver ions are produced at the cathode.

9. Which of the statements about the reaction below is correct?

   $H_2(g) + I_2(g) \Leftrightarrow 2\ HI(g)$

   a. Hydrogen iodide (HI) is a more stable substance than hydrogen or iodine.
   b. At equilibrium, the rate of formation of hydrogen iodide equals the rate of formation of hydrogen and iodine.
   c. All the hydrogen and iodine react to form hydrogen iodide.
   d. The reaction is irreversible.

10. Which statement about nitrogen dioxide ($NO_2$) is true?
    a. N and O both have negative oxidation numbers.
    b. N has a positive oxidation number; O has a negative oxidation number.
    c. N has a negative oxidation number; O has a positive oxidation number.
    d. N and O both have positive oxidation numbers.

## Free Response

Answer each of the following as thoroughly and completely as possible.

11. Use a diagram to show how you would electroplate a metallic spoon with silver metal.

12. Explain why nitrogen is a limiting reactant for plant growth, even though it makes up about 78% of the atmosphere.

13. Using ammonia as an example, describe how the same chemical substance can be used for either constructive or destructive purposes.

14. a. On the diagram of a voltaic cell below, clearly label parts of the cell with the numbers corresponding to the correct choices from the list.

(1) Pb electrode
(2) Au electrode
(3) $Pb(NO_3)_2$ solution
(4) $Au(NO_3)_3$ solution
(5) Anode
(6) Cathode
(7) Salt bridge

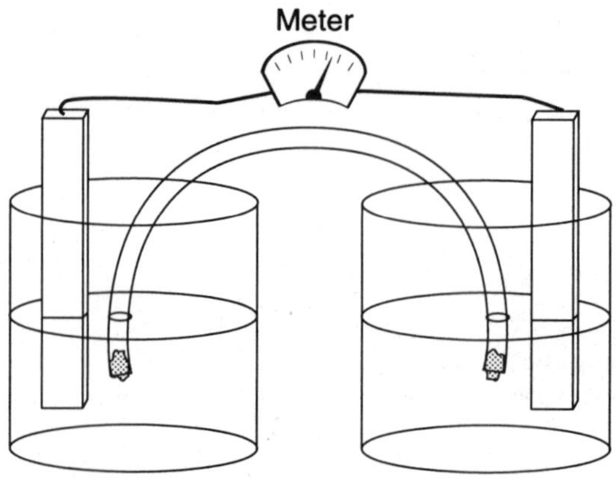


b. On the diagram, indicate the direction electrons will flow in the external wire.
c. Which electrode will eventually lose mass?
d. Which metallic ion in the cell is reduced?
e. Would the cell be able to operate without the salt bridge? Explain why.

15. Aluminum metal is produced industrially using the Hall process. One step in this process involves dissolving aluminum oxide in molten cryolite.
  a. Bauxite (aluminum oxide) is 53% aluminum. Calculate how many grams of aluminum metal can be extracted from 1 metric ton ($10^3$ kg) of bauxite.
  b. How many moles of aluminum metal is that?

16. Select one of the two industries (ammonia-based products and aluminum products) that is interested in locating a plant in Riverwood. Discuss, from a citizen's viewpoint, negative and positive aspects of locating this industry in Riverwood.

## Answers to Questions

**Mutilpe Choice**

| | | | | |
|---|---|---|---|---|
| 1. a | 2. b | 3. b | 4. c | 5. d |
| 6. a | 7. b | 8. c | 9. b | 10. b |

**Free Response**

11.

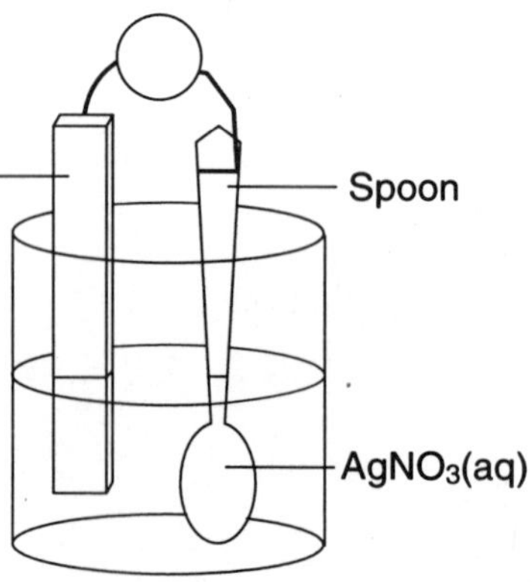


12. Nitrogen is a limiting reactant because plants cannot use nitrogen gas ($N_2$). The nitrogen must be fixed before it is available to plants.

13. Ammonia can be used to synthesize fertilizers (constructive) or to make explosives (destructive.)

14. a.,b.

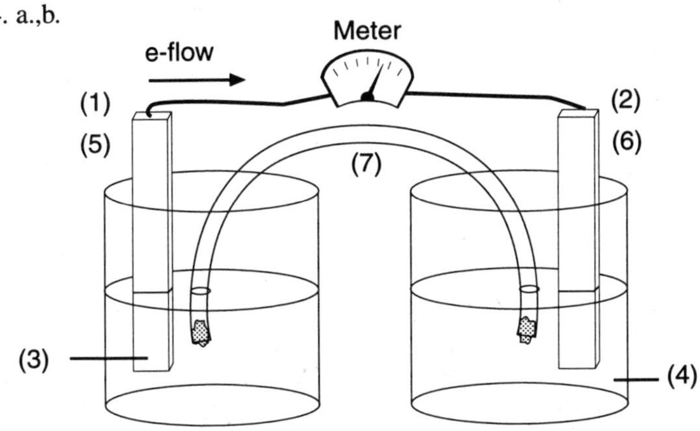


c. Pb electrode
d. $Au^{3+}$ ions
e. No, the circuit is not complete.

15. a. $5.3 \times 10^5$ g Al
b. 20,000 mol Al

16. (Student answers will vary. See Part E.)

# BLACK LINE MASTERS

This section of your Teacher's Guide contains **Black Line Masters** for your use as overhead transparencies.

Permission is hereby granted by the American Chemical Society and Kendall/Hunt Publishing Company for reproduction for classroom use by adopters of *Chemistry in the Community (ChemCom)* of the **Black Line Masters** contained herein.

# SUPPLYING OUR WATER NEEDS

| Data Table 1 | | | ORIGINAL SAMPLES | | | | |
|---|---|---|---|---|---|---|---|
| | | | | | Mixture Classification | | |
| Well | Color | Clarity (Clear or Cloudy?) | Settle Out? | Tyndall Effect | Suspension | Colloid | Solution |
| A1 | | | | | | | |
| A3 | | | | | | | |
| A5 | | | | | | | |
| C1 | | | | | | | |

| Data Table 2 | | | FILTRATES | | | | | |
|---|---|---|---|---|---|---|---|---|
| | | | | | | Mixture Classification | | |
| Well | Color | Clarity (Clear or Cloudy?) | Settle Out? | Tyndall Effect | Cotton Filter | Suspension | Colloid | Solution |
| D2 | | | | | | | | |
| D4 | | | | | | | | |
| D6 | | | | | | | | |
| B6 | | | | | | | | |

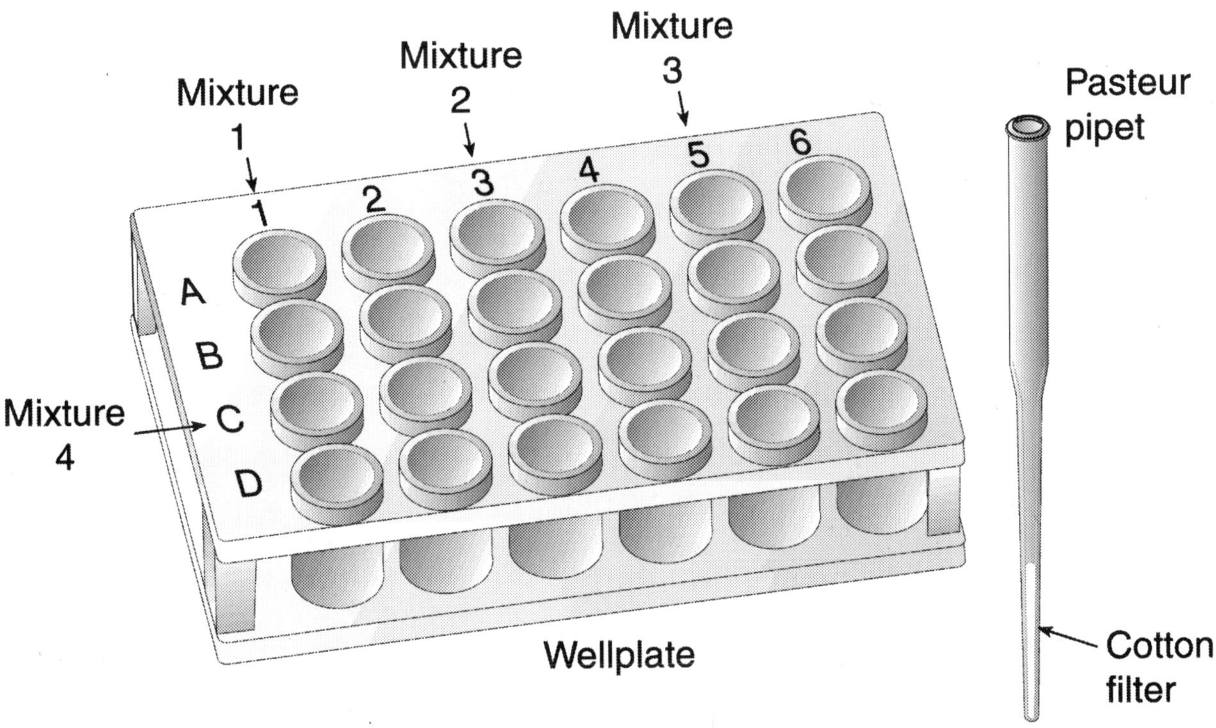


**W1**. Figure 12, page 7

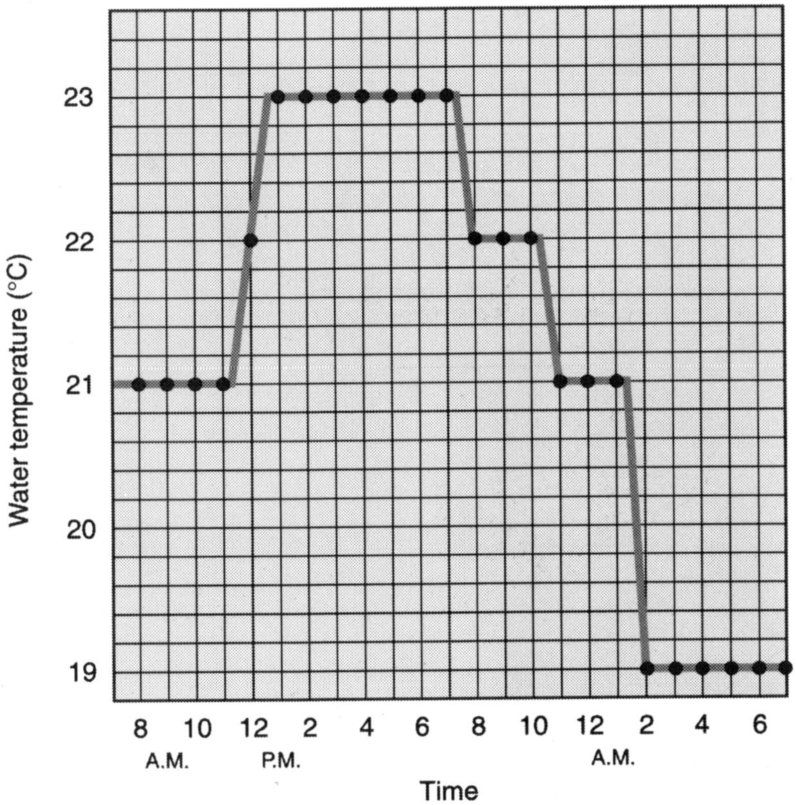


**W2**. Figure 22, page 21

| Table 9 DO Needed | |
|---|---|
| Water °C | 100% Oxygen Saturation (ppm) |
| 0 | 14.6 |
| 1 | 14.2 |
| 2 | 13.9 |
| 3 | 13.5 |
| 4 | 13.2 |
| 5 | 12.8 |
| 6 | 12.5 |
| 7 | 12.2 |
| 8 | 11.9 |
| 9 | 11.6 |
| 10 | 11.3 |
| 11 | 11.1 |
| 12 | 10.8 |
| 13 | 10.6 |
| 14 | 10.4 |
| 15 | 10.2 |
| 16 | 9.9 |
| 17 | 9.7 |
| 18 | 9.5 |
| 19 | 9.3 |
| 20 | 9.2 |
| 21 | 9.0 |
| 22 | 8.8 |
| 23 | 8.7 |
| 24 | 8.5 |
| 25 | 8.4 |

**W4**. Table 9, page 21

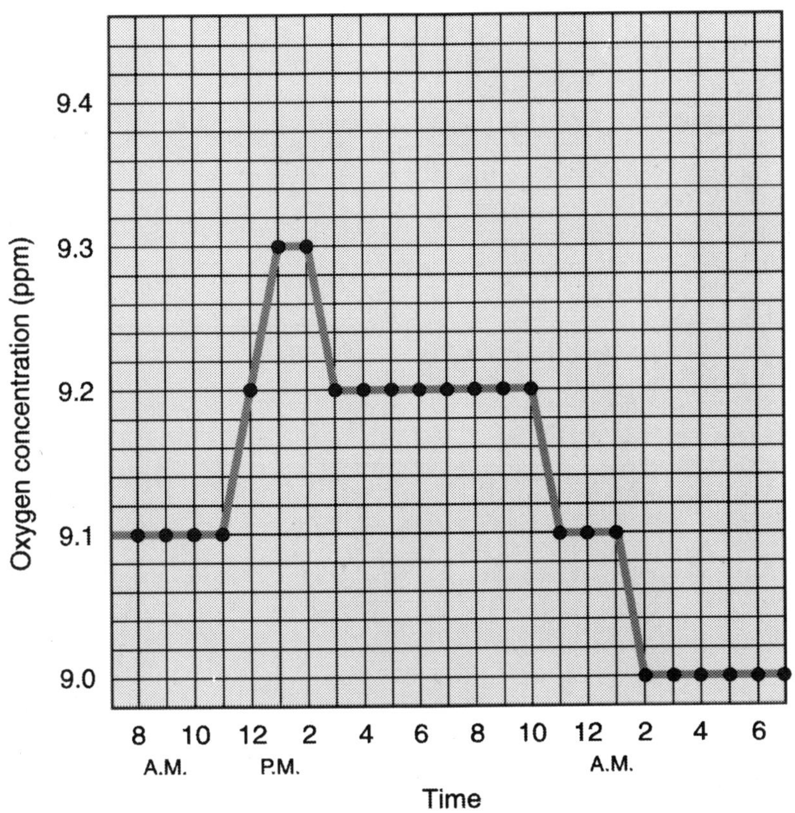


**W3**. Figure 23, page 21

**Table 10 Some Common Acids and Bases**

| Name | Formula | Use |
|---|---|---|
| ***Acids*** | | |
| Acetic acid | $HC_2H_3O_2$ | In vinegar (typically a 5% solution of acetic acid) |
| Carbonic acid | $H_2CO_3$ | In carbonated soft drinks |
| Hydrochloric acid | HCl | Used in removing scale buildup from boilers and for cleaning materials |
| Nitric acid | $HNO_3$ | Used in the manufacture of fertilizers, dyes, and explosives |
| Phosphoric acid | $H_3PO_4$ | Added to some soft drinks to give a tart flavor; also used in the manufacture of fertilizers and detergents |
| Sulfuric acid | $H_2SO_4$ | Largest volume industrial chemical; present in automobile battery fluid |
| ***Bases*** | | |
| Calcium hydroxide | $Ca(OH)_2$ | Present in mortar, plaster, and cement; used in paper pulping and dehairing animal hides |
| Magnesium hydroxide | $Mg(OH)_2$ | Active ingredient in milk of magnesia |
| Potassium hydroxide | KOH | Used in the manufacture of some liquid soaps |
| Sodium hydroxide | NaOH | A major industrial chemical; active ingredient in some drain and oven cleaners; used to convert animal fats into soap |

**W5**. Table 10, page 22

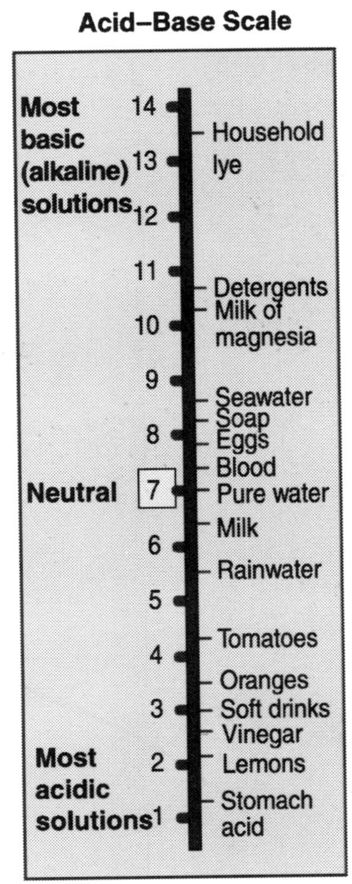


**W6**. Figure 24, page 22

## Table 11 Common Ions

### *Cations*

| +1 Charge Formula | +1 Charge Name | +2 Charge Formula | +2 Charge Name | +3 Charge Formula | +3 Charge Name |
|---|---|---|---|---|---|
| $Na^{+}$ | Sodium | $Mg^{2+}$ | Magnesium | $Al^{3+}$ | Aluminum |
| $K^{+}$ | Potassium | $Ca^{2+}$ | Calcium | $Fe^{3+}$ | Iron(III)* |
| $NH_4^{+}$ | Ammonium | $Ba^{2+}$ | Barium | | |
| | | $Zn^{2+}$ | Zinc | | |
| | | $Cd^{2+}$ | Cadmium | | |
| | | $Hg^{2+}$ | Mercury(II)* | | |
| | | $Cu^{2+}$ | Copper(II)* | | |
| | | $Pb^{2+}$ | Lead(II)* | | |
| | | $Fe^{2+}$ | Iron(II)* | | |

### *Anions*

| −1 Charge Formula | −1 Charge Name | −2 Charge Formula | −2 Charge Name | −3 Charge Formula | −3 Charge Name |
|---|---|---|---|---|---|
| $F^{-}$ | Fluoride | $O^{2-}$ | Oxide | $PO_4^{3-}$ | Phosphate |
| $Cl^{-}$ | Chloride | $S^{2-}$ | Sulfide | | |
| $Br^{-}$ | Bromide | $SO_4^{2-}$ | Sulfate | | |
| $I^{-}$ | Iodide | $SO_3^{2-}$ | Sulfite | | |
| $NO_3^{-}$ | Nitrate | $SeO_4^{2-}$ | Selenate | | |
| $NO_2^{-}$ | Nitrite | $CO_3^{2-}$ | Carbonate | | |
| $OH^{-}$ | Hydroxide | | | | |
| $HCO_3^{-}$ | Hydrogen carbonate (bicarbonate) | | | | |

*Some metals form ions with one charge under certain conditions and a different charge under different conditions. To specify the charge for these metal ions, Roman numerals are used in parentheses after the metal's name.

# CONSERVING CHEMICAL RESOURCES

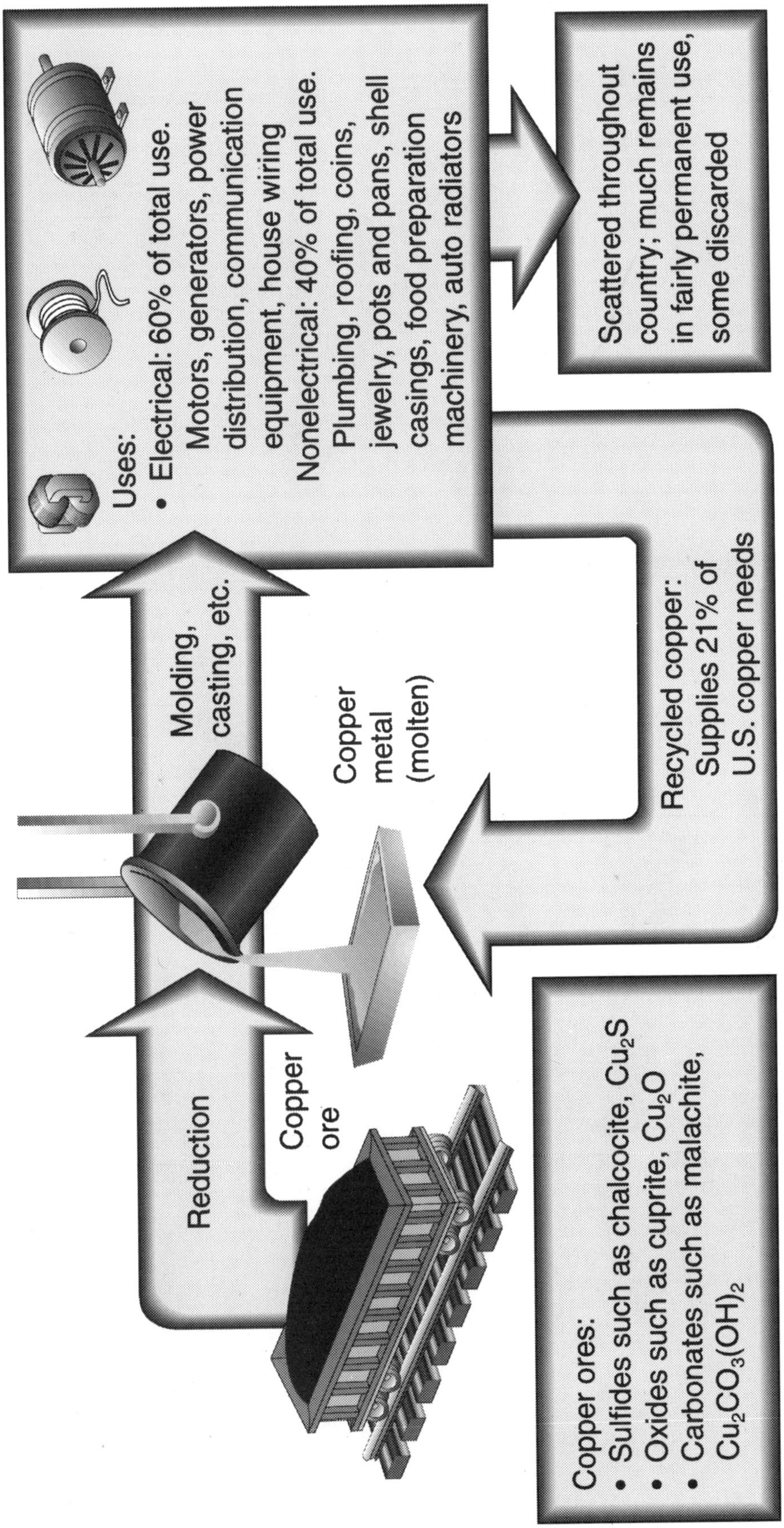


**R1**. Figure 9, page 67

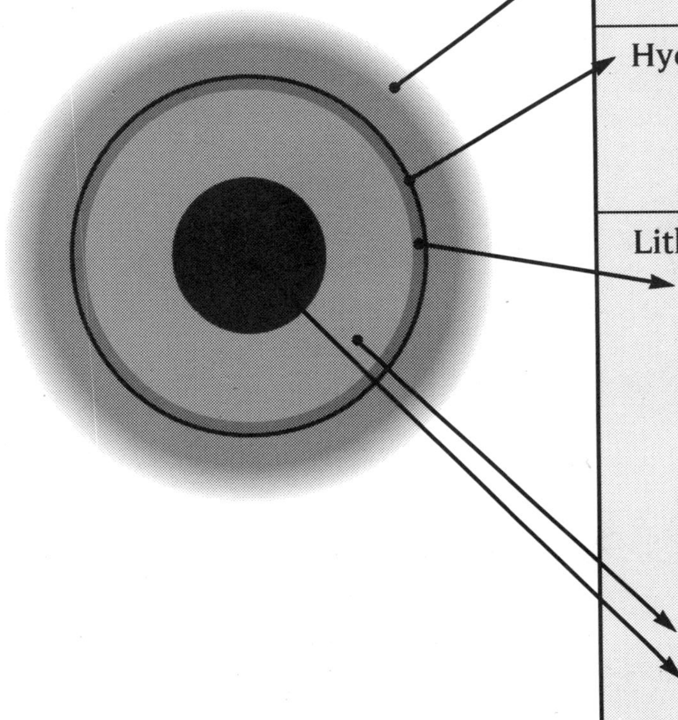

**Table 3** Earth's Composition

| Layer of Planet | Thickness (Average) | Composition (Decreasing Order of Abundance) |
|---|---|---|
| Atmosphere | 100 km | $N_2$ (78%), $O_2$ (21%), Ar (0.9%), He + Ne (<0.01%), variable amounts of $H_2O$, $CO_2$, etc. |
| Hydrosphere | 5 km | $H_2O$, and in the oceans that cover some 71% of Earth's surface, approximately 3.5% NaCl and smaller amounts of Mg, S, Ca, and other elements as ions |
| Lithosphere: Crust | 6400 km; Top 40 km | Silicates (compounds formed of metals, Si, and O atoms). Metals include Al, Na, Fe, Ca, Mg, K, and others<br>Coal, oil, and natural gas<br>Carbonates such as $CaCO_3$<br>Oxides such as $Fe_2O_3$<br>Sulfides such as PbS |
| Mantle | 40–2900 km | Silicates of Mg and Fe |
| Core | 2900 km to the Earth's center | Fe and Ni |

**R2**. Table 3, page 63

## Chemical Elements Crossword Puzzle

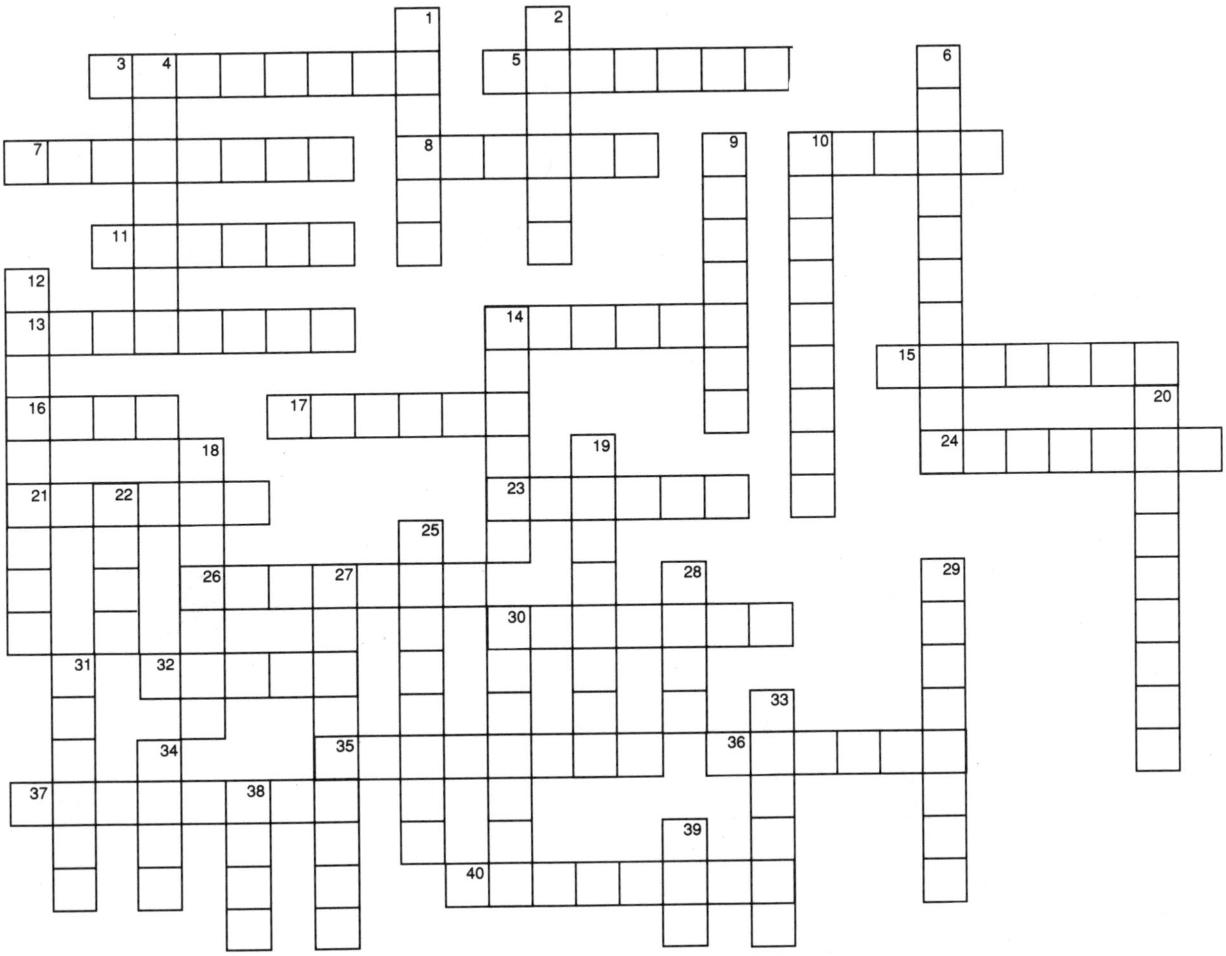

ALUMINUM
FOIL
50

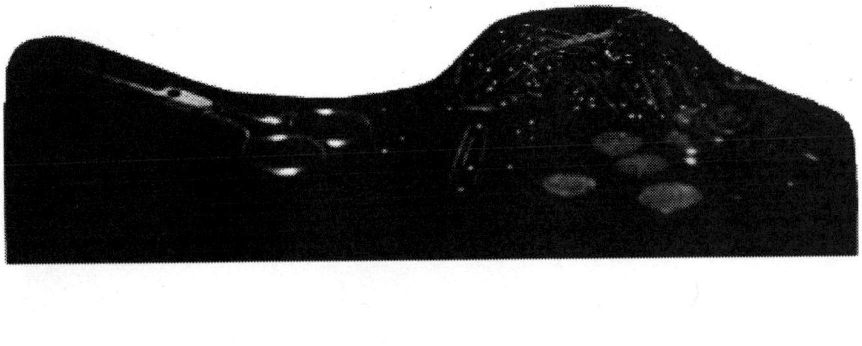

| Table 2 Hydrocarbon Boiling Point | |
|---|---|
| **Hydrocarbon** | **Boiling Point (°C)** |
| Butane | −0.5 |
| Decane | 174.0 |
| Ethane | −88.6 |
| Heptane | 98.4 |
| Hexane | 68.7 |
| Methane | −161.7 |
| Nonane | 150.8 |
| Octane | 125.7 |
| Pentane | 36.1 |
| Propane | −42.1 |

**P2**. Table 2, page 95

| Table 3 Some Members of the Alkane Series | | | |
|---|---|---|---|
| | | **Alkane Molecular Formulas** | |
| **Name** | **Number of Carbons** | **Short Version** | **Long Version** |
| Methane | 1 | $CH_4$ | $CH_4$ |
| Ethane | 2 | $C_2H_6$ | $CH_3CH_3$ |
| Propane | 3 | $C_3H_8$ | $CH_3CH_2CH_3$ |
| Butane | 4 | $C_4H_{10}$ | $CH_3CH_2CH_2CH_3$ |
| Pentane | 5 | $C_5H_{12}$ | $CH_3CH_2CH_2CH_2CH_3$ |
| Hexane | 6 | $C_6H_{14}$ | $CH_3CH_2CH_2CH_2CH_2CH_3$ |
| Heptane | 7 | $C_7H_{16}$ | $CH_3CH_2CH_2CH_2CH_2CH_2CH_3$ |
| Octane | 8 | $C_8H_{18}$ | $CH_3CH_2CH_2CH_2CH_2CH_2CH_2CH_3$ |
| Nonane | 9 | $C_9H_{20}$ | $CH_3CH_2CH_2CH_2CH_2CH_2CH_2CH_2CH_3$ |
| Decane | 10 | $C_{10}H_{22}$ | $CH_3CH_2CH_2CH_2CH_2CH_2CH_2CH_2CH_2CH_3$ |

**P3**. Table 3, page 97

| Table 4 | Alkane Isomers | |
|---|---|---|
| **Alkane** | **Structure** | **Boiling Point (°C)** |
| $C_5H_{12}$ isomers | $CH_3-CH_2-CH_2-CH_2-CH_3$ | 36.1 |
| | 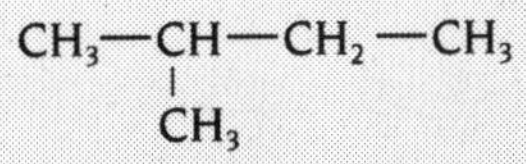 | 27.8 |
| | 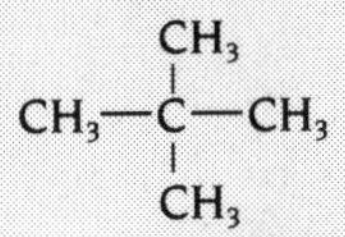 | 9.5 |
| Some $C_8H_{18}$ isomers | $CH_3-CH_2-CH_2-CH_2-CH_2-CH_2-CH_2-CH_3$ | 125.6 |
| | 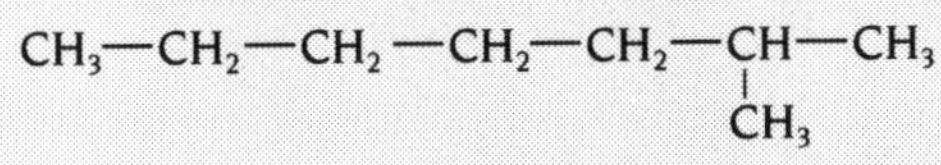 | 117.7 |
| | 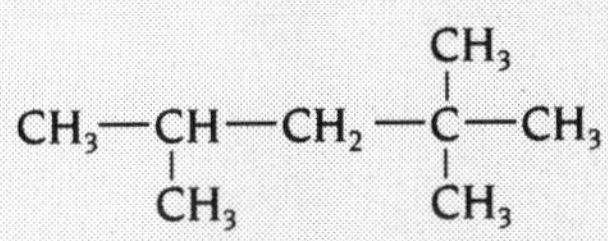 | 99.2 |

**P4**. Table 4, page 98

| Table 5 | Heats of Combustion | | |
|---|---|---|---|
| **Hydrocarbon** | **Formula** | **Heat of Combustion (kJ/g)** | **Molar Heat of Combustion (kJ/mol)** |
| Methane | $CH_4$ | 55.6 | 890 |
| Ethane | $C_2H_6$ | 52.0 | 1560 |
| Propane | $C_3H_8$ | 50.0 | 2200 |
| Butane | $C_4H_{10}$ | 49.3 | 2859 |
| Pentane | $C_5H_{12}$ | 48.8 | 3510 |
| Hexane | $C_6H_{14}$ | 48.2 | 4141 |
| Heptane | $C_7H_{16}$ | 48.2 | 4817 |
| Octane | $C_8H_{18}$ | 47.8 | 5450 |

**P5**. Table 5, page 105

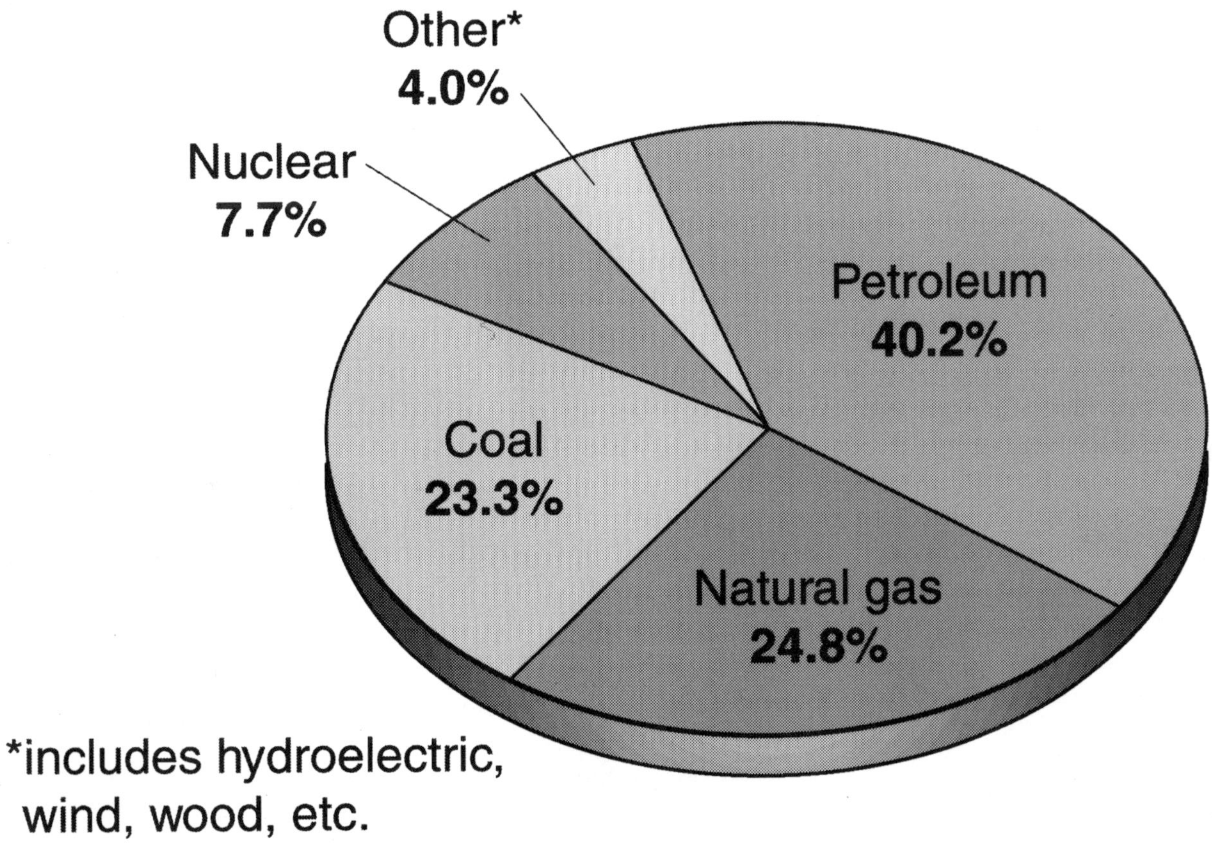


**P6**. Figure 23, page 115

**Table 1**

| **Nutrition Information**<br>**3 items (33 g)** | **Percent Daily Value*** |
|---|---|
| Calories 160<br>Calories from fat 60 | |
| Total fat 7 g<br>Saturated fat 1.5 g<br>Polyunsaturated fat 0.5 g<br>Monounsaturated fat 3 g | 11%<br>8% |
| Cholesterol 0 mg | 0% |
| Sodium 220 mg | 9% |
| Total carbohydrates 23 g<br>Dietary fiber 1 g<br>Sugars 13 g | 8%<br>3% |
| Protein 2 g | |
| | **Percent Daily Value*** |
| Vitamin A | 0% |
| Vitamin C | 0% |
| Calcium | 0% |
| Iron | 4% |

*Percent Daily Values are based on a 2,000-Calorie diet. Your daily values may be higher or lower depending on your Calorie needs.

**F1**. Table 1, page 129

| Table 2 | Energy Equivalents | | |
|---|---|---|---|
| **Activity** | **Energy expended, Cal/min: 120 lb** | **Energy expended, Cal/min: 150 lb** | **Energy expended, Cal/min: 180 lb** |
| Lying down or sleeping | 1.1 | 1.3 | 1.6 |
| Sitting | 1.3 | 1.7 | 2.0 |
| Eating | 2.0 | 2.5 | 3.1 |
| Studying | 1.3 | 1.7 | 2.0 |
| Walking, 3.5 mph | 4.2 | 5.3 | 6.3 |
| Vollyball | 4.7 | 5.9 | 7.0 |
| Tennis, singles | 5.6 | 7.1 | 8.5 |
| Jogging, 9 min per mile | 10.6 | 13.2 | 15.8 |
| Jogging, 7 min per mile | 13.7 | 17.1 | 20.5 |
| Rollerblading, 8 mph | 4.2 | 5.6 | 7.0 |
| Rollerblading, 12 mph | 10.5 | 11.7 | 13.3 |
| Swimming, freestyle | 7.8 | 9.8 | 11.7 |
| Bowling | 3.6 | 4.5 | 5.4 |
| Soccer | 7.2 | 9.0 | 10.8 |
| Basketball, full court | 11.6 | 14.6 | 17.5 |
| Baseball | 3.7 | 4.7 | 5.6 |
| Fishing | 3.4 | 4.2 | 5.0 |
| Stairclimber, 120 steps/min | 5.0 | 6.3 | 7.6 |
| Skiing, water | 6.2 | 7.8 | 9.4 |

**F2**. Table 2, page 133

| Table 3 | Carbohydrates | | |
|---|---|---|---|
| **Classification and Examples** | **Composition** | **Formula** | **Common Name of Source** |
| Monosaccharides | | $C_6H_{12}O_6$ | |
| Glucose | — | | Blood sugar |
| Fructose | — | | Fruit sugar |
| Galactose | — | | — |
| Disaccharides | Monosaccharides | $C_{12}H_{22}O_{11}$ | |
| Sucrose | Fructose + glucose | | Cane sugar |
| Lactose | Galactose + glucose | | Milk sugar |
| Maltose | Glucose + glucose | | Germinating seeds |
| Polysaccharides | Glucose polymers | — | |
| Starch | | | Plants |
| Glycogen | | | Animals |
| Cellulose | | | Plant fibers |

**F3**. Table 3, page 135

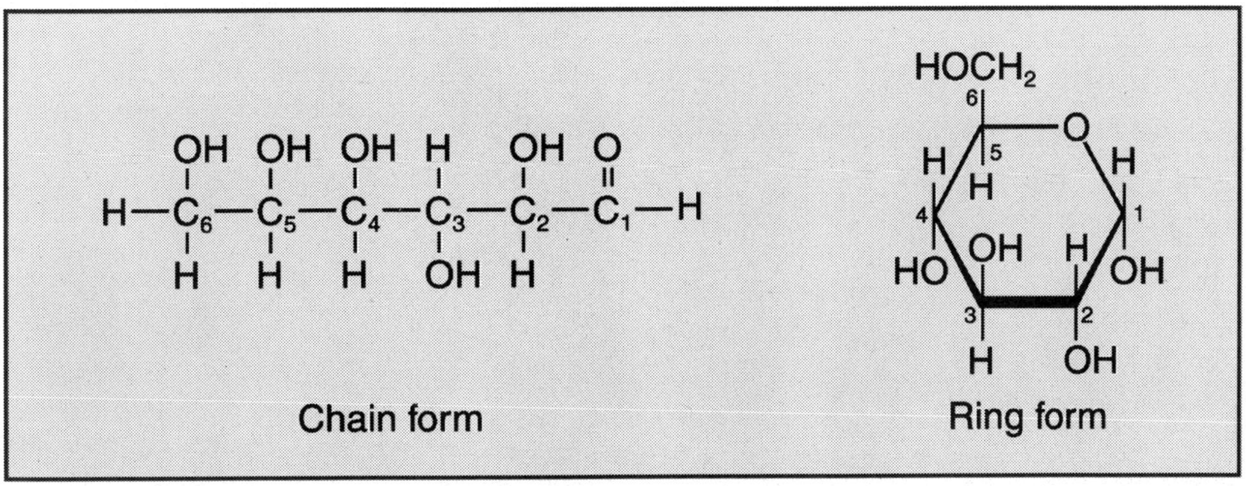


**F4**. Figure 7, page 135

| Table 4 | Elements Needed for Human Nutrition* |
|---|---|
| **In major biomolecules** | Carbon (C)<br>Hydrogen (H)<br>Oxygen (O)<br>Nitrogen (N)<br>Sulfur (S) |
| **Major minerals** | Calcium (Ca)<br>Chlorine (Cl)<br>Magnesium (Mg)<br>Phosphorus (P)<br>Potassium (K)<br>Sodium (Na) |
| **Trace minerals** | Chromium (Cr)<br>Cobalt (Co)<br>Copper (Cu)<br>Fluorine (F)<br>Iodine (I)<br>Iron (Fe)<br>Manganese (Mn)<br>Molybdenum (Mo)<br>Nickel (Ni)<br>Selenium (Se)<br>Zinc (Zn) |

*Essential in animals and very likely essential in humans.

| Table 6 | Proteins in the Body | |
|---|---|---|
| **Type** | **Function** | **Examples** |
| Structural proteins | | |
| Muscle | Contraction, movement | Myosin |
| Connective tissue | Support, protection | Collagen, keratin |
| Chromosomal proteins | Part of chromosome structure | Histones |
| Membranes | Control of influx and outflow, communication | Pore proteins, receptors |
| Transport proteins | Carriers of gases and other substances | Hemoglobin |
| Regulatory proteins | | |
| Fluid balance | Maintenance of pH, water, and salt content of body fluid | Serum albumin |
| Enzymes | Control of metabolism | Proteases |
| Hormones | Regulation of body functions | Insulin |
| Protective proteins | Antibodies | Gamma globulin |

**F6**. Table 6, page 138

| Table 7 | Complementary Proteins |
|---|---|
| **Foods** | **Country** |
| Corn tortillas and dried beans | Mexico |
| Rice and black-eyed peas | United States |
| Peanut butter and bread | United States |
| Rice and bean curd | China and Japan |
| Rice and lentils | India |
| Wheat pasta and cheese | Italy |

**F7**. Table 7, page 138

| Table 8 | RDAs For Protein | | |
|---|---|---|---|
| **Age (yr) or Condition** | **Median Weight (lb)** | **Median Height (in.)** | **RDA (g)** |
| Infants | | | |
| 0–0.5 | 13 | 24 | 13 |
| 0.5–1 | 20 | 28 | 14 |
| Males | | | |
| 11–14 | 99 | 62 | 45 |
| 15–18 | 145 | 58 | 58 |
| 19–24 | 160 | 69 | 58 |
| 25–50 | 174 | 69 | 63 |
| 51+ | 170 | 68 | 63 |
| Females | | | |
| 11–14 | 101 | 62 | 46 |
| 15–18 | 120 | 64 | 44 |
| 19–24 | 128 | 65 | 46 |
| 25–50 | 138 | 64 | 50 |
| 51+ | 143 | 63 | 50 |
| Pregnant | | | 60 |
| Nursing | | | |
| First 6 mo. | | | 65 |
| Second 6 mo. | | | 62 |

**F8**. Table 8, page 139

| Table 9 | Vitamins | |
|---|---|---|
| **Vitamin (Name)** | **Main Sources** | **Deficiency Condition** |
| ***Water-soluble*** | | |
| $B_1$ (Thiamin) | Liver, milk, pasta, bread, wheat germ, lima beans, nuts | Beriberi: nausea, severe exhaustion, paralysis |
| $B_2$ (Riboflavin) | Red meat, milk, eggs, pasta, bread, beans, dark green vegetables, peas, mushrooms | Severe skin problems |
| Niacin | Red meat, poultry, enriched or whole grains, beans, peas | Pellagra: weak muscles, no appetite, diarrhea, skin blotches |
| $B_6$ (Pyridoxine) | Muscle meats, liver, poultry, fish, whole grains | Depression, nausea, vomiting |
| $B_{12}$ (Cobalamin) | Red meat, liver, kidneys, fish, eggs, milk | Pernicious anemia, exhaustion |
| Folic acid | Kidneys, liver, leafy green vegetables, wheat germ, peas, beans | Anemia |
| Pantothenic acid | Plants, animals | Anemia |
| Biotin | Kidneys, liver, egg yolk, yeast, nuts | Dermatitis |
| C (Ascorbic acid) | Citrus fruits, melon, tomatoes, green pepper, strawberries | Scurvy: tender skin; weak, bleeding gums; swollen joints |
| ***Fat-soluble*** | | |
| A (Retinol) | Liver, eggs, butter, cheese, dark green and deep orange vegetables | Inflamed eye membranes, night blindness, scaling of skin, faulty teeth and bones |
| D (Calciferol) | Fish-liver oils, fortified milk | Rickets: soft bones |
| E (Tocopherol) | Liver, wheat germ, whole-grain cereals, margarine, vegetable oil, leafy green vegetables | Breakage of red blood cells in premature infants, oxidation of membranes |
| K (Menaquinone) | Liver, cabbage, potatoes, peas, leafy green vegetables | Hemorrhage in newborns; anemia |

**F9**. Table 9, page 141

| Table 10 | RDAs For Selected Vitamins | | | | | | | |
|---|---|---|---|---|---|---|---|---|
| **Sex and Age** | **A (μg RE)** | **D (μg)** | **C (mg)** | **$B_1$ (mg)** | **$B_2$ (mg)** | **Niacin (mg)** | **$B_{12}$ (μg)** | **K (μg)** |
| **Males** | | | | | | | | |
| 11–14 | 1000 | 10 | 50 | 1.3 | 1.5 | 17 | 2.0 | 45 |
| 15–18 | 1000 | 10 | 60 | 1.5 | 1.8 | 20 | 2.0 | 65 |
| 19–24 | 1000 | 10 | 60 | 1.5 | 1.7 | 19 | 2.0 | 70 |
| 25–50 | 1000 | 10 | 60 | 1.5 | 1.7 | 19 | 2.0 | 80 |
| 51+ | 1000 | 10 | 60 | 1.2 | 1.4 | 15 | 2.0 | 80 |
| **Females** | | | | | | | | |
| 11–14 | 800 | 10 | 50 | 1.1 | 1.3 | 15 | 2.0 | 45 |
| 15–18 | 800 | 10 | 60 | 1.1 | 1.3 | 15 | 2.0 | 55 |
| 19–24 | 800 | 10 | 60 | 1.1 | 1.3 | 15 | 2.0 | 60 |
| 25–50 | 800 | 5 | 60 | 1.1 | 1.3 | 15 | 2.0 | 65 |
| 51+ | 800 | 5 | 60 | 1.0 | 1.2 | 13 | 2.0 | 65 |

Source: Food and Nutrition Board, National Academy of Sciences—National Research Council, Recommended Dietary Allowances, Revised 1989.

**F10**. Table 10, page 141

**Data Table: Drops vit C/mL (Step 1.1) = ________; Drops $I_2$ (Step 1.4) = ________.**

| Beverage | Drops $I_2$ (Step 2.3) × | Conversion Factor (Step 1.5) = | mg Vit C per 25 drops of beverage | Rank |
|---|---|---|---|---|
| 1. | | | | |
| 2. | | | | |
| 3. | | | | |
| 4. | | | | |
| 5. | | | | |

**F11**. Data Table, page 141

| Table 12 | Minerals | |
|---|---|---|
| **Mineral** | **Source** | **Deficiency Condition** |
| **Macrominerals** | | |
| Calcium (Ca) | Canned fish, milk, dairy products | Rickets in children; osteomalacia and osteoporosis in adults |
| Chlorine (Cl) | Meats, salt-processed foods, table salt | — |
| Magnesium (Mg) | Seafoods, cereal grains, nuts, dark green vegetables, cocoa | Heart failure due to spasms |
| Phosphorus (P) | Animal proteins | — |
| Potassium (K) | Orange juice, bananas, dried fruits, potatoes | Poor nerve function, irregular heartbeat, sudden death during fasting |
| Sodium (Na) | Meats, salt-processed foods, table salt | Headache, weakness, thirst, poor memory, appetite loss |
| Sulfur (S) | Proteins | — |
| **Trace minerals** | | |
| Chromium (Cr) | Liver, animal and plant tissue | Loss of insulin efficiency with age |
| Cobalt (Co) | Liver, animal proteins | Anemia |
| Copper (Cu) | Liver, kidney, egg yolk, whole grains | — |
| Fluorine (F) | Seafoods, fluoridated drinking water | Dental decay |
| Iodine (I) | Seafoods, iodized salts | Goiter |
| Iron (Fe) | Liver, meats, green leafy vegetables, whole grains | Anemia; tiredness and apathy |
| Manganese (Mn) | Liver, kidney, wheat germ, legumes, nuts, tea | Weight loss, dermatitis |
| Molybdenum (Mo) | Liver, kidney, whole grains, legumes, leafy vegetables | — |
| Nickel (Ni) | Seafoods, grains, seeds, beans, vegetables | Cirrhosis of liver, kidney failure, stress |
| Selenium (Se) | Liver, organ meats, grains, vegetables | Kashan disease (a heart disease found in China) |
| Zinc (Zn) | Liver, shellfish, meats, wheat germ, legumes | Anemia, stunted growth |

**F12**. Table 12, page 143

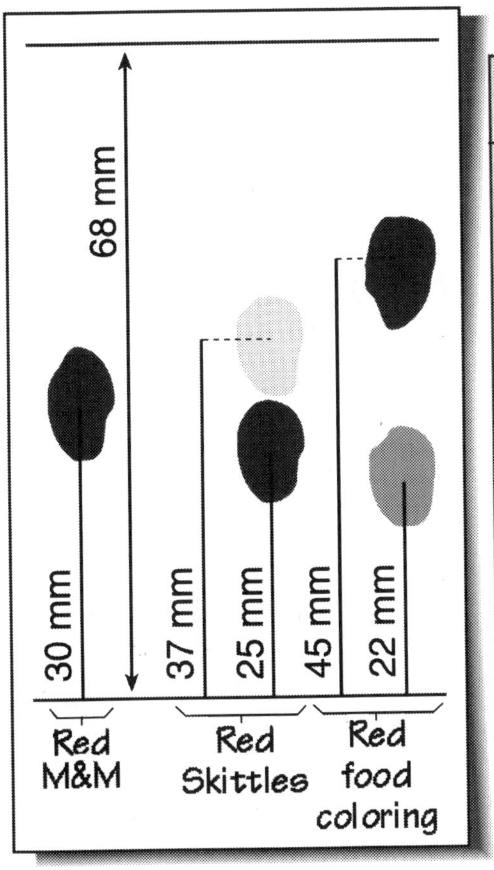


| | $R_f$ values | |
|---|---|---|
| Red M&M | | $R_f = \frac{30\text{ mm}}{68\text{ mm}} = 0.44$ |
| Red Skittles | red spot: | $R_f = \frac{25\text{ mm}}{68\text{ mm}} = 0.37$ |
| | yellow spot: | $R_f = \frac{37\text{ mm}}{68\text{ mm}} = 0.54$ |
| Red food coloring | pink spot: | $R_f = \frac{22\text{ mm}}{68\text{ mm}} = 0.32$ |
| | red spot: | $R_f = \frac{45\text{ mm}}{68\text{ mm}} = 0.66$ |

**F13**. Figure 16, page 145

# NUCLEAR CHEMISTRY IN OUR WORLD

**Table 3 — Some Common Isotopes**

| Symbol | Name | Total Protons (Atomic number) | Total neutrons | Mass number | Total electrons |
|---|---|---|---|---|---|
| $^{7}_{3}Li$ | Lithium-7 | 3 | 4 | 7 | 3 |
| $^{67}_{31}Ga$ | Gallium-67 | 31 | 36 | 67 | 31 |
| $^{201}_{81}Tl$ | Thallium-201 | 81 | 120 | 201 | 81 |
| $^{208}_{82}Pb$ | Lead-208 | 82 | 126 | 208 | 82 |
| $^{208}_{82}Pb^{2+}$ | Lead-208, 2+ ion | 82 | 126 | 208 | 80 |

**N1**. Table 3, page 172

**Table 4 — Some Natural Radioisotopes**

| Name | Symbol | Relative Isotopic Abundance (%) |
|---|---|---|
| Hydrogen-3 | $^{3}_{1}H$ | 0.00013 |
| Carbon-14 | $^{14}_{6}C$ | Trace |
| Potassium-40 | $^{40}_{19}K$ | 0.0012 |
| Rubidium-87 | $^{87}_{37}Rb$ | 27.8 |
| Indium-115 | $^{115}_{49}In$ | 95.8 |
| Lanthanum-138 | $^{138}_{57}La$ | 0.089 |
| Neodymium-144 | $^{144}_{60}Nd$ | 23.9 |
| Samarium-147 | $^{147}_{62}Sm$ | 15.1 |
| Lutetium-176 | $^{176}_{71}Lu$ | 2.60 |
| Rhenium-187 | $^{187}_{75}Re$ | 62.9 |
| Platinum-190 | $^{190}_{78}Pt$ | 0.012 |
| Thorium-232 | $^{232}_{90}Th$ | 100 |
| Uranium-235 | $^{235}_{92}U$ | 0.72 |
| Uranium-238 | $^{238}_{92}U$ | 99.28 |

**N2**. Table 4, page 173

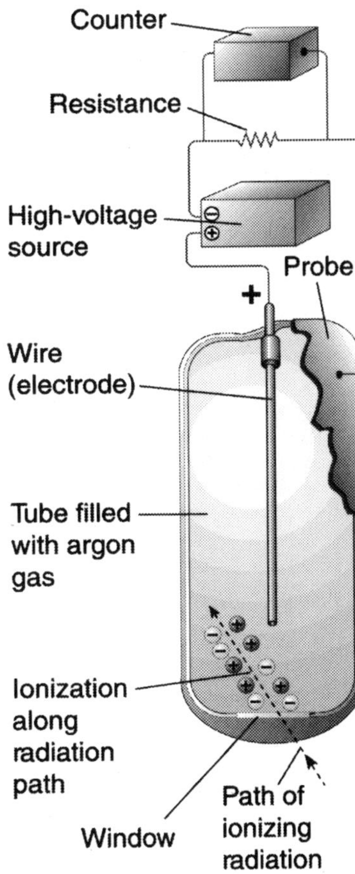


**N3**. Figure 7, page 176

| Data Table | | | | |
|---|---|---|---|---|
| **Radiation** | **No Shielding** | **Counts per Minute (cpm)** Cardboard | **Glass** | **Lead** |
| Gamma | | | | |
| Beta | | | | |
| Alpha | | | | |

**N4**. Data Table, page 176

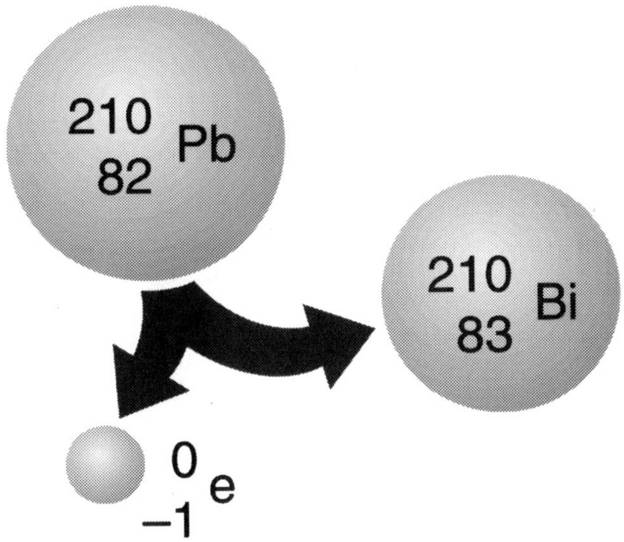


**N5**. Figure 12, page 320

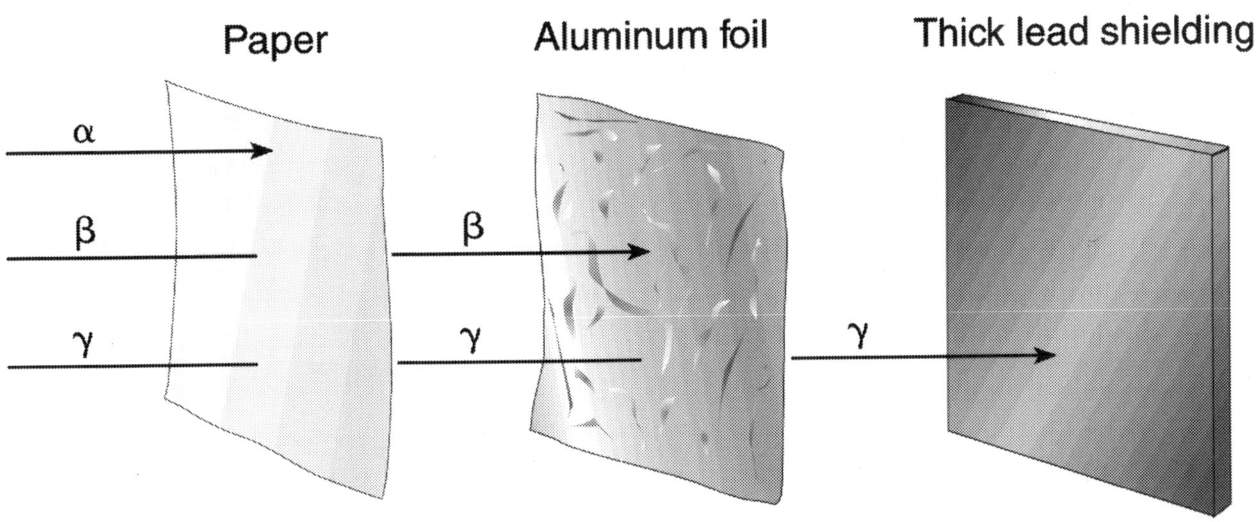


**N6**. Figure 13, page 320

| Table 9 Potassium Content and K-40 Activity of Selected Foods | | | |
|---|---|---|---|
| **Food** | **Quantity** | **Potassium, mg** | **K-40, dps** |
| Hamburger | 4 oz | 960 | 29 |
| Chicken, fried | 1/4 chicken | 240 | 7 |
| Hot dog | 2 regular | 200 | 6 |
| French fries | 3.5 oz | 650 | 19.5 |
| Corn | 1 ear | 200 | 6 |
| Banana | 1 small | 370 | 9 |
| Apple | 1 medium | 160 | 7 |
| Milk, whole | 1 cup | 370 | 11 |
| Cola beverage | 12 oz | 13 | 0.4 |
| Corn flakes | 1 oz | 14 | 0.4 |
| White bread | 1 slice | 30 | 0.9 |
| Peanut butter | 1 tablespoon | 100 | 3 |
| Egg | 1 large | 65 | 2 |
| Oatmeal | 1 cup | 130 | 4 |

**N7**. Table 9, page 179

| Table 11 Selected Medical Radioisotopes | | |
|---|---|---|
| **Radioisotopes** | **Half-Life** | **Use** |
| **Used as tracers** | | |
| Technetium-99m | 6.02 h | Measure cardiac output; locate strokes and brain and bone tumors. |
| Gallium-67 | 78 h | Diagnosis of Hodgkin's disease |
| Iron-59 | 45.1 d | Determine rate of red blood cell formation (these contain iron); anemia assessment |
| Chromium-51 | 27.8 d | Determine blood volume and lifespan of red blood cells |
| Hydrogen-3 (tritium) | 12.3 y | Determine volume of body's water; assess vitamin D usage in body |
| Thallium-201 | 74 h | Cardiac assessment |
| Iodine-123 | 13.3 h | Thyroid function diagnosis |
| **Used for irradiation therapy** | | |
| Cesium-137 | 30.0 y | Treat shallow tumors (external source) |
| Phosphorus-32 | 14.3 d | Treat leukemia, a bone cancer affecting white blood cells (internal source) |
| Iodine-131 | 8 d | Treat thyroid cancer (external source) |
| Cobalt-60 | 5.3 y | Treat shallow tumors (external source) |
| Yttrium-90 | 64 h | Treat pituitary gland cancer internally with ceramic beads |

**N8**. Table 11, page 183

| Table 1 | Tropospheric Air | |
|---|---|---|
| **Substance** | **Formula** | **Percent of Gas Molecules** |
| Major components | | |
| Nitrogen | $N_2$ | 78.08 |
| Oxygen | $O_2$ | 20.95 |
| Minor components | | |
| Argon | Ar | 0.93 |
| Carbon dioxide | $CO_2$ | 0.033 |
| Trace components | | |
| Neon | Ne | 0.0018 |
| Ammonia | $NH_3$ | 0.0010 |
| Helium | He | 0.0005 |
| Methane | $CH_4$ | 0.0002 |
| Krypton | Kr | 0.00011 |
| Others | see text | <0.0001 each |

**A1**. Table 1, page 210

| Table 2 | | Atmospheric Data | | |
|---|---|---|---|---|
| **Altitude (km)** | **Temp. (°C)** | **Pressure (mmHg)** | **Mass (g) of 1-L Sample** | **Total Molecules in 1-L Sample** |
| 0 | 20 | 760 | 1.20 | $250 \times 10^{20}$ |
| 5 | −12 | 407 | 0.73 | $150 \times 10^{20}$ |
| 10 | −45 | 218 | 0.41 | $90 \times 10^{20}$ |
| 12 | −60 | 170 | 0.37 | $77 \times 10^{20}$ |
| 20 | −53 | 62 | 0.13 | $27 \times 10^{20}$ |
| 30 | −38 | 18 | 0.035 | $7 \times 10^{20}$ |
| 40 | −18 | 5.1 | 0.009 | $2 \times 10^{20}$ |
| 50 | 2 | 1.5 | 0.003 | $0.5 \times 10^{20}$ |
| 60 | −26 | 0.42 | 0.0007 | $0.2 \times 10^{20}$ |
| 80 | −87 | 0.03 | 0.00007 | $0.02 \times 10^{20}$ |

**A2**. Table 2, page 211

| Table 3<br>$CO_2$ in the Atmosphere | |
|---|---|
| **Year** | **Approximate $CO_2$ Level (ppm by volume)** |
| 1900 | 287 |
| 1920 | 303 |
| 1930 | 310 |
| 1960 | 317 |
| 1965 | 320 |
| 1970 | 325 |
| 1972 | 328 |
| 1974 | 330 |
| 1976 | 332 |
| 1978 | 335 |
| 1980 | 338 |
| 1982 | 341 |
| 1984 | 344 |
| 1986 | 347 |
| 1988 | 351 |
| 1990 | 354 |
| 1992 | 356 |
| 1994 | 358 |

**A3**. Table 3, page 218

| Table 6 | U.S. Pollutant Standards Index | | | | | | |
|---|---|---|---|---|---|---|---|
| Air Pollutant Levels (micrograms per cubic meter) | | | | | | | |
| Air Quality Index Value | Air Quality Description | Total Suspended Particulate Matter (24 Hours) | Sulfur Dioxide (8 Hours) | Carbon Monoxide (1 Hour) | Ozone (1 Hour) | Nitrogen Dioxide (1 Hour) | Effects and Suggested Actions |
| 500 | Hazardous | 1,000 | 2,620 | 57,000 | 1,200 | 3,750 | Normal activity impossible. All should remain indoors with windows and doors closed. Fatal for some in high-risk group.* |
| 400 | Hazardous | 875 | 2,100 | 46,000 | 1,000 | 3,000 | High-risk group should stay quietly indoors. Others should avoid outdoor activity. |
| 300 | Very unhealthful | 625 | 1,600 | 34,000 | 800 | 2,260 | High-risk group has more symptoms, and should stay indoors and reduce physical activity. All persons notice lung irritation. |
| 200 | Unhealthful | 375 | 800 | 17,000 | 400 | 1,130 | Those with lung or heart disease should reduce physical exertion. Healthy persons notice irritations. |
| 100 | Moderate | 260 | 365 | 10,000 | 235 | Not reported | Some damage to materials and plants. Human health not affected unless level continues for many days. |
| 50 | Good | 75 | 80 | 5,000 | 118 | Not reported | No significant effects. |

*High-risk group includes elderly people and those with heart or lung diseases.

**A4**. Table 6, page 223

| Table 7 Changes in Average Concentrations of Air Pollutants in the United States, 1975–1991 | | | |
|---|---|---|---|
| **Pollutant** | **1975** | **1991** | **Change** |
| Sulfur dioxide | 0.0132 ppm | 0.0075 ppm | 43% decrease |
| Nitrogen oxides | 0.021ppm | 0.021 ppm | 1% decrease |
| Carbon monoxide | 10 ppm | 6 ppm | 40% decrease |
| Ozone | 0.147 ppm | 0.115 ppm | 22% decrease |
| Particulates | 63 μg/m$^3$ | 47 μg/m$^3$ | 22% decrease |
| Lead | 0.68 μg/m$^3$ | 0.048 μg/m$^3$ | 93% decrease |

**A5.** Table 7, page 223

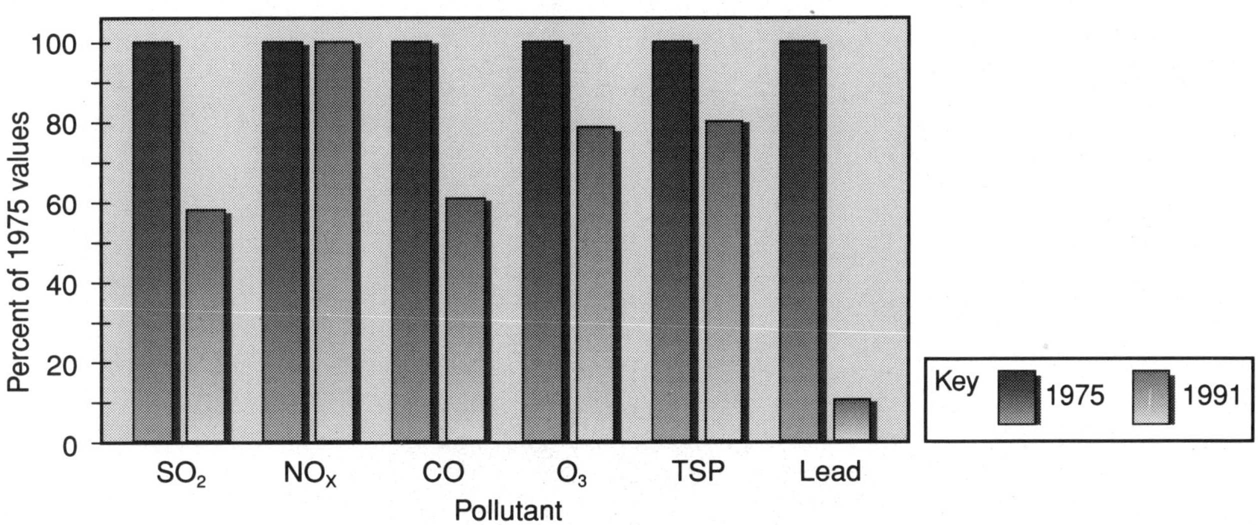


**A6.** Figure 17, page 223

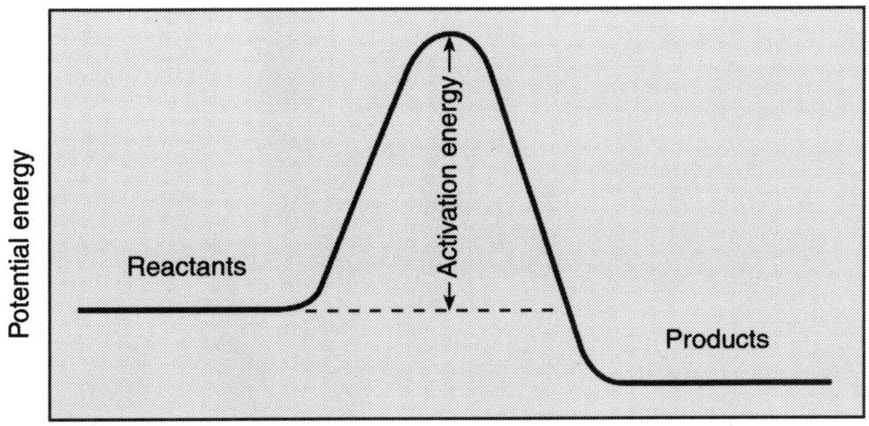


**A7**. Figure 21, page 225

**Table 9 Some Indoor Pollutants and Their Sources**

| Phase of Matter | Source | Pollutant |
|---|---|---|
| Solid/particulate | Plants | Molds, mildew, bacteria, viruses |
| | Pets | Pet dander, dust, flea feces |
| | Floor tile | Asbestos |
| Liquids/gases | Clothes | Dry-cleaning fluid (1,1,2 trichlorotrifluoroethane), moth balls (*p*-dichlorobenzene) |
| | Furniture | Formaldehyde |
| | Particle board | Formaldehyde |
| | Carpet | Styrene |
| | Paint and paint thinners | Methanol, methylene chloride |
| | Glues and solvents | Xylene, acetone, toluene |
| | Cigarette smoke | Formaldehyde, carbon monoxide, benzo[*a*]pyrene |
| | Faulty furnaces and/or kerosene space heaters | Carbon monoxide |
| | Rocks under house | Radon from decay of uranium |
| | Water from showers | Chloroform |
| | Electric arcing | Ozone |

**A8**. Table 9, page 230

# PERSONAL CHEMISTRY AND CHOICES

**Some Compounds in Cigarette Smoke in Addition to Carbon Monoxide and Nicotine**

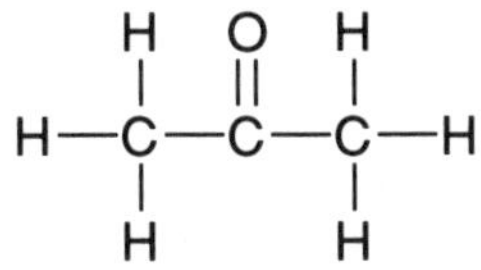

Acetone, $C_3H_6O$

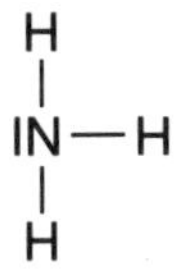

Ammonia, $NH_3$

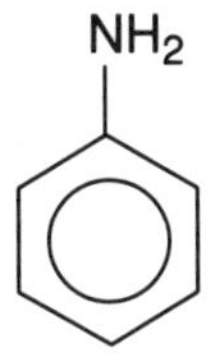

Aniline, $C_6H_5NH_2$

Benzene, $C_6H_6$

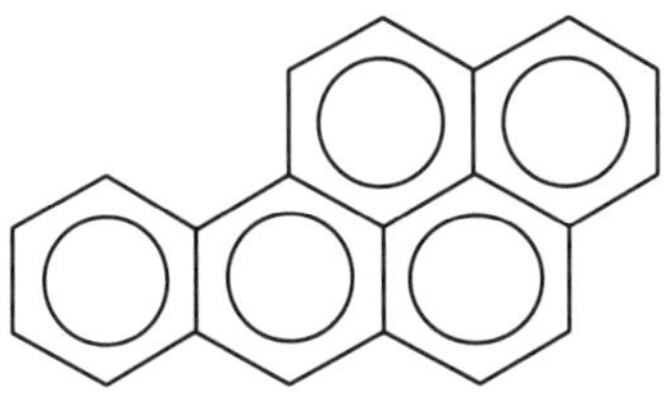

Benzo (a) pyrene, $C_{20}H_{12}$

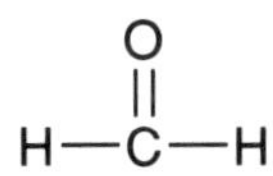

Formaldehyde, $H_2CO$

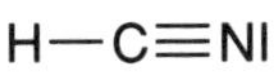

Hydrogen cyanide, HCN

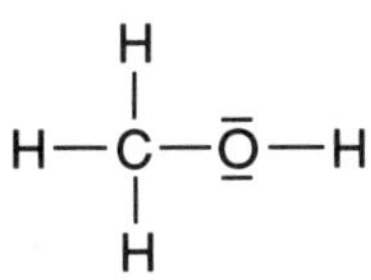

Methanol, $CH_3OH$

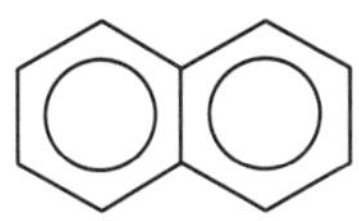

Naphthalene, $C_{10}H_8$

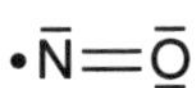

Nitric oxide, NO

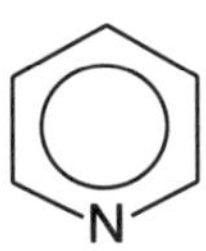

Pyridine, $C_5H_5N$

**C1.** page 241

**Data Table**

| Step number | Starting material | Added | Color | Approximate pH |
|---|---|---|---|---|
| 1 | Distilled water | Universal indicator | | |
| 3 | Buffer solution | Universal indicator | | |
| 4 | Distilled water + universal indicator | 0.1 M NaOH | | |
| 5 | Buffer solution + universal indicator | 0.1 M NaOH | | |
| 7 | Distilled water + universal indicator | 0.1 M HCl | | |
| 8 | Bufffer solution + universal indicator | 0.1 M HCl | | |

**C2.** Data Table, page 251

| Table 4 | UV Index | |
|---|---|---|
| **Index** | **Minutes for some damage to occur in light-skinned people who "never tan"** | **Minutes for some damage to occur in dark-skinned people who "never tan"** |
| 0–2<br>Minimal | 30 minutes | More than 120 minutes |
| 3–4<br>Low | 15–20 minutes | 75–90 minutes |
| 5–6<br>Moderate | 10–12 minutes | 50–60 minutes |
| 7–9<br>High | 7–8.5 minutes | 33–40 minutes |
| 10–15<br>Very High | Less than 4–6 minutes | 20–30 minutes |

**C3**. Table 4, page 254

| Table 5 | Sunscreen Protection Factor (SPF) | |
|---|---|---|
| **Rating of Sun Protection Product** | **SPF Values** | **Protection Level** |
| Minimal | 2–4 | Least protection; permits tanning. Recommended for people who rarely burn, but tan profusely. |
| Moderate | 4–6 | Moderate protection from sunburning; permits some suntanning.<br>Recommended for people who burn minimally and tan well. |
| Extra | 6–8 | Extra protection from sunburn; permits limited tanning.<br>Recommended for people who burn moderately and tan gradually. |
| Maximal | 8–15 | High protection from sunburn; permits little or no tanning.<br>Recommended for people who burn easily and tan minimally. |
| Ultra | 15 or more | Most protection from sunburn; permits no tanning.<br>Recommended for people who burn easily and rarely tan. |

**C4**. Table 5, page 254

| Table 6 | Alcohol in the Blood | | | | | | |
|---|---|---|---|---|---|---|---|
| | Blood Alcohol Level (Percent)* Body Weight (lbs) | | | | | | |
| No. of Drinks | 100 | 125 | 150 | 175 | 200 | 225 | 250 |
| 1 | 0.03 | 0.03 | 0.02 | 0.02 | 0.01 | 0.01 | 0.01 |
| 2 | 0.06 | 0.05 | 0.04 | 0.04 | 0.03 | 0.03 | 0.03 |
| 3 | 0.10 | 0.08 | 0.06 | 0.06 | 0.05 | 0.04 | 0.04 |
| 4 | 0.13 | 0.10 | 0.09 | 0.07 | 0.06 | 0.06 | 0.05 |
| 5 | 0.16 | 0.13 | 0.11 | 0.09 | 0.08 | 0.07 | 0.06 |
| 6 | 0.19 | 0.16 | 0.13 | 0.11 | 0.10 | 0.09 | 0.08 |
| 7 | 0.22 | 0.18 | 0.15 | 0.13 | 0.11 | 0.10 | 0.09 |
| 8 | 0.26 | 0.21 | 0.17 | 0.15 | 0.13 | 0.11 | 0.10 |
| 9 | 0.29 | 0.24 | 0.19 | 0.17 | 0.14 | 0.13 | 0.12 |
| 10 | 0.33 | 0.26 | 0.22 | 0.18 | 0.16 | 0.14 | 0.13 |
| 11 | 0.36 | 0.29 | 0.24 | 0.20 | 0.18 | 0.16 | 0.14 |
| 12 | 0.39 | 0.31 | 0.26 | 0.22 | 0.19 | 0.17 | 0.16 |

**Effect on Behavior**

| Blood Alcohol Level (%) | Behavior |
|---|---|
| 0.05 | Lowered alertness; reduced coordination |
| 0.10 | Reaction time slowed 15–25%; visual sensitivity reduced up to 32%; headlight recovery 7–32 seconds longer |
| 0.25 | Severe disturbance to coordination; dizziness; staggering; senses impaired |
| 0.35 | Surgical anaesthesia; reduced body temperature |
| 0.40 | 50% of people die of alcohol poisoning |

*Data from MADD

**C5**. Table 6, page 260

**Data Table—Ion Tests**

| | Known Cations | | | Known Anions | | | Unknown |
|---|---|---|---|---|---|---|---|
| Ion Tests | Ammonium $NH_4^+$ | Iron(III) $Fe^{3+}$ | Potassium $K^+$ | Nitrate $NO_3^-$ | Phosphate $PO_4^{3-}$ | Sulfate $SO_4^{2-}$ | # ______ |
| 1. Color | | | | | | | |
| 2a. NaOH | | | | | | | |
| 2b. $BaCl_2$ | | | | | | | |
| 2c. $BaCl_2$ & HCl | | | | | | | |
| 3. Brown ring test | × | × | × | | × | × | |
| 4. NaOH & litmus | | | | × | × | × | |
| 5c. Flame test | × | × | | × | × | × | |
| 5d. Flame test & cobalt glass | × | × | | × | × | × | |
| 6. KSCN test | × | | × | × | × | × | |

**I1**. Data Table , page 278

| Table 1 | Industrial Nitrogen Compounds | |
|---|---|---|
| **Compound** | **Structure** | **Use** |
| Ammonia, $NH_3$ | $NH_3$ | Fertilizer, explosives, fibers, plastics |
| Nitric Acid, $HNO_3$ | HO—$NO_2$ | Fertilizer, explosives, manufacturing |
| Hydrazine, $N_2H_4$ | $H_2N$—$NH_2$ | Rocket fuel, plastics |
| Amines, R—$NH_2$ (e.g., propylamine) | $CH_3CH_2CH_2NH_2$ | Intermediate |
| Nitrates, $MNO_3$ (e.g., sodium nitrate) | $Na^+NO_3^-$ | Fertilizer, explosives, food preservation |
| Nitrites, $MNO_2$ (e.g., sodium nitrite) | $Na^+NO_2^-$ | Textile bleach, manufacturing, food preservation |
| Hydroxylamine, $NH_2OH$ | H—$\ddot{N}$(H)—OH | Reducing agent |
| Phenylhydrazine, $C_6H_5N_2H_3$ | 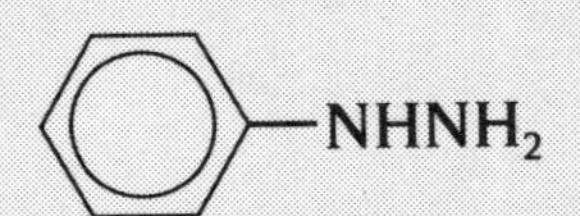 | Dye intermediate |
| Urea, $(NH_2)_2CO$ | $H_2NC(=O)NH_2$ | Fertilizer, plastics |

**I2.** Table 1, page 281

**Table 3 Activity Series of Common Metals**

| Metal | | Products of Metal Reactivity | | |
|---|---|---|---|---|
| Li(s) | $\rightarrow$ | $Li^{+}(aq)$ | + | $e^{-}$ |
| Na(s) | $\rightarrow$ | $Na^{+}(aq)$ | + | $e^{-}$ |
| Mg(s) | $\rightarrow$ | $Mg^{2+}(aq)$ | + | $2\ e^{-}$ |
| Al(s) | $\rightarrow$ | $Al^{3+}(aq)$ | + | $3\ e^{-}$ |
| Mn(s) | $\rightarrow$ | $Mn^{2+}(aq)$ | + | $2\ e^{-}$ |
| Zn(s) | $\rightarrow$ | $Zn^{2+}(aq)$ | + | $2\ e^{-}$ |
| Cr(s) | $\rightarrow$ | $Cr^{3+}(aq)$ | + | $3\ e^{-}$ |
| Fe(s) | $\rightarrow$ | $Fe^{2+}(aq)$ | + | $2\ e^{-}$ |
| Ni(s) | $\rightarrow$ | $Ni^{2+}(aq)$ | + | $2\ e^{-}$ |
| Sn(s) | $\rightarrow$ | $Sn^{2+}(aq)$ | + | $2\ e^{-}$ |
| Pb(s) | $\rightarrow$ | $Pb^{2+}(aq)$ | + | $2\ e^{-}$ |
| Cu(s) | $\rightarrow$ | $Cu^{2+}(aq)$ | + | $2\ e^{-}$ |
| Ag(s) | $\rightarrow$ | $Ag^{+}(aq)$ | + | $e^{-}$ |
| Au(s) | $\rightarrow$ | $Au^{3+}(aq)$ | + | $3\ e^{-}$ |

**I3.** Table 3, page 285

| Table 4 Top 25 Chemicals Produced in the United States in 1994 | | | |
|---|---|---|---|
| **Rank** | **Name** | **Formula** | **Billions of Pounds** |
| 1 | Sulfuric acid | $H_2SO_4$ | 89.2 |
| 2 | Nitrogen | $N_2$ | 67.5 |
| 3 | Oxygen | $O_2$ | 49.7 |
| 4 | *Ethene (ethylene) | $C_2H_4$ | 48.5 |
| 5 | Calcium oxide (lime) | CaO | 38.4 |
| 6 | Ammonia | $NH_3$ | 37.9 |
| 7 | *Propene (propylene) | $C_3H_6$ | 28.8 |
| 8 | Sodium hydroxide | NaOH | 25.8 |
| 9 | Phosphoric acid | $H_3PO_4$ | 25.3 |
| 10 | Chlorine | $Cl_2$ | 24.2 |
| 11 | Sodium carbonate | $Na_2CO_3$ | 20.6 |
| 12 | *1,2-Dichloroethane (ethylene dichloride) | $C_2H_4Cl_2$ | 18.7 |
| 13 | Nitric acid | $HNO_3$ | 17.6 |
| 14 | Ammonium nitrate | $NH_4NO_3$ | 17.6 |
| 15 | *Urea | $(NH_2)_2CO$ | 16.1 |
| 16 | *Vinyl chloride | $C_2H_3Cl$ | 14.8 |
| 17 | *Benzene | $C_6H_6$ | 14.7 |
| 18 | *Methyl *tert*-butyl ether (MTBE) | $C_5H_{12}O$ | 12.3 |
| 19 | *Ethylbenzene | $C_8H_{10}$ | 11.9 |
| 20 | *Styrene | $C_8H_8$ | 11.3 |
| 21 | Carbon dioxide | $CO_2$ | 11.0 |
| 22 | *Methanol | $CH_3OH$ | 10.8 |
| 23 | *Xylene | $C_8H_{10}$ | 9.1 |
| 24 | *Terphthalic acid | $C_8H_6O_4$ | 8.6 |
| 25 | *Formaldehyde | $H_2CO$ | 7.9 |

| | |
|---|---|
| Total Organics | 279.2 |
| Total Inorganics | 450.2 |
| Grand Total | 729.4 |



*Organic

**I4.** Table 4, page 289

# Periodic Table of the Elements

**Key**

| |
|---|
| 1 — Atomic Number<br>Hydrogen — Name<br>H — Symbol<br>1.0079 — Atomic Weight |

- Metals
- Metalloids
- Nonmetals

| Period | Group 1A | 2A | 3B | 4B | 5B | 6B | 7B | 8B | 8B | 8B | 1B | 2B | 3A | 4A | 5A | 6A | 7A | 8A |
|---|---|---|---|---|---|---|---|---|---|---|---|---|---|---|---|---|---|---|
| 1 | 1<br>Hydrogen<br>**H**<br>1.0079 | | | | | | | | | | | | | | | | | 2<br>Helium<br>**He**<br>4.0026 |
| 2 | 3<br>Lithium<br>**Li**<br>6.941 | 4<br>Beryllium<br>**Be**<br>9.0122 | | | | | | | | | | | 5<br>Boron<br>**B**<br>10.811 | 6<br>Carbon<br>**C**<br>12.0112 | 7<br>Nitrogen<br>**N**<br>14.0067 | 8<br>Oxygen<br>**O**<br>15.9994 | 9<br>Fluorine<br>**F**<br>18.9984 | 10<br>Neon<br>**Ne**<br>20.179 |
| 3 | 11<br>Sodium<br>**Na**<br>22.989 | 12<br>Magnesium<br>**Mg**<br>24.305 | | | | | | | | | | | 13<br>Aluminum<br>**Al**<br>26.9815 | 14<br>Silicon<br>**Si**<br>28.086 | 15<br>Phosphorus<br>**P**<br>30.9738 | 16<br>Sulfur<br>**S**<br>32.064 | 17<br>Chlorine<br>**Cl**<br>35.453 | 18<br>Argon<br>**Ar**<br>39.948 |
| 4 | 19<br>Potassium<br>**K**<br>39.098 | 20<br>Calcium<br>**Ca**<br>40.08 | 21<br>Scandium<br>**Sc**<br>44.956 | 22<br>Titanium<br>**Ti**<br>47.90 | 23<br>Vanadium<br>**V**<br>50.942 | 24<br>Chromium<br>**Cr**<br>51.996 | 25<br>Manganese<br>**Mn**<br>54.938 | 26<br>Iron<br>**Fe**<br>55.847 | 27<br>Cobalt<br>**Co**<br>58.933 | 28<br>Nickel<br>**Ni**<br>58.71 | 29<br>Copper<br>**Cu**<br>63.546 | 30<br>Zinc<br>**Zn**<br>65.38 | 31<br>Gallium<br>**Ga**<br>69.723 | 32<br>Germanium<br>**Ge**<br>72.59 | 33<br>Arsenic<br>**As**<br>74.922 | 34<br>Selenium<br>**Se**<br>78.96 | 35<br>Bromine<br>**Br**<br>79.904 | 36<br>Krypton<br>**Kr**<br>83.80 |
| 5 | 37<br>Rubidium<br>**Rb**<br>85.468 | 38<br>Strontium<br>**Sr**<br>87.62 | 39<br>Yttrium<br>**Y**<br>88.905 | 40<br>Zirconium<br>**Zr**<br>91.22 | 41<br>Niobium<br>**Nb**<br>92.906 | 42<br>Molybdenum<br>**Mo**<br>95.94 | 43<br>Technetium<br>**Tc**<br>(99) | 44<br>Ruthenium<br>**Ru**<br>101.07 | 45<br>Rhodium<br>**Rh**<br>102.905 | 46<br>Palladium<br>**Pd**<br>106.4 | 47<br>Silver<br>**Ag**<br>107.868 | 48<br>Cadmium<br>**Cd**<br>112.40 | 49<br>Indium<br>**In**<br>114.82 | 50<br>Tin<br>**Sn**<br>118.69 | 51<br>Antimony<br>**Sb**<br>121.75 | 52<br>Tellurium<br>**Te**<br>127.60 | 53<br>Iodine<br>**I**<br>126.904 | 54<br>Xenon<br>**Xe**<br>131.30 |
| 6 | 55<br>Cesium<br>**Cs**<br>132.905 | 56<br>Barium<br>**Ba**<br>137.34 | *57<br>Lanthanum<br>**La**<br>138.91 | 72<br>Hafnium<br>**Hf**<br>178.49 | 73<br>Tantalum<br>**Ta**<br>180.948 | 74<br>Tungsten<br>**W**<br>183.85 | 75<br>Rhenium<br>**Re**<br>186.2 | 76<br>Osmium<br>**Os**<br>190.2 | 77<br>Iridium<br>**Ir**<br>192.2 | 78<br>Platinum<br>**Pt**<br>195.09 | 79<br>Gold<br>**Au**<br>196.967 | 80<br>Mercury<br>**Hg**<br>200.59 | 81<br>Thallium<br>**Tl**<br>204.37 | 82<br>Lead<br>**Pb**<br>207.19 | 83<br>Bismuth<br>**Bi**<br>208.980 | 84<br>Polonium<br>**Po**<br>(209) | 85<br>Astatine<br>**At**<br>(210) | 86<br>Radon<br>**Rn**<br>(222) |
| 7 | 87<br>Francium<br>**Fr**<br>(223) | 88<br>Radium<br>**Ra**<br>(226) | **89<br>Actinium<br>**Ac**<br>(227) | 104<br>Rutherfordium<br>**Rf**<br>(261) | 105<br>Hahnium<br>**Ha**<br>(262) | 106<br>Seaborgium<br>**Sg**<br>(263) | 107<br>Nielsbohrium<br>**Ns**<br>(261) | 108<br>Hassium<br>**Hs**<br>(265) | 109<br>Meitnerium<br>**Mt**<br>(266) | 110 | 111 | 112 | | | | | | |

| | | | | | | | | | | | | | | |
|---|---|---|---|---|---|---|---|---|---|---|---|---|---|---|
| *Lanthanides 6 | 58<br>Cerium<br>**Ce**<br>140.12 | 59<br>Praseodymium<br>**Pr**<br>140.907 | 60<br>Neodymium<br>**Nd**<br>144.24 | 61<br>Promethium<br>**Pm**<br>144.913 | 62<br>Samarium<br>**Sm**<br>150.35 | 63<br>Europium<br>**Eu**<br>151.96 | 64<br>Gadolinium<br>**Gd**<br>157.25 | 65<br>Terbium<br>**Tb**<br>158.925 | 66<br>Dysprosium<br>**Dy**<br>162.50 | 67<br>Holmium<br>**Ho**<br>164.930 | 68<br>Erbium<br>**Er**<br>167.26 | 69<br>Thulium<br>**Tm**<br>168.934 | 70<br>Ytterbium<br>**Yb**<br>173.04 | 71<br>Lutetium<br>**Lu**<br>174.97 |
| **Actinides 7 | 90<br>Thorium<br>**Th**<br>232.038 | 91<br>Protactinium<br>**Pa**<br>(231) | 92<br>Uranium<br>**U**<br>238.03 | 93<br>Neptunium<br>**Np**<br>(237) | 94<br>Plutonium<br>**Pu**<br>244.064 | 95<br>Americium<br>**Am**<br>(243) | 96<br>Curium<br>**Cm**<br>(247) | 97<br>Berkelium<br>**Bk**<br>(247) | 98<br>Californium<br>**Cf**<br>242.058 | 99<br>Einsteinium<br>**Es**<br>(254) | 100<br>Fermium<br>**Fm**<br>257.095 | 101<br>Mendelevium<br>**Md**<br>258.10 | 102<br>Nobelium<br>**No**<br>259.101 | 103<br>Lawrencium<br>**Lr**<br>260.105 |